VOLUME FOUR

METHODS IN CHEMICAL PROCESS SAFETY

Advanced Methods of Risk Assessment and Management

VOLUME FOUR

METHODS IN CHEMICAL PROCESS SAFETY

Advanced Methods of Risk Assessment and Management

Edited by

FAISAL I. KHAN

Centre for Risk, Integrity and Safety Engineering (C-RISE)
Faculty of Engineering & Applied Science
Memorial University of Newfoundland
St. John's, NL, Canada

PAUL R. AMYOTTE

Department of Process Engineering & Applied Science
Dalhousie University
Halifax, NS, Canada

ELSEVIER

ACADEMIC PRESS
An imprint of Elsevier

Academic Press is an imprint of Elsevier
50 Hampshire Street, 5th Floor, Cambridge, MA 02139, United States
525 B Street, Suite 1650, San Diego, CA 92101, United States
The Boulevard, Langford Lane, Kidlington, Oxford OX5 1GB, United Kingdom
125 London Wall, London, EC2Y 5AS, United Kingdom

First edition 2020

ISBN: 978-0-12-821824-2
ISSN: 2468-6514

For information on all Academic Press publications
visit our website at https://www.elsevier.com/books-and-journals

Publisher: Zoe Kruze
Acquisitions Editor: Jason Mitchell
Editorial Project Manager: Leticia M. Lima
Production Project Manager: Denny Mansingh
Cover Designer: Miles Hitchen

Typeset by SPi Global, India

Contents

Contributors

Rouzbeh Abbassi
Faculty of Science and Engineering, School of Engineering, Macquarie University, Sydney, NSW, Australia

Albert Addo
Department of Process Engineering & Applied Science, Dalhousie University, Halifax, NS, Canada

Md. Tanjin Amin
Centre for Risk, Integrity and Safety Engineering (C-RISE), Faculty of Engineering & Applied Science, Memorial University of Newfoundland, St. John's, NL, Canada

Paul R. Amyotte
Department of Process Engineering & Applied Science, Dalhousie University, Halifax, NS, Canada

Ehsan Arzaghi
Faculty of Science and Engineering, School of Mechanical, Medical & Process Engineering, Queensland University of Technology, Brisbane, QLD, Australia

Terje Aven
University of Stavanger, Stavanger, Norway

Ezzeddin Bakhtavar
School of Engineering, University of British Columbia (Okanagan Campus), Kelowna, BC, Canada; Department of Mining and Material Engineering, Urmia University of Technology, Urmia, Iran

Kayleigh Rayner Brown
Department of Process Engineering & Applied Science, Dalhousie University, Halifax, NS, Canada

Chao Chen
Safety and Security Science Group, Faculty of Technology, Policy and Management, TU Delft, Delft, The Netherlands

Feifan Cheng
Department of Chemical Engineering, Tsinghua University, Beijing, PR China

Gyan Chhipi-Shrestha
School of Engineering, University of British Columbia (Okanagan Campus), Kelowna, BC, Canada

Yiyang Dai
School of Chemical Engineering, Sichuan University, Chengdu, Sichuan, PR China

Kasun Hewage
School of Engineering, University of British Columbia (Okanagan Campus), Kelowna, BC, Canada

Hirushie Karunathilake
School of Engineering, University of British Columbia (Okanagan Campus), Kelowna, BC, Canada; Department of Mechanical Engineering, University of Moratuwa, Moratuwa, Sri Lanka

Nima Khakzad
School of Occupational and Public Health, Ryerson University, Toronto, ON, Canada

Faisal I. Khan
Centre for Risk, Integrity and Safety Engineering (C-RISE), Faculty of Engineering & Applied Science, Memorial University of Newfoundland, St. John's, NL, Canada

Haroon R. Mian
School of Engineering, University of British Columbia (Okanagan Campus), Kelowna, BC, Canada

Norafneeza Binti Norazahar
Centre of Hydrogen Energy, Institute of Future Energy; School of Chemical and Energy Engineering, Faculty of Engineering, Universiti Teknologi Malaysia, Johor Bahru, Malaysia

Nicola Paltrinieri
Department of Mechanical and Industrial Engineering, Norwegian University of Science and Technology NTNU, Trondheim, Norway

Hans J. Pasman
TEES Mary Kay O'Connor Process Safety Center, Texas A&M University System, College Station, TX, United States

Howard Pike
Centre for Risk, Integrity and Safety Engineering (C-RISE), Faculty of Engineering, Memorial University, St. John's, NL, Canada

Genserik Reniers
Safety and Security Science Group, Faculty of Technology, Policy and Management, TU Delft, Delft, The Netherlands; Faculty of Applied Economics, Antwerp Research Group on Safety and Security (ARGoSS), University Antwerp, Antwerp; CEDON, KULeuven, Campus Brussels, Brussels, Belgium

William J. Rogers
Artie McFerrin Department of Chemical Engineering, Texas A&M University, College Station, TX, United States

Rehan Sadiq
School of Engineering, University of British Columbia (Okanagan Campus), Kelowna, BC, Canada

Zaman Sajid
Faculty of Engineering & Applied Science, Centre for Risk, Integrity and Safety Engineering, Memorial University of Newfoundland, St. John's, NL, Canada

Deyang Wu
Department of Chemical Engineering, Tsinghua University, Beijing, PR China

Hao Wu
Department of Chemical Engineering, Tsinghua University, Beijing, PR China

Jinsong Zhao
Department of Chemical Engineering, Tsinghua University, Beijing, PR China

Preface

Chemical process operations are hazardous by nature. These operations cause events leading to adverse consequences resulting in significant financial losses, injuries, fatalities, adverse environmental impact, and reputation damage. These undesired events can be predicted and prevented by utilizing the advanced methods of process safety and risk management.

This book is the fourth volume of the *Methods in Chemical Process Safety* book series. The book series intends to serve as a unified source of information for students, academic researchers, and professional practitioners. It aims to publish the fundamentals of process safety science and leading state-of-the-art advancements in the field. The book series pays a special focus on maintaining a practical approach to process safety methods and their application to industry. An international editorial board and authorship ensures that the series captures the global expertise and the latest developments. Each volume is dedicated to a specific thematic area of process safety to ensure detailed coverage of the methodological development.

This fourth volume presents the risk assessment and management methods used in the chemical industries over 12 distinct chapters: (i) a brief overview of advanced risk assessment and management methods and their evolution, (ii) fundamentals of dynamic risk assessment, (iii) logic based techniques for dynamic risk assessment, (iv) Bayesian methods for assessing dynamic risk, (v) data-driven methods used in risk assessment, (vi) uncertainty analysis during risk assessment of rare events, (vii) basics of risk management and multicriteria risk-based and multiobjective decision making, (viii) importance of inclusion of human factors in risk assessment and utilization of virtual technology in testing and verification of human factor risk, (ix) risk acceptance benchmarks such as "as low as reasonably practicable" (ALARP) and the precautionary principle (PP), (x) advanced methods used for environmental risk modeling and management, (xi) process economic risk management and techniques to model loss and integrate them in risk assessment, and (xii) implementation of inherently safer design (ISD) principles in risk management.

We wholeheartedly thank all the authors for their precious contributions to this volume on risk assessment and management. We also acknowledge the efforts provided by the Elsevier editorial team—Jason Mitchell, Acquisitions Editor and Leticia Lima, Editorial Project Manager.

Catastrophic accidents are often considered as unforeseeable or unpreventable. However, these can be avoided by employing a proper risk assessment and management method. We hope readers will perceive the appropriateness of this statement by reading this book and reflecting on the content's application to their own work. Also, it is expected that this fourth volume of the *Methods in Chemical Process Safety* book series will help academic researchers to become familiar with the existing methods used in chemical process system risk assessment and management, and industrial safety practitioners to upgrade their existing safety technology.

Faisal I. Khan
Centre for Risk, Integrity and Safety Engineering (C-RISE)
Faculty of Engineering & Applied Science
Memorial University of Newfoundland, St. John's, NL, Canada

Paul R. Amyotte
Department of Process Engineering & Applied Science
Dalhousie University, Halifax, NS, Canada

CHAPTER ONE

Advanced methods of risk assessment and management: An overview

Faisal I. Khan[a,*], Paul R. Amyotte[b], Md. Tanjin Amin[a]
[a]Centre for Risk, Integrity and Safety Engineering (C-RISE), Faculty of Engineering & Applied Science, Memorial University of Newfoundland, St. John's, NL, Canada
[b]Department of Process Engineering & Applied Science, Dalhousie University, Halifax, NS, Canada
*Corresponding author: e-mail address: fikhan@mun.ca

Contents

1. Background

1.1 Are we comfortably safe?

In recent years, human civilization has been experiencing unprecedented growth. It is widely agreed that technological advancements have a significant influence on this observed growth. Chemical process industries are also impacted by the increasing demand for a wider range of products, higher product quality, and higher production efficiency. These are consequently resulting in a transformation of simple process operations to complex ones. Also, the trends mentioned above have forced to extract the natural

Methods in Chemical Process Safety, Volume 4
ISSN 2468-6514
https://doi.org/10.1016/bs.mcps.2020.03.002

resources including hydrocarbons from challenging locations, as these continue to be the most reliable and readily available sources of energy and raw materials that can meet up industrial requirements. Advancements in science and technology have provided us with a comfort that the production and processing of natural resources are safer. However, are we comfortably safe? Do we have enough understanding and methods to ensure that we are maintaining an adequate level of safety? The straightforward answer to both these questions will be "No." This answer may sound astonishing, especially considering the safety technologies that are being used in modern process industries. However, the answer remains the same. A number of recent catastrophic accidents is a clear indication that "We are still not comfortably safe." This milestone is yet to be achieved. Maybe we are relatively safer, compared to the past. However, not to a level, we can feel comfortably safe.

1.2 Who should be accountable for the accidents?

The answer to this question entails detail causation analysis of past accidents. Although several crucial factors can be identified, human factors are the most frequently noticed causes behind many catastrophic accidents (Coco, 2003). After reviewing the Marsh report, Crowl and Louvar (2011) explicitly stated that human factors are the major reasons behind the studied cases. Even, the causes that are often termed as technical failures can be related to human error. If an accident occurs due to the failure of an aging component, how appropriate it would be to blame this component? It is a human failure as well. Perhaps, a proper and careful inspection could have avoided the unwanted situation.

Crowl and Louvar (2011) also opined that all accidents could be ascribed by human factors, apart from those caused by natural disasters. However, human factors were also prevalent in the accidents that were commenced by natural causes. The Fukushima Daiichi nuclear power plant accident, caused by a massive tsunami, is a glaring example. Although a magnitude 9 earthquake accompanied by the tsunami initiated a series of failures, a careful examination of the possible causes revealed that significant human errors were prevalent. The safety technology was not up-to-the-date; even, the management was aware that a massive tsunami might lead to a disaster, and the plant was at considerable risk (Amano, 2015). Later, several studies reflected that the magnitude of the Fukushima Daiichi power plant accident could have been mitigated by taking the right action when required (Acton & Hibbs, 2012; Synolakis & Kânoğlu, 2015).

The catastrophic BP Texas Refinery explosion (2005) and Deepwater Horizon offshore accidents (2010) are other notable cases where initial blame went to equipment failure. However, it was not the exact reason behind these catastrophes. The BP Texas Refinery accident was highly attributed to the company's managerial decision, while the latter one happened due to a flawed well plan (CSB, 2014; Holmstrom et al., 2006). Even, if the blame goes to the system failure, a series of questions may arise. Who has designed the system? Who is maintaining the system? Who is responsible for operating the process? Who is taking the managerial decisions? Who has produced the equipment? The answer to all these questions is "We, the human beings!" It is utterly unjust to blame the lifeless stuff that we have created to serve the purpose.

1.3 Can we predict the major accidents?

Major accidents are indeed infrequent. The process industries suffer from near-misses and minor accidents frequently. Therefore, the decision-makers are often reluctant to take safeguards against the major ones and somehow miss the critical issues that may lead to a catastrophic outcome. The cost of implementation of safety is another vital factor. Many industries often consider this as redundant. However, the cost of a devastating accident is way more than that is used to avoid it (Hopkins, 1999). There is no space to be contented with the previous successes on preventing the near-misses. The vital factors need to be taken care of, and safety technology should be designed to tackle every possible outcome.

But, do we know why major accidents are still happening? Do we have the understanding of the mechanism of these catastrophes? Admittedly, we have. Even, if we do not have the complete picture, at least, we have the partial one. Detail analysis of the accidents with catastrophic consequences are published in public domain so that crucial safety lessons can be learned, and future safeguards can be designed. These informative analyses are pillars for the future accident models (Lundberg, Rollenhagen, & Hollnagel, 2010). The available data in many well-known databases such as the Major Accident Reporting System (MARS), Process Safety Incident Database (PSID), Pondicherry University Process-Industry Accident Database (PUPAD), and Relational Information System for Chemical Accidents Database (RISCAD) may reflect that industrial safety has been improved by a significant margin. However, this may not depict the real picture, as many industries are often naturally reluctant to provide the actual information from the fear of reputation loss (Kletz, 2004).

Even, if the industries share accurate information, can we predict the rare accidents? Anything rare contains a higher level of uncertainty. It may be an arduous task to demystify them. However, it is not an impossible thing to be done. We are the deviser of the process systems, and we do an extensive brainstorming on what can go wrong in the process and subsequently, ensure the adequate system safety. We should be able to predict the major accidents, given that we have an understanding of the operational mechanism and analyzed reports on catastrophic accidents suffered by a similar kind of industries.

1.4 Do we have the solution?

The answer to this question is "Yes, we have." Risk assessment is considered as one of the reliable ways that can aid in preventing major accidents (Meng, Chen, Zhu, & Zhu, 2018; Rathnayaka, Khan, & Amayotte, 2013). Risk is an alternative expression of safety. Safety and risk are inversely related to each other, as shown in Fig. 1. Lower the risk, higher the safety and vice versa. Risk can be defined by answering "what can go wrong?," "how bad an accident could be?," and "how often could it happen?." For quantitative analysis, two terms: the probability of failure and severity of the failure are used to get the numerical value of risk (CCPS, 2007; Kletz, 1992). Can we measure safety directly? Unfortunately, no. Suppose people are attacked by fever. When they visit the doctor, their body temperature is measured, and it is decided whether they have a fever or not, based on the temperature measurement. Similarly, risk can be measured, and the level of safety is determined from it.

Although the terms: risk assessment and management are closer, there is a clear distinction between them. Risk assessment mainly focuses on the identification of hazard and analysis of the possible loss due to this hazard; it starts with the hazard identification and ends up with the loss estimation due to an accident initiated by this hazard. On the contrary, risk management (shown in Fig. 2) is a continuous process that aims to identify,

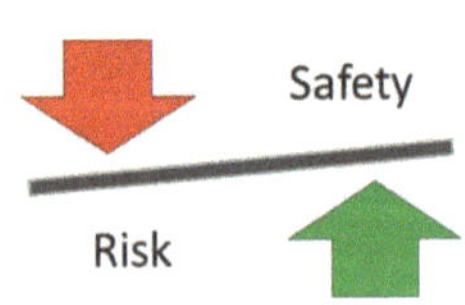

Fig. 1 Relationship between safety and risk.

Fig. 2 A general risk management cycle.

analyze, evaluate, and treat loss exposures. It also monitors the risk control and financial resources to diminish the undesired effects (Aven, 2016). Risk assessment is the core element of the risk management process. Assessing and managing the risk provides a robust solution to prevent catastrophic accidents and ensure process safety (Kleindorfer, Oktem, Pariyani, & Seider, 2012). According to the Center for Chemical Process Safety (CCPS), risk assessment plays a pivotal role in safety and risk management (CCPS, 2007). Hence, there is no alternative to risk assessment to prevent the occurrence of "rare" events.

1.5 A brief history of risk assessment and management

It may sound surprising; however, the concept of assessing and managing the risk is somewhat an ancient practice. At about 3200 BC, a group of people called the Asipu from the Tigris-Euphrates valley used to conduct the risk assessment and suggest their clients how to manage risk. This is the first simplified form of risk assessment that has been recorded (Covello & Mumpower, 1985; Grier, 1980, 1981). People used to contact one of the members from Asipu before making crucial decisions such as risky ventures, engineering projects, and familial issues to get their advice. How did they use to do the risk assessment? They tried to analyze the problems from every possible perspective and create an alternative solution for each view. After collecting data, they used to assign the possible outcomes such as success and failure for each alternative. They would create a register book with spaces and used to put a plus if the result was in favor; otherwise, a minus would have been put. Once the analysis was conducted, the Asipu used to recommend the best alternative and possible ways to mitigate the risks associated with the suggested alternative to their clients and provided them with a detail report, written in a clay tablet (Trimpop, 1994).

There is a clear distinction between the risk assessment techniques that have been utilized by the Asipu and modern days' ones due to the application of probability theories. Although the Arab mathematicians are the first to be considered who developed the probability theory, Blaise Pascal, a French mathematician, is popularly credited as the inventor of the modern probability theory (Broemeling, 2011; Ore, 1960). In recent years, most of the risk assessment frameworks are probabilistic by nature. On the contrary, the people of Asipu used to claim that they received the message from God. Therefore, their assessment was deterministic. The concept of probabilistic risk assessment (PRA) (often termed as the quantitative risk assessment (QRA)) emerged from the aviation industry. Later, it became popular in the nuclear and chemical industries (Bedford & Cooke, 2001; Guymer, Kaiser, Mc Kelvey, & Hannaman, 1987). Kirchsteiger (1999) differentiated the probabilistic and deterministic approaches with examples from nuclear and chemical industries and concluded that the PRA is a more cost-effective approach.

A careful examination of the available process safety and risk analysis literature suggests that the growth of this field is greatly influenced by the lessons learned from the catastrophic accidents. The Flixborough disaster (1974), an explosion in a chemical plant increased the attention to risk assessment. It introduced the Management of Change (MOC), as well (Turney, 2014). The Seveso disaster (1976) accelerated the emphasis of inherently safer design and improvement to emergency and response planning to mitigate the loss (Jain, Pasman, Waldram, Rogers, & Mannan, 2017; Khan & Amyotte, 2008). The BP Texas Refinery accident (2005) showed the necessity of improved risk assessment and MOC, adequate safety technology and equipment, and human factor analysis. The Deepwater Horizon offshore accident (2010) and West Fertilizer explosion (2013) speeded up the significance to enhance safety culture and understand the correlation among the system, operation, and risk (Antonsen, Nilsen, & Almklov, 2017; Dahle et al., 2012; Mannan, Reyes-Valdes, Jain, Tamim, & Ahammad, 2016).

Even, the first document in this field was published on the statistics of industrial accidents, according to a bibliometric review by Amin, Khan, and Amyotte (2019). The authors also stated that the dissemination of crucial research findings in the public domain became regular only after 1965. Although the risk-based safety analysis approaches started at the beginning of the 1980s, these became frequent in about the mid of 1990s. Legislative authorities played a pivotal role in the development of risk management as well. The European Commission inaugurated the Seveso Directive in 1982 after the Seveso disaster (1976), which was later substituted by Seveso II and Seveso III Directives in 1996 and 2012, respectively. The Canadian Chemical Producers' Association (CCPA) introduced the "Responsible Care" initiative that urged the Canadian chemical industries to go through their existing safety practices so that they can report the potential weakness and relevant findings from the review. The CCPA also provided the guidelines to transport the hazardous chemical after the Mississauga train derailment (1979) (Belanger, Topalovic, Krantzberg, & West, 2009). The Occupational Health and Safety Administration (OSHA) introduced 29CFR 1910.119 in 1992 for safety management of highly hazardous chemicals. Although this standard was developed for the US industries, later, it was adopted by the industries around the world. After a couple of years, the Environmental Protection Agency (EPA) inaugurated the 40 CFR Part 68 for chemical accident prevention and risk management (Macza, 2008). The CCPS has taken one of the recent initiatives; they have issued a detailed guideline for the risk-based process safety analysis and management. Commit to process safety, understand hazards and risk, manage risk, and learn from experience are the four functional elements, as described in this guideline (CCPS, 2007).

Is the industry at the same pace as academia? According to Klein (2009), the DuPont company, a gunpowder producer in the United States, implemented process safety management principles from the beginning of the 1800s. The Imperial Chemical Industries (ICI) is another name that is found whenever the evolution of process safety and risk analysis is investigated since it is the first chemical industry where the hazard and operability study (HAZOP) was implemented (Kletz, 1983). As a result, chemical process risk management experienced a paradigm shift in the following decades. It is worth mentioning that Professor Kletz first implemented the HAZOP in the ICI and trained a team to perform it.

2. Evolving methods

2.1 Classification

As risk assessment and management has a lengthy historical background, there are abundant documents on this field. However, the focus of this chapter is concentrating solely on the materials that are mainly concerned with the chemical processes. Several techniques have been found in the existing publicized documents where the outcome (risk) have been obtained by following distinct methods. According to Khan, Rathnayaka, and Ahmed (2015), perhaps, the simplest and easiest way to classify the existing approaches in the qualitative, semi-quantitative, quantitative, and hybrid types. The quantification of risk is the basis of this classification, as shown in Fig. 3.

2.2 Qualitative methods

The qualitative methods do not provide a numerical representation and description based on the observed data, studied process flow diagram, characteristic graphs, and other risk indicators. The output of these methods is purely non-numerical; however, provides a strong perception of risk that can be further utilized for risk mitigation. Consider a reactor where 10 °C increase in temperature means that there is a high risk of a runaway reaction. By monitoring the temperature profile, the operational risk (i.e., high, medium, and low) can be predicted; nevertheless, no quantification of risk is obtained from such procedure.

2.2.1 Risk matrix

The idea of a risk matrix was first developed in the Electronic System Centre, US Air force (Garvey & Lansdowne, 1998). Since its inception, the risk matrix has become one of the most widely used qualitative risk

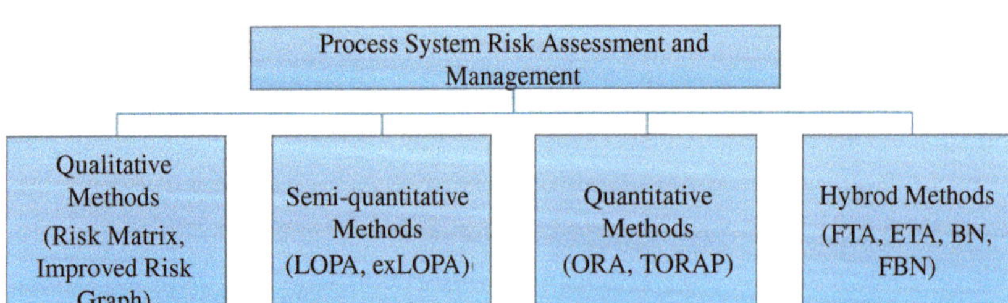

Fig. 3 A classification of risk assessment and management methods.

assessment technique that has been well adopted by the industries for its simplicity, yet, effectiveness. The risk matrix has been described as a semi-quantitative approach by many scholars (Ni, Chen, & Chen, 2010; Ruge, 2004). However, if both the fundamentals of risk: the probability of failure and severity of consequences are expressed in a qualitative manner, the risk matrix becomes a purely qualitative approach. Fig. 4 shows an illustrative risk matrix where the probability of failure is shown on the vertical axis, and the severity of consequences is displayed on the horizontal axis. Qualitative outcomes are separated from each other by distinct colors which make it easier to visualize. The major advantage of the risk matrix is that it allows representing complex risk data in a compressed form. However, there are several significant disadvantages, as well. Firstly, it is a user perspective approach; the same matrix can be interpreted differently by another user unless each condition is clearly described. Secondly, it is often found oversimplified which cannot capture the detail process risk profile. Finally, it cannot model complex dependency and uncertainty.

2.2.2 Risk graph

The risk graph has its origin in Germany in 1994 (Din, 1994). It is one of the most frequently used technique to measure the safety integrity level (SIL) of process systems. A risk graph is usually comprised of four parameters: consequences of a hazard, frequency of being in the hazardous zone and operating time, probability of avoiding the hazard, and the probability of the undesired consequences. The risk is estimated from the combination of these parameters. Although it is a simple and effective method, it is often found confusing that makes it strenuous to be interpreted. It sometimes provides a measure of over risk, which results in impractical management action. Besides, defining all the parameters and condition is a tedious and laborious task that needs significant workforce and time. To overcome some of the limitations of conventional risk graphs, an improved version was

		Severity of Consequences			
		Negligible	Minor	Moderate	Severe
Probability of Failure	Highly Frequent	Low	Medium	Medium High	High
	Moderately Frequent	Low	Low Medium	Medium	High
	Less Frequent	Low	Low Medium	Medium	Medium High
	Rarely Frequent	Low	Low	Low Medium	Medium

Fig. 4 A simple risk matrix.

proposed by Baybutt (2007). This advanced version was developed based on a strong theoretical foundation. The author clarified the existed ambiguities by displaying a more robust way to define the parameters and operating conditions.

2.2.3 Qualitative accident models

The accidents models are of paramount significance in process risk management, as these provide a deeper understanding of the accident scenarios; more importantly, the dependence between the cause and effects. Several accident models have been developed to describe an accident. Although many of these models were constructed for distinct industries, later, process industries also adopted them. A pioneering work of sequential Domino accident model by Heinrich (1941) was one of the earliest sequential models that introduced a new era in safety and risk management. This model describes an accident as a chain of sequential events. Heinrich proposed a total of five metaphorical dominoes: social environment and ancestry, fault of the person, unsafe act or mechanical or physical hazard (unsafe condition), accident, and injury. He defined each of these dominoes explicitly. According to Heinrich's Domino model, when all these dominoes are prevalent, we are likely to experience an unwanted consequence. Hence, action needs to be taken to minimize or eliminate their presence in the sequence. An updated version of the original Domino model was presented in the "Loss Causation Model" (Bird, Cecchi, Tilche, & Mata-Alvarez, 1974). It prioritized the management and organizational factors while analyzing an accident. Johnson (1973) introduced the Management Oversight and Risk Tree (MORT) to investigate the system safety and demystify the innate relationships between the management and organizational factors and plant operational conditions.

Professor Kletz proposed a model-oriented methodology for accident investigation (Kletz, 1991). In this work, the author placed an accident on the top, and the causes and leading events were put under this accident. The Swiss cheese model by Professor James Reason is another notable work in sequential accident analysis. It contains a series of barriers, and each barrier has many holes (latent errors that may lead to accident). When holes from all the barriers are aligned, an accident occurs (Reason, 1990). Svenson (1991) described an accident in terms of a series of interactions between human and process systems and developed the Accident Evaluation and Barrier (AEB) model. The Uncontrolled Flow of Energy (UFOE) model was proposed by Rasmussen and Grønberg (1997). The authors used the uncontrolled flow of energy to explain an accident and stated that an uncontrolled flow or transfer of energy because of loss of containment causes an accident. Almost 5 years later, Svedung and Rasmussen (2002) came up with the idea of representing an accident model by mapping the system structure and causation factors. This work emphasized the revelation of relationship between hazards and socio-technological factors. Also, it depicted the necessity of translating an accident to hazard analysis. The major contribution of this work was the introduction of a set of pictorial formats to explain the distinct causation factors.

An automated tool, called yAGAS (Yonsei Automatic Generator of Accident Scenarios) was developed by Kim, Moon, Lee, and Yoon (2003) to generate possible accident scenarios. The prior operational experience was the basic building block of this software. Leveson (2004) stated the deviation of safety as a pressing issue that significantly affects the control structure of adaptive socio-technological systems. She viewed an accident as a consequence of external disturbances, component failures, and dysfunctional interaction among system components. A conceptual offshore accident model was demonstrated by Kujath, Amyotte, and Khan (2010). This model allowed to build the sub-barriers under the main safety barrier. The authors identified the required safety barriers for accident prevention and risk minimization. Another work on safety barrier identification was presented by Chen, Moan, and Verhoeven (2008). They described that three main safety barriers: position loss prevention mechanism, vessel movement arrest barrier, and well integrity loss prevention barrier are required to ensure the safety of the dynamic positioning of a mobile drilling unit.

2.2.4 *Qualitative inherent safety analysis*

The Flixborough accident (1974) was the major catastrophic accident that inspired the application of inherent safety principles in process industries. Professor Kletz, the father of inherent safety, delivered an outstanding speech at the annual jubilee lecture to the Society of Chemical Industry in London; he told, "What you don't have, can't leak" (Srinivasan & Natarajan, 2012). This was the formal introduction of inherent safety which has a vital role in minimizing the risk and safety management. However, is it possible to disappear the chemicals from the industries? Surely, not. However, the batch size can be reduced to mitigate the risk. Professor Kletz proposed four key principles: minimization, simplification, substitution, and moderation that are essential parts of an inherently safer design (Kletz, 1985). Suppose a chemical plant is dealing with highly hazardous materials, and the stakeholders want to apply the concept of inherent safety to reduce the risk. Table 1 shows a simple way of doing this.

Table 1 Inherent safety design example.

Principle	Action
Minimize	Use the chemicals in a lesser quantity
Substitute	Replace the chemical with a less hazardous one if it serves the purpose
Moderate	Use the material at a form that provides the least hazard. Ambient condition may need to be modified to ensure that the hazard strength has been reduced
Simplify	Simplify the process flow diagram and equipment

Since the inception of inherent safety, it has been used as a reliable mechanism to operate a plant more efficiently and risk management. Gupta and Edwards (2003) proposed a graphical method to measure inherent safety. This work mainly focused on finding the best process route. A qualitative approach, Qualitative Assessment for Inherently Safer Design (QAISD), was proposed by Rusli and Mohd Shariff (2010). The aim of developing QASID was to apply it at the early process design stages. The authors also stated that an amended version of the developed method could be utilized in the later stages of the process life cycle.

2.3 Semi-quantitative methods

Semi-quantitative approaches generate the results taking the advantages of both qualitative and quantitative methods. Although it can provide numerical outputs, the obtained result is an approximate estimate rather than an actual one. It often provides the estimation of risk by means of scores. Still, these methods are useful, especially when the straightforward measurement of risk is unavailable. The risk matrix shown in Fig. 4 can be translated to a semi-quantitative one by assigning risk values for different outcomes such as low, medium, low medium, medium high, and high.

2.3.1 Layer of protection analysis (LOPA)

The LOPA is the most dominant member of the semi-quantitative analysis-based methods; it is widely used in risk management in the chemical plants. It has been considered as an important technique for highly hazardous process industries from the late 1990s (Amin et al., 2019). Bridges and Williams (1997) and Dowell (1997) described how the LOPA was implemented in their plants and consultancies, as it could capture the independent protective safety systems without an accurate quantitative analysis. The CCPS issued a

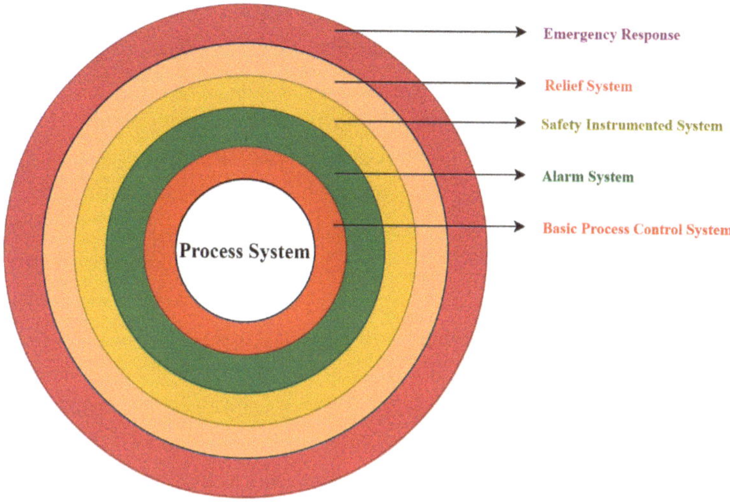

Fig. 5 A pictorial illustration of LOPA.

detailed guideline on LOPA when it started becoming popular in the United States and elsewhere in the world (Summers, 2003). The concept of LOPA is illustrated in Fig. 5. A process system is surrounded by five layers named as the basic process control system, alarm system, safety instrumented system, relief system, and emergency response. Among these, the first four are the accident prevention safeguards while the last one is the mitigation safeguard. It should be noted that the number of layers is expected to vary for different industries. Despite being a strong risk assessment tool, the LOPA cannot deal with uncertainty prevailed in input data.

In the last 20 years, many derivatives of LOPA have been developed to capture different issues that may question the efficacy of this technique. While conducting the layer of protection analysis, human factors are often excluded, which may result in a flaw in the analysis. To overcome this issue, Baybutt (2002) came up with an idea of incorporating relevant human factors to hazards while performing LOPA; it was named as the LOPA-HF. Markowski (2007) demonstrated an application of LOPA in the explosive atmospheric conditions. This methodology was named "exLOPA." Later, Markowski and Mannan (2010) improved this method and proposed a new technique called "ExSys-LOPA." This work aimed at reducing the hazard identification time significantly by using the available database and expert opinion. Another application of LOPA to the reactive chemicals has been displayed by Wei, Rogers, and Mannan (2008).

2.3.2 Index-based approaches

A fatality index-based model for assessing the consequences of ammonia and chlorine release was proposed by Brockhoff, Petersen, and Haastrup (1992). Historical data were used to estimate the fatality index. Khan and Abbasi (1997a) developed an index-based impact assessment technique that included several surrounding factors such as the population, asset, and ecosystem. The runaway risk index (RRI) methodology was proposed by Kao, Duh, Chen, and Yu (2002) to predict the risk of an exothermic runaway reaction. This work used a simplified mathematical and tabular method for assessing the risk, based on calculated hazard index. Al-Sharrah, Edwards, and Hankinson (2007) proposed an index-based approach for petrochemical industries. A total of four terms: frequency of accidents, hazardous effect of the chemical, inventory of the chemical released, and size of the plant were used to determine the risk index. However, unlike the traditional concept of risk assessment, this work expressed the risk in terms of number of people affected per year.

A semi-quantitative short-cut risk analysis method (SCRAM) was presented by Davis, Hinze, Hansen, and Van Wingerden (2011). This methodology was applied to deal with the dust explosion problems; the probability that the cloud will reach to its flammability limit and ignition probability to were used to estimate the likelihood of failure. A risk matrix was utilized to express the assessed risk. Zadakbar, Imtiaz, and Khan (2012) proposed a risk-based fault detection methodology which considered a number of indices for severity estimation. Later, a similar concept of severity estimation was used in the works by Zadakbar, Imtiaz, and Khan (2013) and Yu, Khan, and Garaniya (2015). However, detecting the process faults by monitoring the risk profile was the main contribution of these works.

2.3.3 Other semi-quantitative methods

Prioritizing the risk is of paramount significance in risk management. Long and Fischhoff (2000) proposed a risk ranking model; its efficacy was validated by examining both the sequential and simultaneous risk rankings. Aven (2008) argued that the semi-quantitative methods could be considered as the superior techniques over the quantitative ones, as the calculations in quantitative risk assessment (QRA) are often tedious. He developed a semi-quantitative method using the hazards and barriers, risk influencing factors (RIFs), and safety improvement measures. Khan, Abunada, John, and Benmosbah (2010) proposed a technique to measure the process safety performance using the risk and named it as SPI. This methodology used a risk metric to assess process safety. A total of three design elements: operational, design, and mechanical integrity were used to analyze both the leading and lagging indicators. Each element's risk was calculated from the risk factors associated with the performance indicators, and then, operational risk was calculated from the element risk.

Deacon, Amyotte, and Khan (2010) developed a human error risk analysis method during offshore emergencies. The level of risk and its tolerability were determined based on consequences from past incidents in the offshore industry and the probabilities of human error. A convention to select and evaluate the safety measures to mitigate the risk was developed from the ARAMIS (accidental risk assessment methodology for industries) approach. The authors used the Ocean Odyssey Blowout (1988) as the case study to validate their model. Tugnoli, Landucci, Salzano, and Cozzani (2012) developed the Inherent Safety Key Performance Indicators (IS-KPIs) methodology; it provided a flexible procedure for hazard identification and a sound consequence-based quantitative safety performance.

2.4 Quantitative methods

The quantitative methods provide exact and practical quantification of risk on contrary to the semi-quantitative techniques. This intrinsic nature of these methodologies gives them an edge over the qualitative and semi-quantitative counterparts from the decision-making perspective and safety analysis.

2.4.1 Probabilistic risk assessment

Khan and Abbasi (1998a) developed a software called DOMIFFECT (DOMIno eFFECT). It was applicable to domino effect analysis. The uniqueness of this computer program was that it could capture the secondary accident, initiated from the primary one. This tool provided a new direction in domino effect related risk assessment. The HAZDIG, another software package for the accidental release of toxic chemicals' risks assessment, was developed by Khan and Abbasi (1999b). The data, release scenario generation, dispersion, characteristics estimation, and graphic were the five main elements of this unique software. The tool could provide a flexible output which could have been utilized for reporting without further editing.

A probabilistic fault tree analysis (PROFAT) tool was developed by Khan and Abbasi (1999d). It was able to find out the initiating events, which may eventually lead to a major accident and shortest routes (minimum cut sets). Besides, it could estimate the probabilities of occurrence of the initiating events, relative contribution of each of the initiating event, and initiating events with the highest potential to cause the major accidents. Later, this tool was advanced by Khan and Abbasi (2000) on the basis of AS-II algorithm. A rapid risk assessment tool, MAXCRED (MAXimum CREDible accident analysis), was developed by Khan and Abbasi (1997d). This tool was converted to software by Khan and Abbasi (1999c). It could generate a wide range of accident scenarios and provide the results rapidly. Two other developed versions of MAXCRED are also available in the existing literature (Khan & Abbasi, 1997c, 1999a).

An optimal risk assessment method, named as the ORA, was proposed by Khan and Abbasi (1998c). It used the Hazard Identification and Ranking (HIRA) (Khan & Abbasi, 1998b), optHAZOP (Khan & Abbasi, 1997b), and TOPHAZOP (Khan & Abbasi, 1997e) for hazard identification. The PROFAT was used for probabilistic hazard assessment. The consequences were modeled using the HAZDIG and DOMIFFECT. Another computer-aided automated tool for quantitative risk assessment, called the TORAP (Tool for Rapid Risk Assessment in the Petroleum Refinery and Petrochemical Industries), was developed by Khan and Abbasi (1999e). Later, this tool was applied to a fertilizer plant's risk assessment in Mumbai, India (Khan, Iqbal, & Abbasi, 2001).

Antonioni, Spadoni, and Cozzani (2009) emphasized the requirement of QRA for domino effect analysis in the chemical industrial plants and developed a methodology for risk assessment in the presence of the domino effect. Baesi, Abdolhamidzadeh, Hassan, Hamid, and Reniers (2013) proposed a QRA-based decision-making strategy for establishing a new plant, adjacent to an old one. The past experience data were used to estimate the likelihood of domino effect. Many articles have been published on QRA considering natural events such as the lightning, floods, and earthquakes, as well (Antonioni, Landucci, Necci, Gheorghiu, & Cozzani, 2015; Borghetti et al., 2017).

2.4.2 Uncertainty treatment

Uncertainty is an inevitable part of risk assessment. Lower the uncertainty, better the confidence over the obtained result. Theofanous (1996) developed a framework called the Risk Oriented Accident Analysis Methodology (ROAAM) to assess the accidents and associated uncertainties. This methodology was developed using six phases named as scoping, analytical, quantitative, maturation, extension, and defense-in-depth. Each phase is composed of a number of steps. The ROAAM was successfully applied to the advanced passive plants under the Advanced Reactor Severe Accident Program (ARSAP) of the US Department of Energy (DOE). Later, Scobel, Theofanous, and Sorrell (1998) applied the ROAAM methodology for managing accidents in the AP600, an advanced light water reactor. With the help of the Monte Carlo method, Rai and Krewski (1998) proposed an uncertainty analysis technique for the multiplicative risk models. Markowski, Mannan, and Bigoszewska (2009) applied fuzzy set theory to the bow-tie (BT) approach for accident risk assessment in the process industries.

Markowski, Mannan, Bigoszewska, and Siuta (2010) proposed a risk assessment technique to estimate uncertainty. This method comprised of the traditional risk assessment approach and fuzzy logic. Integrating the analytical hierarchy process (AHP) with fuzzy logic, Miri Lavasani, Yang, Finlay, and Wang (2011) developed a risk assessment framework for quantifying the risk in offshore oil and gas wells. Ferdous, Khan, Sadiq, Amyotte, and Veitch (2011) applied fuzzy logic to minimize uncertainty in QRA. The Dempster–Shafer theory (DST) was also used to handle uncertainty in the FTAs (Curcurù, Galante, & La Fata, 2012). Abdo and Flaus (2015) developed an uncertainty analysis technique for the dynamic process systems using the fuzzy set theory and Monte Carlo simulation. This work considered both the stochastic and subjective uncertainties in risk calculation.

2.4.3 Recent trend in QRA

One of the most recent trends in process safety is the shift from traditional QRA to dynamic risk assessment (DRA). The inability of capturing the dynamic nature of risk by static QRA techniques is the pivotal reason behind this shift (Villa, Paltrinieri, Khan, & Cozzani, 2016). Although the concept of dynamic safety and risk analysis dates back to the 1990s, it only became a widespread practice from the late 2000s (Marseguerra, Nutini, & Zio, 1994; Marseguerra, Ricotti, & Zio, 1995; Marseguerra & Zio, 1996). A Bayes' theorem based failure updating mechanism was proposed by Shafaghi (2008). Using the accident precursor data and Bayesian updating mechanism, Kalantarnia, Khan, and Hawboldt (2009) proposed a DRA methodology. Potential accident scenarios were modeled using the event tree analysis (ETA), and safety barriers' failure probability were updated from the current accident data; subsequently, the risk profile was updated. Most of the current literature deal with the DRA-based methodologies, which will be discussed in the latter part of this chapter.

2.5 Hybrid methods

Hybrid methods are constructed using qualitative and quantitative approaches. It may seem to be analogous to the semi-quantitative techniques; however, there is a clear difference between these two. The semi-quantitative ones often provide a vague numerical value while the quantitative methods usually give a more accurate and realistic risk measurement. Also, the semi-quantitative approaches are often found without any qualitative part.

2.5.1 Fault tree analysis (FTA)

Perhaps, the FTA is the most widely used technique in the domain of safety and risk analysis. It is a top-down deductive technique that shows both the failure paths and logical relationships among the top event (unwanted event) and bottom events (basic failure cause). These two are the qualitative aspects of an FTA. In addition to these, using the Boolean logic and probability theory, it can also provide an accurate quantitative estimate. Fig. 6 shows a simple FTA. It can be seen that two types of gates: AND and OR are used to depict the relationship between the causes and effects. The nodes that represent the effects are always placed on top of the causes.

The FTA was first developed in 1961 by H. Watson and A. Mearns in the Bell Labs. The initial purpose was to evaluate the Minuteman I Intercontinental Ballistic Missile (ICBM) Launch Control System. Soon,

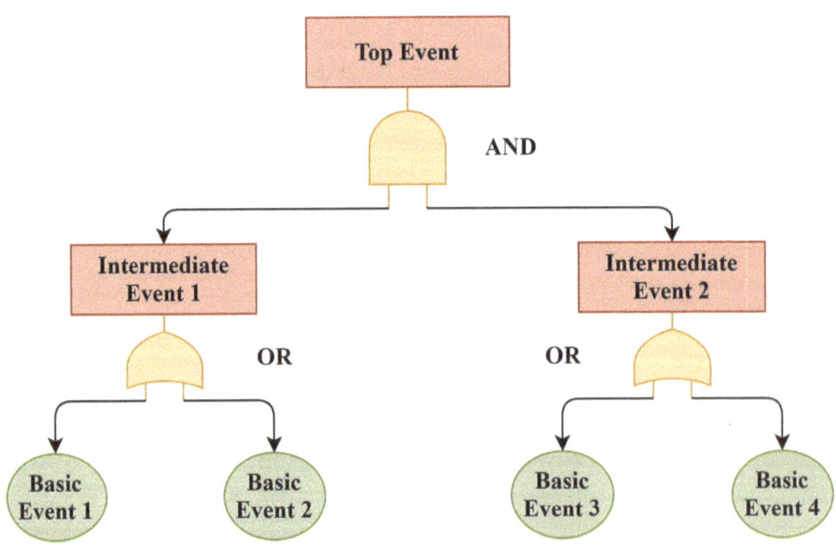

Fig. 6 A simple fault tree.

it was adopted by Boeing as a remarkable safety analysis tool (Ericson, 1999). The first technical paper on FTA was presented by Haasl (1965) at a symposium in Seattle, jointly organized by the University of Washington and Boeing. Although the FTA is popular in distinct industrial settings, it cannot model complex interdependency among the variables. Besides, its ability to handle uncertainty and multi-state variables is limited to an extent.

Even though the development of a fault tree for a simple system is easier, it becomes complex with the increase in system complexity. Often, the minimal cut set (MCS) analysis provides a more relaxed and straight forward way to determine the set of basic events that may lead to the top event. Rauzy (1993) proposed a combination of the binary decision diagram (BDD) and Shannon's decomposition theorem to find the MCSs. Reay and Andrews (2002) claimed that the then research failed to identify any scheme that could produce the BDDs for all the fault trees, and they proposed a new algorithm. This algorithm could reorganize the fault trees into the smaller sub-trees using the Faunet reduction technique, linear-time algorithm, and BDD. In addition to the BDD-based construction procedures, Remenyte-Prescott and Andrews (2008) proposed a ternary decision diagram (TDD)-based fault tree construction methodology; it was found that the proposed method could provide a more efficient and faster processing, compared to the existing methodologies.

As mentioned earlier, the FTA is vulnerable to uncertainty. Tanaka, Fan, Lai, and Toguchi (1983) incorporated the fuzzy theory to handle uncertainty. Yuhua and Datao (2005) displayed the fuzzy-FTA-based safety assessment technique for an oil and gas transmission pipeline. Lavasani, Zendegani, and Celik (2015) showed how fuzzy-FTA could be applied to the petrochemical process industries in the presence of limited data. The D-S theory was also adopted to minimize uncertainty (Limbourg, Savić, Petersen, & Kochs, 2007, 2008). Other notable works to reduce uncertainty in FTA are (Brissaud, Barros, Bérenguer, & Troffaes, 2010; Choi & Cho, 2007).

By nature, a conventional fault tree is a static tool; however, the process systems are dynamic. Hence, the traditional FTA becomes less suitable when time dependency is required. Using the house event matrix and time dependent probabilistic models, a dynamic version of the traditional fault tree was used in safety analysis at the nuclear industries (Čepin & Mavko, 2002). A Bayesian network (BN)-based DFT modeling technique has been demonstrated by Boudali and Dugan (2005). Additionally, the dynamic fault tree (DFT) can be constructed using four gates: priority AND (PAND), sequence enforcing (SEQ), standby or spare (SPARE), and functional dependency (FDEP). Durga Rao et al. (2009) proposed a Monte Carlo simulation-based framework to solve these gates and construct the DFT. Recently, Kabir, Walker, Papadopoulos, Rüde, and Securius (2016) proposed a new extension of the FTA, named Pandora to handle the dynamic process behavior. A combination of expert elicitation and fuzzy set theory are the basic building blocks of this extension.

2.5.2 Event tree analysis (ETA)

The ETA (Fig. 7) is an inductive technique that is frequently used in consequence modeling in context to the risk assessment. It is a graphical representation of sequential events. The probability of a hazard is shown on the left of the tree while the failure paths are shown on the right. On the basis of the performance the safety barriers employed to prevent the accidents, several consequences can be obtained as the outputs of an event tree. The performance of the safety barriers is often estimated using the FTA and BN.

The inception of the ETA in process safety dates back to 1975 during the WASH 1400 nuclear safety study. The WASH team, led by Professor Norman Rasmussen, was trying to assess the accident risk of the US commercial atomic industries. However, as the size of the fault trees was larger, they developed the event tree as an alternative method for quantifying the failure risk (Commission, 1975). Over the years, several research articles have utilized the event trees to build the risk assessment and accident

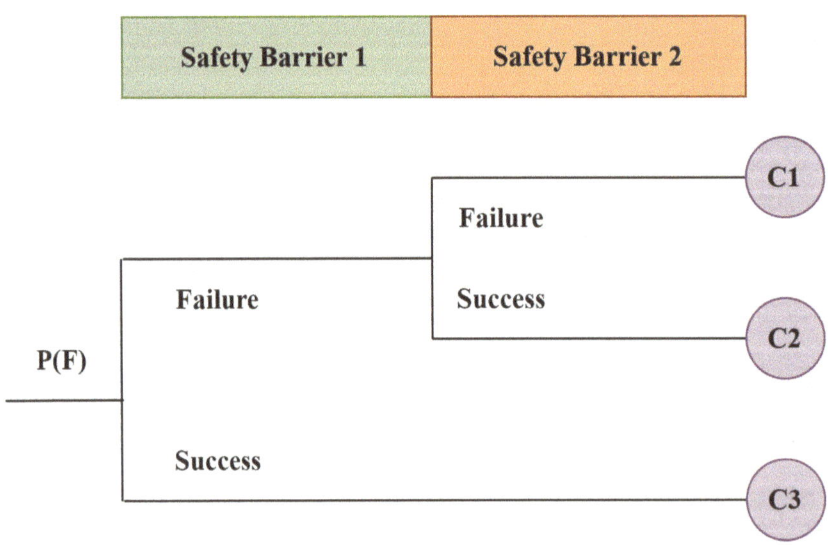

Fig. 7 An illustrative event tree.

analysis models (Adedigba, Khan, & Yang, 2016b; Alileche, Olivier, Estel, & Cozzani, 2017; Hashemi, Ahmed, & Khan, 2014b; Renni, Krausmann, & Cozzani, 2010). However, similar to the FTA, the ETA suffers from the same disadvantages as mentioned in the previous section.

Although the fuzzy set theory is widely used in epistemic uncertainty analysis (Huang, Chen, & Wang, 2001), Baraldi and Zio (2008) stated that such technique might not provide a better representation of uncertainty. They proposed a hybrid Monte Carlo and possibilistic method in this context. The application to a nuclear power plant suggested that the developed approach outperformed the conventional approaches. Ferdous, Khan, Sadiq, Amyotte, and Veitch (2009) utilized the fuzzy and evidence theories to assess the uncertainty associated with the initiating event. The fuzzy theory was used to tackle the imprecision and subjectivity, and evidence theory was used to handle the inconsistent, incomplete, and conflicting data. You and Tonon (2012) developed a novel methodology to feed the event tree with the imprecise probability. The authors focused on the calculation of upper and lower bounds of the probability.

2.5.3 Bow-tie analysis

The bow-tie analysis is another powerful tool in quantitative risk assessment. It is used to depict the cause-consequences relationship. The works by Dan Nielsen founded the underlying formalism of the bow-tie approach (Nielsen, 1971, 1974; Nielsen, Platz, & Runge, 1975). It is structured by putting thc FTA and ETA on the left and right, respectively (Fig. 8).

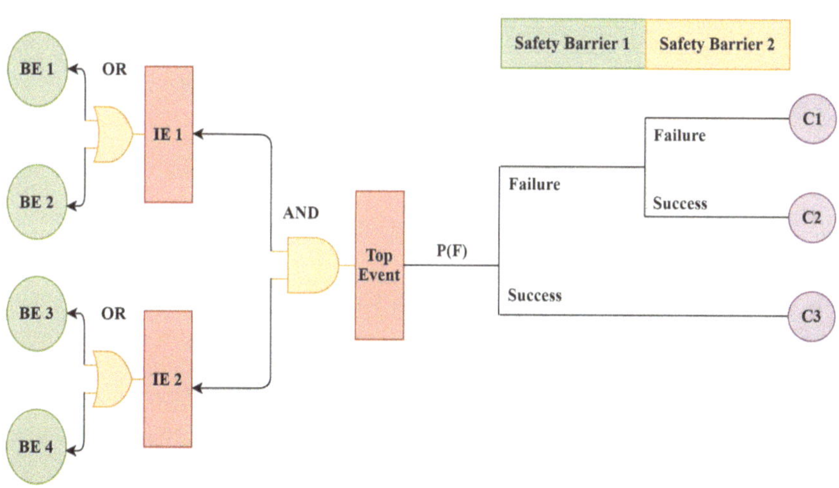

Fig. 8 The bow-tie diagram.

As the bow-tie is a combination of the fault tree and event tree, it suffers from the same limitations that are faced by the FTA and ETA.

The ARAMIS project, a scheme by the European Commission, considered the bow-tie diagram as an integrated part of their proposed risk assessment methodology (De Dianous & Fiévez, 2006). Khakzad, Khan, and Amyotte (2012) demonstrated an application of how the bow-tie could be efficiently utilized in dynamic risk assessment. This methodology considered updating the safety barriers using the Bayes' theorem, upon the availability of the new information which resulted in the dynamically updated risk information as the outputs. Saud, Israni, and Goddard (2014) applied the bow-tie to the downstream and chemical process industries in the United States; the authors stated that the bow-tie could serve as a robust risk management tool. A Bayesian approach-based construction mechanism for the bow-tie diagram was proposed by Badreddine and Amor (2013). The authors concluded that bow-tie could be used to capture real-time system behavior. As previously mentioned, the results of the bow-tie analysis also highly suffer from uncertainty. Many articles have been published that demonstrated how to deal with the uncertainty while conducting the risk assessment and management using the bow-tie approach (Ferdous, Khan, Sadiq, Amyotte, & Veitch, 2013; Lu et al., 2015; Mokhtari, Ren, Roberts, & Wang, 2011). A brief review of the bow-tie method can be found in de Ruijter and Guldenmund (2016).

2.5.4 Bayesian network (BN)

The BN is the most widely adopted tool in the current risk assessment literature due to its robustness to represent the multi-state variables, update upon receiving new information, and represent uncertainty. Another notable feature of a BN is the inference mechanism which enables it to refine its belief. These properties have made the BN more suitable over the FTA, ETA, and bow-tie for dynamic quantitative risk analysis.

Bobbio, Portinale, Minichino, and Ciancamerla (2001) proposed an algorithm to map a fault tree to a BN. Khakzad, Khan, and Amyotte (2011) claimed that the ability of this algorithm is limited while incorporating the dependent failure and functional uncertainty in the BN, and they proposed a new mapping technique to overcome the stated limitations. Later, Khakzad, Khan, and Amyotte (2013) proposed a method to transform a bow-tie into a BN. Rather than probability updating, this work introduced the probability adaption technique to assess the dynamic safety and risk of process operations. Marsh and Bearfield (2008) proposed an algorithm to develop the BN form the ETA. This work demonstrated that multiple event trees could be modeled as a single generalized BN.

In spite of being a robust tool, the traditional BN lacks in capturing the dynamic process behavior and nonlinear dependence among the process variables. According to Murphy (2002), a dynamic Bayesian network can be utilized to model the time dependency. Cai et al. (2013) used the DBN to quantify the human factors during the offshore blowout situations. The pseudo-fault trees were used to structuralize the human factors, and these were translated into the BN and DBN. A DBN-based DRA methodology for domino effects was proposed by Khakzad (2015). Barua, Gao, Pasman, and Mannan (2016) proposed an algorithm to map the DFT to BN and subsequently DBN and introduced a dynamic operational risk assessment methodology. Halabi, Kenett, and Sacerdote (2017) demonstrated how a DBN could be utilized to manage technical risks.

Adedigba, Khan, and Yang (2016a) used the Noisy-OR and Leaky Noisy-OR gates to handle the nonlinear dependence. The authors proposed a nonlinear and non-sequential accident model for process operations using the BN. The sequential nature of an accident was replaced by the non-sequential relationships among the consequences (i.e., safe, accident, mishap, near-miss, incident, and catastrophe). The copula theory was incorporated with the BN by Hashemi, Khan, and Ahmed (2016) which was named as CBN; it could represent the unknown complex nonlinear dependence structures and accommodate noisy process variables. An offshore managed pressure drilling (MPD) was used to demonstrate the superiority of the proposed CBN over the traditional BN. Further applications of copula integrated BN can be found in Ghosh, Ahmed, Khan, and Rusli (2020) and Guo, Khan, and Imtiaz (2019).

2.5.5 Other hybrid approaches

In most of the existing methods, the loss due to a failure is either assumed constant or ignored; more priority is given to assess the probability of failure. However, the loss may vary with time, and the loss functions (LFs) are widely used to model the time-variant losses due to process deviations. Hashemi, Ahmed, and Khan (2014a) developed a risk-based warning generation scheme. The LFs were used to measure the dynamic losses. Zadakbar, Khan, and Imtiaz (2015) proposed an LF-based process risk assessment methodology. Different LFs were used based on the observed fault signals (i.e., step, ramp, and random variation).

Even though the BN is robust to uncertainty, a suitable plugin method is required to handle the events that may create uncertainty. Several articles are published on the integrated fuzzy-BN model to overcome the above-stated problem. Ren, Jenkinson, Wang, Xu, and Yang (2009) proposed a fuzzy-BN (FBN) approach for offshore risk analysis. This method allowed multiple forms of information to be used to quantify model relationships and expert opinion, which provided better flexibility. Zarei, Khakzad, Cozzani, and Reniers (2019) used the fuzzy theory and subject matter specialists' opinion to estimate the probabilities and minimize uncertainty.

The Petri net (PN) and its derivative are also being used in risk assessment. Kadri and Châtelet (2013) proposed a QRA framework for domino effect analysis using the colored stochastic Petri net (CSPN). Zhou, Reniers, and Zhang (2017) proposed a fuzzy-weighted PN-based framework for security risk assessment in the process facilities. This approach could model several relationships between the risk factors and their importance. Recently, Taleb-Berrouane, Khan, and Amyotte (2020) have integrated the BN with the stochastic Petri net (SPN). The SPN captures the dynamic behavior of complex systems while the BN enables to habituate to changes in the data and operating conditions. This hybrid framework has been named as Bayesian Stochastic Petri Nets (BSPN). Like the DBN, it provides a flexible mechanism for safety and reliability assessment.

3. Conclusion

The main aim of this chapter is to present an overview of advanced methods used in chemical process risk assessment and management. This brief review includes documents that are available publicly. It gives the readers an idea of how this field is evolving and what are the main reasons for the growing popularity of risk assessment in recent decades. One of the salient features of this chapter is that it displays the evolution of significant methods and tools used for risk assessment and management in chemical process industries.

Although risk assessment and management were a part of industrial practice from the inception of industrialization, it only became popular after the 1970s due to a number of catastrophic accidents such as the Flixborough explosion in North Lincolnshire, England (1974), Seveso disaster in Seveso, Italy (1976), and Union Carbide plant gas leakage in Bhopal, India (1984). The global regulatory bodies responded to these catastrophes by introducing many regulations to improve industrial safety practices to prevent such events. However, catastrophes kept occurring, which imply that we are still far away from ensuring a comfortable level of safety.

The research community also contributed by developing methods for risk assessment, as it is the most reliable technique to measure the level of safety. These methods are divided into qualitative, semi-quantitative, quantitative, and hybrid approaches for the sake of simplicity. After reviewing several published documents, it can be inferred that the LOPA, risk matrix, risk graph, ETA, FTA, BT, and BN are the vital tools that are widely utilized by researchers to devise new methods. Although the techniques are mostly qualitative and semi-quantitative till the mid of 1990s, the QRA approaches became popular from the late 1990s. It is also noticed that the current research is heavily focusing on developing hybrid methods due to their added advantages over the other approaches. Dynamic risk assessment methods cover most of the contemporary literature, as these techniques can provide an estimation of real-time risk which can be further utilized for risk management.

References

Abdo, H., & Flaus, J. M. (2015). A mixed fuzzy probabilistic approach for risk assessment of dynamic systems. *IFAC-PapersOnLine*, *48*(3), 960–965. https://doi.org/10.1016/j.ifacol.2015.06.207.

Acton, J. M., & Hibbs, M. (2012). *Why Fukushima was preventable*. Washington, DC: Carnegie Endowment for International Peace.

Adedigba, S. A., Khan, F., & Yang, M. (2016a). Dynamic safety analysis of process systems using nonlinear and non-sequential accident model. *Chemical Engineering Research and Design*, *111*, 169–183. https://doi.org/10.1016/j.cherd.2016.04.013.

Adedigba, S. A., Khan, F., & Yang, M. (2016b). Process accident model considering dependency among contributory factors. *Process Safety and Environmental Protection*, *102*, 633–647. https://doi.org/10.1016/j.psep.2016.05.004.

Alileche, N., Olivier, D., Estel, L., & Cozzani, V. (2017). Analysis of domino effect in the process industry using the event tree method. *Safety Science*, *97*, 10–19. https://doi.org/10.1016/j.ssci.2015.12.028.

Al-Sharrah, G. K., Edwards, O., & Hankinson, G. (2007). A new safety risk index for use in petrochemical planning. *Process Safety and Environmental Protection*, *85*(6), 533–540. https://doi.org/10.1205/psep06039.

Amano, Y. (2015). *The Fukushima Daiichi accident*. IAEA.

Amin, M. T., Khan, F., & Amyotte, P. (2019). A bibliometric review of process safety and risk analysis. *Process Safety and Environmental Protection*, *126*, 366–381. https://doi.org/10.1016/j.psep.2019.04.015.

Antonioni, G., Landucci, G., Necci, A., Gheorghiu, D., & Cozzani, V. (2015). Quantitative assessment of risk due to NaTech scenarios caused by floods. *Reliability Engineering and System Safety*, *142*, 334–345. https://doi.org/10.1016/j.ress.2015.05.020.

Antonioni, G., Spadoni, G., & Cozzani, V. (2009). Application of domino effect quantitative risk assessment to an extended industrial area. *Journal of Loss Prevention in the Process Industries*, *22*(5), 614–624. https://doi.org/10.1016/j.jlp.2009.02.012.

Antonsen, S., Nilsen, M., & Almklov, P. G. (2017). Regulating the intangible. Searching for safety culture in the Norwegian petroleum industry. *Safety Science, 92*, 232–240. https://doi.org/10.1016/j.ssci.2016.10.013.

Aven, T. (2008). A semi-quantitative approach to risk analysis, as an alternative to QRAs. *Reliability Engineering and System Safety, 93*(6), 790–797. https://doi.org/10.1016/j.ress.2007.03.025.

Aven, T. (2016). Risk assessment and risk management: Review of recent advances on their foundation. *European Journal of Operational Research, 253*(1), 1–13. https://doi.org/10.1016/j.ejor.2015.12.023.

Badreddine, A., & Amor, N. B. (2013). A Bayesian approach to construct bow tie diagrams for risk evaluation. *Process Safety and Environmental Protection, 91*(3), 159–171. https://doi.org/10.1016/j.psep.2012.03.005.

Baesi, S., Abdolhamidzadeh, B., Hassan, C. R. C., Hamid, M. D., & Reniers, G. (2013). Application of a multi-plant QRA: A case study investigating the risk impact of the construction of a new plant on an existing chemical plant's risk levels. *Journal of Loss Prevention in the Process Industries, 26*(5), 895–903. https://doi.org/10.1016/j.jlp.2012.11.005.

Baraldi, P., & Zio, E. (2008). A combined Monte Carlo and possibilistic approach to uncertainty propagation in event tree analysis. *Risk Analysis, 28*(5), 1309–1326. https://doi.org/10.1111/j.1539-6924.2008.01085.x.

Barua, S., Gao, X., Pasman, H., & Mannan, M. S. (2016). Bayesian network based dynamic operational risk assessment. *Journal of Loss Prevention in the Process Industries, 41*, 399–410. https://doi.org/10.1016/j.jlp.2015.11.024.

Baybutt, P. (2002). Layers of protection analysis for human factors (LOPA-HF). *Process Safety Progress, 21*(2), 119–129. https://doi.org/10.1002/prs.680210208.

Baybutt, P. (2007). An improved risk graph approach for determination of safety integrity levels (SILs). *Process Safety Progress, 26*(1), 66–76. https://doi.org/10.1002/prs.10172.

Bedford, T., & Cooke, R. (2001). *Probabilistic risk analysis: Foundations and methods*. Cambridge University Press.

Belanger, J., Topalovic, P., Krantzberg, G., & West, J. (2009). *Responsible care: History & development*. Canada, ON: McMaster University.

Bird, F. E., Cecchi, F., Tilche, A., & Mata-Alvarez, J. (1974). *Management guide to loss control*. Institute Press.

Bobbio, A., Portinale, L., Minichino, M., & Ciancamerla, E. (2001). Improving the analysis of dependable systems by mapping fault trees into Bayesian networks. *Reliability Engineering & System Safety, 71*(3), 249–260. https://doi.org/10.1016/S0951-8320(00)00077-6.

Borghetti, A., Cozzani, V., Mazzetti, C., Nucci, C. A., Paolone, M., & Renni, E. (2017). *Monte Carlo based lightning risk assessment in oil plant tank farms. In 2010 30th international conference on lightning protection, ICLP 2010* (pp. 1–7). https://doi.org/10.1109/ICLP.2010.7845977.

Boudali, H., & Dugan, J. B. (2005). *A new Bayesian network approach to solve dynamic fault trees. In Proceedings—Annual reliability and maintainability symposium* (pp. 451–456). https://doi.org/10.1109/rams.2005.1408404.

Bridges, W. G., & Williams, T. R. (1997). Risk acceptance criteria and risk judgment tools (LOPA) applied worldwide within a chemical company. In *International conference and workshop on risk analysis in process safety* (pp. 545–557).

Brissaud, F., Barros, A., Bérenguer, C., & Troffaes, M. (2010). Handling parameter and model uncertainties by continuous gates in fault tree analyses. *Proceedings of the Institution of Mechanical Engineers, Part O: Journal of Risk and Reliability, 224*(4), 253–265. https://doi.org/10.1243/1748006XJRR313.

Brockhoff, L., Petersen, H. J. S., & Haastrup, P. (1992). A consequence model for chlorine and ammonia based on a fatality index approach. *Journal of Hazardous Materials*, *29*(3), 405–425. https://doi.org/10.1016/0304-3894(92)85044-2.

Broemeling, L. D. (2011). An account of early statistical inference in Arab cryptology. *American Statistician*, *65*(4), 255–257. https://doi.org/10.1198/tas.2011.10191.

Cai, B., Liu, Y., Zhang, Y., Fan, Q., Liu, Z., & Tian, X. (2013). A dynamic Bayesian networks modeling of human factors on offshore blowouts. *Journal of Loss Prevention in the Process Industries*, *26*(4), 639–649. https://doi.org/10.1016/j.jlp.2013.01.001.

CCPS. (2007). *Guidelines for risk based process safety*. Center for Chemical Process Safety and John Wiley & Sons, Inc.

Čepin, M., & Mavko, B. (2002). A dynamic fault tree. *Reliability Engineering & System Safety*, *75*(1), 83–91. https://doi.org/10.1016/S0951-8320(01)00121-1.

Chen, H., Moan, T., & Verhoeven, H. (2008). Safety of dynamic positioning operations on mobile offshore drilling units. *Reliability Engineering and System Safety*, *93*(7), 1072–1090. https://doi.org/10.1016/j.ress.2007.04.003.

Choi, J. S., & Cho, N. Z. (2007). A practical method for accurate quantification of large fault trees. *Reliability Engineering and System Safety*, *92*(7), 971–982. https://doi.org/10.1016/j.ress.2006.07.005.

Coco, J. C. (2003). The 100 largest losses 1972-2001: Large property damage losses in the hydrocarbon-chemical industries. *Marsh Risk Consulting*.

Covello, V. T., & Mumpower, J. (1985). Risk analysis and risk management: An historical perspective. *Risk Analysis*, *5*(2), 103–120. https://doi.org/10.1111/j.1539-6924.1985.tb00159.x.

Crowl, D. A., & Louvar, J. F. (2011). *Chemical process safety: Fundamentals with applications* (3rd ed.). Prentice Hall.

CSB. (2014). Investigation report volume 1: Explosion and fire at the Macondo Well. In *Vol. 1*. The U.S. Chemical Safety Board.

Curcurù, G., Galante, G. M., & La Fata, C. M. (2012). Epistemic uncertainty in fault tree analysis approached by the evidence theory. *Journal of Loss Prevention in the Process Industries*, *25*(4), 667–676. https://doi.org/10.1016/j.jlp.2012.02.003.

Dahle, I. B., Dybvig, G., Ersdal, G., Guldbrandsen, T., Hanson, B. A., Tharaldsen, J. E., et al. (2012). Major accidents and their consequences for risk regulation. *Advances in Safety, Reliability and Risk Management: ERSEL, 2011*, 33–41.

Davis, S. G., Hinze, P. C., Hansen, O. R., & Van Wingerden, K. (2011). Does your facility have a dust problem: Methods for evaluating dust explosion hazards. *Journal of Loss Prevention in the Process Industries*, *24*(6), 837–846. https://doi.org/10.1016/j.jlp.2011.06.010.

De Dianous, V., & Fiévez, C. (2006). ARAMIS project: A more explicit demonstration of risk control through the use of bow-tie diagrams and the evaluation of safety barrier performance. *Journal of Hazardous Materials*, *130*(3), 220–233. https://doi.org/10.1016/j.jhazmat.2005.07.010.

de Ruijter, A., & Guldenmund, F. (2016). The bowtie method: A review. *Safety Science*, *88*, 211–218. https://doi.org/211-218.

Deacon, T., Amyotte, P. R., & Khan, F. I. (2010). Human error risk analysis in offshore emergencies. *Safety Science*, *48*(6), 803–818. https://doi.org/10.1016/j.ssci.2010.02.013.

Din, V. (1994). *19250. Control technology: Fundamental safety aspects to be considered for measurement and control equipment*. Berlin, Germany: German Institute for Standardization (DIN).

Dowell, A. M. (1997). Layer of protection analysis: A new PHA tool after HAZOP, before fault tree analysis. In *1997 International conference and workshop on risk analysis in process safety*.

Durga Rao, K., Gopika, V., Sanyasi Rao, V. V. S., Kushwaha, H. S., Verma, A. K., & Srividya, A. (2009). Dynamic fault tree analysis using Monte Carlo simulation in probabilistic safety assessment. *Reliability Engineering and System Safety*, *94*(4), 872–883. https://doi.org/10.1016/j.ress.2008.09.007.

Ericson, C. A. (1999). Fault tree analysis—A history. In *The 17th international system safety conference* (pp. 1–9).

Ferdous, R., Khan, F., Sadiq, R., Amyotte, P., & Veitch, B. (2009). Handling data uncertainties in event tree analysis. *Process Safety and Environmental Protection*, *87*(5), 283–292. https://doi.org/10.1016/j.psep.2009.07.003.

Ferdous, R., Khan, F., Sadiq, R., Amyotte, P., & Veitch, B. (2011). Fault and event tree analyses for process systems risk analysis: Uncertainty handling formulations. *Risk Analysis*, *31*(1), 86–107. https://doi.org/10.1111/j.1539-6924.2010.01475.x.

Ferdous, R., Khan, F., Sadiq, R., Amyotte, P., & Veitch, B. (2013). Analyzing system safety and risks under uncertainty using a bow-tie diagram: An innovative approach. *Process Safety and Environmental Protection*, *91*(1–2), 1–18. https://doi.org/10.1016/j.psep.2011.08.010.

Garvey, P. R., & Lansdowne, Z. F. (1998). Risk matrix: An approach for identifying, assessing, and ranking program risks. *Air Force Journal of Logistics*, *22*, 18–21.

Ghosh, A., Ahmed, S., Khan, F., & Rusli, R. (2020). Process safety assessment considering multivariate non-linear dependence among process variables. *Process Safety and Environmental Protection*, *135*, 70–80. https://doi.org/10.1016/j.psep.2019.12.006.

Grier, B. (1980). *One thousand years of mathematical psychology*. Madison, Wis: Society for Mathematical Psychology Convention; Society for Mathematical Psychology Convention.

Grier, B. (1981). In *The early history of the theory and management of risk. Judgment and decision making group meeting, Philadelphia, Pa*.

Guo, C., Khan, F., & Imtiaz, S. (2019). Copula-based Bayesian network model for process system risk assessment. *Process Safety and Environmental Protection*, *123*, 317–326. https://doi.org/10.1016/j.psep.2019.01.022.

Gupta, J. P., & Edwards, D. W. (2003). A simple graphical method for measuring inherent safety. *Journal of Hazardous Materials*, *104*(1–3), 15–30. https://doi.org/10.1016/S0304-3894(03)00231-0.

Guymer, P., Kaiser, G. D., Mc Kelvey, T. C., & Hannaman, G. W. (1987). Probabilistic risk assessment in the CPI. *Chemical Engineering Progress*, *83*(1), 37–45.

Haasl, D. F. (1965). Advanced concepts in fault tree analysis (fault tree construction and probability evaluation techniques used in safety evaluation of rocket launching device). In *System safety symposium, Seattle, Washington, 1965*.

Halabi, A., Kenett, R. S., & Sacerdote, L. (2017). Using dynamic Bayesian networks to model technical risk management efficiency. *Quality and Reliability Engineering International*, *33*(6), 1179–1196. https://doi.org/10.1002/qre.2186.

Hashemi, S. J., Ahmed, S., & Khan, F. (2014a). Loss functions and their applications in process safety assessment. *Process Safety Progress*, 285–291. https://doi.org/10.1002/prs.11659.

Hashemi, S. J., Ahmed, S., & Khan, F. I. (2014b). Risk-based operational performance analysis using loss functions. *Chemical Engineering Science*, *116*, 99–108. https://doi.org/10.1016/j.ces.2014.04.042.

Hashemi, S. J., Khan, F., & Ahmed, S. (2016). Multivariate probabilistic safety analysis of process facilities using the Copula Bayesian Network model. *Computers and Chemical Engineering*, *93*(4), 128–142. https://doi.org/10.1016/j.compchemeng.2016.06.011.

Heinrich, H. W. (1941). Industrial accident prevention. A scientific approach. In *Industrial accident prevention. A scientific approach* (2nd ed.). New York & London: McGraw-Hill Book Company, Inc.

Holmstrom, D., Altamirano, F., Banks, J., Joseph, G., Kaszniak, M., Mackenzie, C., et al. (2006). CSB investigation of the explosions and fire at the BP Texas City refinery on March 23, 2005. *Process Safety Progress*, *25*(4), 345–349. https://doi.org/10.1002/prs.10158.

Hopkins, A. (1999). For whom does safety pay? The case of major accidents. *Safety Science*, *32*(2–3), 143–153. https://doi.org/10.1016/S0925-7535(99)00017-X.

Huang, D., Chen, T., & Wang, M. J. J. (2001). A fuzzy set approach for event tree analysis. *Fuzzy Sets and Systems*, *118*(1), 153–165. https://doi.org/10.1016/S0165-0114(98)00288-7.

Jain, P., Pasman, H. J., Waldram, S. P., Rogers, W. J., & Mannan, M. S. (2017). Did we learn about risk control since Seveso? Yes, we surely did, but is it enough? An historical brief and problem analysis. *Journal of Loss Prevention in the Process Industries*, *49*(Part A), 5–17. https://doi.org/10.1016/j.jlp.2016.09.023.

Johnson, W. G. (1973). The Management Oversight And Risk Tree-MORT. In *Vol.* 7. US Atomic Energy Commission.

Kabir, S., Walker, M., Papadopoulos, Y., Rüde, E., & Securius, P. (2016). Fuzzy temporal fault tree analysis of dynamic systems. *International Journal of Approximate Reasoning*, 77(C), 20–37. https://doi.org/10.1016/j.ijar.2016.05.006.

Kadri, F., & Châtelet, E. (2013). Domino effect analysis and assessment of industrial sites: A review of methodologies and software tools. *International Journal of Computers and Distributed Systems*, *2*(III), 1–10.

Kalantarnia, M., Khan, F., & Hawboldt, K. (2009). Dynamic risk assessment using failure assessment and Bayesian theory. *Journal of Loss Prevention in the Process Industries*, *22*(5), 600–606. https://doi.org/10.1016/j.jlp.2009.04.006.

Kao, C. S., Duh, Y. S., Chen, T. J. H., & Yu, S. W. (2002). An index-based method for assessing exothermic runaway risk. *Process Safety Progress*, *21*(4), 294–304. https://doi.org/10.1002/prs.680210406.

Khakzad, N. (2015). Application of dynamic Bayesian network to risk analysis of domino effects in chemical infrastructures. *Reliability Engineering and System Safety*, *138*, 263–272. https://doi.org/10.1016/j.ress.2015.02.007.

Khakzad, N., Khan, F., & Amyotte, P. (2011). Safety analysis in process facilities: Comparison of fault tree and Bayesian network approaches. *Reliability Engineering & System Safety*, *96*(8), 925–932.

Khakzad, N., Khan, F., & Amyotte, P. (2012). Dynamic risk analysis using bow-tie approach. *Reliability Engineering and System Safety*, *104*, 36–44. https://doi.org/10.1016/j.ress.2012.04.003.

Khakzad, N., Khan, F., & Amyotte, P. (2013). Dynamic safety analysis of process systems by mapping bow-tie into Bayesian network. *Process Safety and Environmental Protection*, *91*(1–2), 46–53. https://doi.org/10.1016/j.psep.2012.01.005.

Khan, F. I., & Abbasi, S. A. (1997a). Accident hazard index: A multi-attribute method for process industry hazard rating. *Process Safety and Environmental Protection*, *75*(4), 217–224. https://doi.org/10.1205/095758297529093.

Khan, F. I., & Abbasi, S. A. (1997b). OptHAZOP—An effective and optimum approach for HAZOP study. *Journal of Loss Prevention in the Process Industries*, *10*(3), 191–204. https://doi.org/10.1016/S0950-4230(97)00002-8.

Khan, F. I., & Abbasi, S. A. (1997c). Risk analysis of a chloralkali industry situated in a populated area using the software package MAXCRED-II. *Process Safety Progress*, *16*(3), 172–184. https://doi.org/10.1002/prs.680160312.

Khan, F. I., & Abbasi, S. A. (1997d). Risk analysis of an epichlorohydrin manufacturing industry using the new computer automated tool MAXCRED. *Journal of Loss Prevention in the Process Industries*, *10*(2), 91–100. https://doi.org/10.1016/S0950-4230(96)00042-3.

Khan, F. I., & Abbasi, S. A. (1997e). TOPHAZOP: A knowledge-based software tool for conducting HAZOP in a rapid, efficient yet inexpensive manner. *Journal of Loss Prevention in the Process Industries*, *10*, 333–343. https://doi.org/10.1016/S0950-4230(97)00023-5.

Khan, F. I., & Abbasi, S. A. (1998a). DOMIFFECT (DOMIno eFFECT): User-friendly software for domino effect analysis. *Environmental Modelling and Software*, *13*(2), 163–177. https://doi.org/10.1016/S1364-8152(98)00018-8.

Khan, F. I., & Abbasi, S. A. (1998b). Multivariate hazard identification and ranking system. *Process Safety Progress, 17*(3), 157–170. https://doi.org/10.1002/prs.680170303.
Khan, F. I., & Abbasi, S. A. (1998c). Techniques and methodologies for risk analysis in chemical process industries. *Journal of Loss Prevention in the Process Industries, 11*(4), 261–277. https://doi.org/10.1016/S0950-4230(97)00051-X.
Khan, F. I., & Abbasi, S. A. (1999a). Assessment of risks posed by chemical industries—Application of a new computer automated tool MAXCRED-III. *Journal of Loss Prevention in the Process Industries, 12*(6), 455–469. https://doi.org/10.1016/S0950-4230(98)00064-3.
Khan, F. I., & Abbasi, S. A. (1999b). HAZDIG: A new software package for assessing the risks of accidental release of toxic chemicals. *Journal of Loss Prevention in the Process Industries, 12*, 167–181. https://doi.org/10.1016/S0950-4230(97)00043-0.
Khan, F. I., & Abbasi, S. A. (1999c). MAXCRED—A new software package for rapid risk assessment in chemical process industries. *Environmental Modelling & Software, 14*(1), 11–25.
Khan, F. I., & Abbasi, S. A. (1999d). PROFAT: A user friendly system for probabilistic fault tree analysis. *Process Safety Progress, 18*, 42–49. https://doi.org/10.1002/prs.680180109.
Khan, F. I., & Abbasi, S. A. (1999e). TORAP—A new tool for conducting rapid risk assessment in petroleum refineries and petrochemical industries. *Journal of Loss Prevention in the Process Industries, 12*(4), 299–313. https://doi.org/10.1016/S0950-4230(98)00063-1.
Khan, F. I., & Abbasi, S. A. (2000). Analytical simulation and PROFAT II: A new methodology and a computer automated tool for fault tree analysis in chemical process industries. *Journal of Hazardous Materials, 75*, 1–27. https://doi.org/10.1016/S0304-3894(00)00169-2.
Khan, F., Abunada, H., John, D., & Benmosbah, T. (2010). Development of risk-based process safety indicators. *Process Safety Progress, 29*(2), 133–143. https://doi.org/10.1002/prs.10354.
Khan, F. I., & Amyotte, P. R. (2008). How to make inherent safety practice a reality. *The Canadian Journal of Chemical Engineering, 81*(1), 2–16. https://doi.org/10.1002/cjce.5450810101.
Khan, F. I., Iqbal, A., & Abbasi, S. A. (2001). Rapid risk assessment of a fertilizer industry using recently developed computer-automated tool TORAP. *Journal of Loss Prevention in the Process Industries, 14*(5), 413–427. https://doi.org/10.1016/S0950-4230(00)00055-3.
Khan, F., Rathnayaka, S., & Ahmed, S. (2015). Methods and models in process safety and risk management: Past, present and future. *Process Safety and Environmental Protection, 98*, 116–147. https://doi.org/10.1016/j.psep.2015.07.005.
Kim, D., Moon, I., Lee, Y., & Yoon, D. (2003). Automatic generation of accident scenarios in domain specific chemical plants. *Journal of Loss Prevention in the Process Industries, 16*(2), 121–132. https://doi.org/10.1016/S0950-4230(02)00111-0.
Kirchsteiger, C. (1999). On the use of probabilistic and deterministic methods in risk analysis. *Journal of Loss Prevention in the Process Industries, 12*(5), 399–419. https://doi.org/10.1016/S0950-4230(99)00012-1.
Klein, J. A. (2009). Two centuries of process safety at DuPont. *Process Safety Progress, 28*(2), 114–122. https://doi.org/10.1002/prs.10309.
Kleindorfer, P., Oktem, U. G., Pariyani, A., & Seider, W. D. (2012). Assessment of catastrophe risk and potential losses in industry. *Computers and Chemical Engineering, 47*, 85–96. https://doi.org/10.1016/j.compchemeng.2012.06.033.
Kletz, T. A. (1983). *HAZOP & HAZAN: Identifying and assessing process industry hazards* (1st ed.). Hardeover.
Kletz, T. A. (1985). Inherently safer plants. *Plant/Operations Progress, 4*(3), 164–167. https://doi.org/10.1002/prsb.720040311.

Kletz, T. A. (1991). Piper alpha: Latest chapter in a long story. *Journal of Loss Prevention in the Process Industries*, *4*(4), 277–278. https://doi.org/10.1016/0950-4230(91)85013-J.

Kletz, T. A. (1992). *HAZOP and HAZAN: Identifying and assessing process industry standards* (3rd ed.). Hemisphere.

Kletz, T. A. (2004). Safety recommendations for small businesses and/or plants. *Process Safety Progress*, *23*(2), 83. https://doi.org/10.1002/prs.10016.

Kujath, M. F., Amyotte, P. R., & Khan, F. I. (2010). A conceptual offshore oil and gas process accident model. *Journal of Loss Prevention in the Process Industries*, *23*(2), 323–330. https://doi.org/10.1016/j.jlp.2009.12.003.

Lavasani, S. M., Zendegani, A., & Celik, M. (2015). An extension to Fuzzy Fault Tree Analysis (FFTA) application in petrochemical process industry. *Process Safety and Environmental Protection*, *93*, 75–88. https://doi.org/10.1016/j.psep.2014.05.001.

Leveson, N. (2004). A new accident model for engineering safer systems. *Safety Science*, *42*(4), 237–270. https://doi.org/10.1016/S0925-7535(03)00047-X.

Limbourg, P., Savić, R., Petersen, J., & Kochs, H. D. (2007). Fault tree analysis in an early design stage using the Dempster-Shafer theory of evidence. In *Proceedings of the European safety and reliability conference 2007, ESREL 2007—Risk, reliability and societal safety* (pp. 713–722).

Limbourg, P., Savić, R., Petersen, J., & Kochs, H. D. (2008). Modelling uncertainty in fault tree analyses using evidence theory. *Proceedings of the Institution of Mechanical Engineers, Part O: Journal of Risk and Reliability*, *222*(3), 291–302. https://doi.org/10.1243/1748006XJRR142.

Long, J., & Fischhoff, B. (2000). Setting risk priorities: A formal model. *Risk Analysis*, *20*(3), 339–352. https://doi.org/10.1111/0272-4332.203033.

Lu, L., Liang, W., Zhang, L., Zhang, H., Lu, Z., & Shan, J. (2015). A comprehensive risk evaluation method for natural gas pipelines by combining a risk matrix with a bow-tie model. *Journal of Natural Gas Science and Engineering*, *25*, 124–133. https://doi.org/10.1016/j.jngse.2015.04.029.

Lundberg, J., Rollenhagen, C., & Hollnagel, E. (2010). What you find is not always what you fix-how other aspects than causes of accidents decide recommendations for remedial actions. *Accident Analysis and Prevention*, *42*(6), 2132–2139. https://doi.org/10.1016/j.aap.2010.07.003.

Macza, M. (2008). A Canadian perspective of the history of process safety management legislation. *Proceedings of the 8th Internationale Symposium Programmable Electronic System in Safety-Related Applications*. 2–3.

Mannan, M. S., Reyes-Valdes, O., Jain, P., Tamim, N., & Ahammad, M. (2016). The evolution of process safety: Current status and future direction. *Annual Review of Chemical and Biomolecular Engineering*, 7, 135–162. https://doi.org/10.1146/annurev-chembioeng-080615-033640.

Markowski, A. S. (2007). exLOPA for explosion risks assessment. *Journal of Hazardous Materials*, *142*(3), 669–676. https://doi.org/10.1016/j.jhazmat.2006.06.070.

Markowski, A. S., & Mannan, M. S. (2010). ExSys-LOPA for the chemical process industry. *Journal of Loss Prevention in the Process Industries*, *23*, 688–696. https://doi.org/10.1016/j.jlp.2010.05.011.

Markowski, A. S., Mannan, M. S., & Bigoszewska, A. (2009). Fuzzy logic for process safety analysis. *Journal of Loss Prevention in the Process Industries*, *22*(6), 695–702. https://doi.org/10.1016/j.jlp.2008.11.011.

Markowski, A. S., Mannan, M. S., Bigoszewska, A., & Siuta, D. (2010). Uncertainty aspects in process safety analysis. *Journal of Loss Prevention in the Process Industries*, *23*(3), 446–454. https://doi.org/10.1016/j.jlp.2010.02.005.

Marseguerra, M., Nutini, M., & Zio, E. (1994). Approximate physical modelling in dynamic PSA using artificial neural networks. *Reliability Engineering and System Safety*, *45*(1–2), 47–56. https://doi.org/10.1016/0951-8320(94)90075-2.

Marseguerra, M., Ricotti, M., & Zio, E. (1995). Approaching system evolution in dynamic PSA by neural networks. *Reliability Engineering and System Safety*, *49*(1), 91–99. https://doi.org/10.1016/0951-8320(95)00039-5.

Marseguerra, M., & Zio, E. (1996). Monte Carlo approach to PSA for dynamic process systems. *Reliability Engineering and System Safety*, *52*(3), 227–241. https://doi.org/10.1016/0951-8320(95)00131-X.

Marsh, D. W. R., & Bearfield, G. (2008). Generalizing event trees using Bayesian networks. *Proceedings of the Institution of Mechanical Engineers, Part O: Journal of Risk and Reliability*, *222*(2), 105–114. https://doi.org/10.1243/1748006XJRR131.

Meng, X., Chen, G., Zhu, G., & Zhu, Y. (2018). Dynamic quantitative risk assessment of accidents induced by leakage on offshore platforms using DEMATEL-BN. *International Journal of Naval Architecture and Ocean Engineering*, *11*(1), 22–32. https://doi.org/10.1016/j.ijnaoe.2017.12.001.

Miri Lavasani, S. M., Yang, Z., Finlay, J., & Wang, J. (2011). Fuzzy risk assessment of oil and gas offshore wells. *Process Safety and Environmental Protection*, *89*(5), 277–294. https://doi.org/10.1016/j.psep.2011.06.006.

Mokhtari, K., Ren, J., Roberts, C., & Wang, J. (2011). Application of a generic bow-tie based risk analysis framework on risk management of sea ports and offshore terminals. *Journal of Hazardous Materials*, *192*(2), 465–475. https://doi.org/10.1016/j.jhazmat.2011.05.035.

Murphy, K. P. (2002). *Dynamic Bayesian networks: Representation, inference and learning*. PhD Thesis, (pp. 1–281). https://doi.org/10.1.1.129.7714.

Ni, H., Chen, A., & Chen, N. (2010). Some extensions on risk matrix approach. *Safety Science*, *48*(10), 1269–1278. https://doi.org/10.1016/j.ssci.2010.04.005.

Nielsen, D. S. (1971). *The cause/consequence diagram method as a basis for quantitative accident analysis*. Risø National Laboratory.

Nielsen, D. S. (1974). *Use of cause-consequence charts in practical systems analysis*. Risø National Laboratory.

Nielsen, D. S., Platz, O., & Runge, B. (1975). A cause-consequence chart of a redundant protection system. *IEEE Transactions on Reliability*, *24*(1), 8–13. https://doi.org/10.1109/TR.1975.5215314.

Ore, O. (1960). Pascal and the invention of probability theory. *The American Mathematical Monthly*, *67*(5), 409–419. https://doi.org/10.2307/2309286.

Rai, S. N., & Krewski, D. (1998). Uncertainty and variability analysis in multiplicative risk models. *Risk Analysis*, *18*(1), 37–45. https://doi.org/10.1111/j.1539-6924.1998.tb00914.x.

Rasmussen, B., & Grønberg, C. D. (1997). Accidents and risk control. *Journal of Loss Prevention in the Process Industries*, *10*(5–6), 325–332. https://doi.org/10.1016/S0950-4230(97)00022-3.

Rathnayaka, S., Khan, F., & Amayotte, P. (2013). Accident modeling and risk assessment framework for safety critical decision-making: Application to deepwater drilling operation. *Proceedings of the Institution of Mechanical Engineers, Part O: Journal of Risk and Reliability*, *227*(1), 86–105. https://doi.org/10.1177/1748006X12472158.

Rauzy, A. (1993). New algorithms for fault trees analysis. *Reliability Engineering and System Safety*, *40*(3), 203–211. https://doi.org/10.1016/0951-8320(93)90060-C.

Reason, J. T. (1990). The contribution of latent human failures to the breakdown of complex systems. *Philosophical Transactions of the Royal Society B*, *327*(1241), 475–484. https://doi.org/10.1098/rstb.1990.0090.

Reay, K. A., & Andrews, J. D. (2002). A fault tree analysis strategy using binary decision diagrams. *Reliability Engineering and System Safety, 78*(1), 45–56. https://doi.org/10.1016/S0951-8320(02)00107-2.
Remenyte-Prescott, R., & Andrews, J. (2008). Analysis of non-coherent fault trees using ternary decision diagrams. *Proceedings of the Institution of Mechanical Engineers, Part O: Journal of Risk and Reliability, 222*(2), 127–138. https://doi.org/10.1243/1748006XJRR154.
Ren, J., Jenkinson, I., Wang, J., Xu, D. L., & Yang, J. B. (2009). An offshore risk analysis method using fuzzy Bayesian network. *Journal of Offshore Mechanics and Arctic Engineering, 131*(4), 041101-1–041101–041112. https://doi.org/10.1115/1.3124123.
Renni, E., Krausmann, E., & Cozzani, V. (2010). Industrial accidents triggered by lightning. *Journal of Hazardous Materials, 184*(1–3), 42–48. https://doi.org/10.1016/j.jhazmat.2010.07.118.
Ruge, B. (2004). Risk matrix as tool for risk assessment in the chemical process industries. *Probabilistic Safety Assessment and Management*, 2693–2698. https://doi.org/10.1007/978-0-85729-410-4_431.
Rusli, R., & Mohd Shariff, A. (2010). Qualitative assessment for inherently safer design (QAISD) at preliminary design stage. *Journal of Loss Prevention in the Process Industries, 23*(1), 157–165. https://doi.org/10.1016/j.jlp.2009.07.005.
Saud, Y. E., Israni, K. C., & Goddard, J. (2014). Bow-tie diagrams in downstream hazard identification and risk assessment. *Process Safety Progress, 33*(1), 26–35. https://doi.org/10.1002/prs.11576.
Scobel, J. H., Theofanous, T. G., & Sorrell, S. W. (1998). Application of the risk oriented accident analysis methodology (ROAAM) to severe accident management in the AP600 advanced light water reactor. *Reliability Engineering and System Safety, 62*(1–2), 51–58. https://doi.org/10.1016/S0951-8320(97)00170-1.
Shafaghi, A. (2008). Equipment failure rate updating-Bayesian estimation. *Journal of Hazardous Materials, 159*(1), 87–91. https://doi.org/10.1016/j.jhazmat.2008.01.042.
Srinivasan, R., & Natarajan, S. (2012). Developments in inherent safety: A review of the progress during 2001-2011 and opportunities ahead. *Process Safety and Environmental Protection, 90*(5), 389–403. https://doi.org/10.1016/j.psep.2012.06.001.
Summers, A. E. (2003). Introduction to layers of protection analysis. Journal of Hazardous Materials, 104(1–3), 163–168. doi: https://doi.org/10.1016/S0304-3894(03)00242-5
Svedung, I., & Rasmussen, J. (2002). Graphic representation of accident scenarios: Mapping system structure and the causation of accidents. *Safety Science, 40*(5), 397–417. https://doi.org/10.1016/S0925-7535(00)00036-9.
Svenson, O. (1991). The accident evolution and barrier function (AEB) model applied to incident analysis in the processing industries. *Risk Analysis, 11*(3), 499–507. https://doi.org/10.1111/j.1539-6924.1991.tb00635.x.
Synolakis, C., & Kânoğlu, U. (2015). The Fukushima accident was preventable. *Philosophical Transactions of the Royal Society A: Mathematical, Physical and Engineering Sciences, 373*(2053), 20140379. https://doi.org/10.1098/rsta.2014.0379.
Taleb-Berrouane, M., Khan, F., & Amyotte, P. (2020). Bayesian Stochastic Petri Nets (BSPN)—A new modelling tool for dynamic safety and reliability analysis. *Reliability Engineering and System Safety, 193*, 106587. https://doi.org/10.1016/j.ress.2019.106587.
Tanaka, H., Fan, L. T., Lai, F. S., & Toguchi, K. (1983). Fault-tree analysis by fuzzy probability. *IEEE Transactions on Reliability, R-32*(5), 453–457. https://doi.org/10.1109/TR.1983.5221727.
Theofanous, T. G. (1996). On the proper formulation of safety goals and assessment of safety margins for rare and high-consequence hazards. *Reliability Engineering and System Safety, 54*(2–3), 243–257. https://doi.org/10.1016/S0951-8320(96)00079-8.
Trimpop, R. M. (1994). *The psychology of risk taking behavior*. Elsevier.

Tugnoli, A., Landucci, G., Salzano, E., & Cozzani, V. (2012). Supporting the selection of process and plant design options by inherent safety KPIs. *Journal of Loss Prevention in the Process Industries*, *25*(5), 830–842. https://doi.org/10.1016/j.jlp.2012.03.008.

Turney, R. (2014). Flixborough: Lessons which are still relevant today. In *Institution of chemical engineers symposium series* (pp. 1–7).

United States Nuclear Regulatory Commission. (1975). *Reactor safety study: An assessment of accident risks in US commercial nuclear power plants*. Vol. 88, NUREG-75/014(WASH-1400).

Villa, V., Paltrinieri, N., Khan, F., & Cozzani, V. (2016). Towards dynamic risk analysis: A review of the risk assessment approach and its limitations in the chemical process industry. *Safety Science*, *89*, 77–93. https://doi.org/10.1016/j.ssci.2016.06.002.

Wei, C., Rogers, W. J., & Mannan, M. S. (2008). Layer of protection analysis for reactive chemical risk assessment. *Journal of Hazardous Materials*, *159*(1), 19–24. https://doi.org/10.1016/j.jhazmat.2008.06.105.

You, X., & Tonon, F. (2012). Event-tree analysis with imprecise probabilities. *Risk Analysis*, *32*(2), 330–344. https://doi.org/10.1111/j.1539-6924.2011.01721.x.

Yu, H., Khan, F., & Garaniya, V. (2015). Risk-based fault detection using self-organizing map. *Reliability Engineering and System Safety*, *139*(82–96), 82–96. https://doi.org/10.1016/j.ress.2015.02.011.

Yuhua, D., & Datao, Y. (2005). Estimation of failure probability of oil and gas transmission pipelines by fuzzy fault tree analysis. *Journal of Loss Prevention in the Process Industries*, *18*(2), 83–88. https://doi.org/10.1016/j.jlp.2004.12.003.

Zadakbar, O., Imtiaz, S., & Khan, F. (2012). Dynamic risk assessment and fault detection using principal component analysis. *Industrial & Engineering Chemistry Research*, *52*, 809–816. https://doi.org/10.1021/ie202880w.

Zadakbar, O., Imtiaz, S., & Khan, F. (2013). Dynamic risk assessment and fault detection using a multivariate technique. *Process Safety Progress*, *32*(4), 365–375. https://doi.org/10.1002/prs.11609.

Zadakbar, O., Khan, F., & Imtiaz, S. (2015). Development of economic consequence methodology for process risk analysis. *Risk Analysis*, *35*(4), 713–731. https://doi.org/10.1111/risa.12313.

Zarei, E., Khakzad, N., Cozzani, V., & Reniers, G. (2019). Safety analysis of process systems using Fuzzy Bayesian Network (FBN). *Journal of Loss Prevention in the Process Industries*, *57*, 7–16. https://doi.org/10.1016/j.jlp.2018.10.011.

Zhou, J., Reniers, G., & Zhang, L. (2017). A weighted fuzzy Petri-net based approach for security risk assessment in the chemical industry. *Chemical Engineering Science*, *174*, 136–145. https://doi.org/10.1016/j.ces.2017.09.002.

CHAPTER TWO

Dynamic risk analysis—Fundamentals

Nicola Paltrinieri[a,*], Faisal I. Khan[b]
[a]Department of Mechanical and Industrial Engineering, Norwegian University of Science and Technology NTNU, Trondheim, Norway
[b]Centre for Risk, Integrity and Safety Engineering (C-RISE), Faculty of Engineering & Applied Science, Memorial University of Newfoundland, St John's, NL, Canada
*Corresponding author: e-mail address: nicola.paltrinieri@ntnu.no

Contents

1. Introduction

One of the most renowned definitions of risk was given by Kaplan and Garrick (1981). It states that risk (R) can be expressed by what can go wrong (scenario s), what likelihood it will have (probability p), and how severe consequences will be (consequence c):

$$R = f(s, p, c) \tag{1}$$

Numerous revisions have been suggested since this fundamental definition. Aven (2012) and Villa, Paltrinieri, Khan, and Cozzani (2016a) provide a thorough review of the risk concept definitions available in literature. Villa et al. (2016a); Villa, Paltrinieri, Khan, and Cozzani (2016b) demonstrate how different risk definitions may ultimately shift the assessment and

Methods in Chemical Process Safety, Volume 4
ISSN 2468-6514
https://doi.org/10.1016/bs.mcps.2020.04.001

management focus. Some authors focus on loss (Willis, 2007), disutility (Campbell, 2005), effect (ISO, 2009, 2018) and outcome (Knight, 1921), and this might give relatively more emphasis on the consequence variable while defining priorities for the application of safety measures (Villa et al., 2016a). On the other hand, other authors stress the terms of likelihood (Kirchsteiger, 2002), potential or possibility (Rowe, 1977), which might give relatively more emphasis on the probability variable while defining priorities for safety measures (Villa et al., 2016a). The different perspectives on the definition of risk and related risk strategies suggest that there is not a single approach, but several paths leading to relatively different results, which may be all beneficial but intrinsically incomplete. This is demonstrated by the occurrence of major accidents whose scenarios were disregarded by safety reports because deemed improbable (Paltrinieri, Dechy, Salzano, Wardman, & Cozzani, 2012).

The issue of realistic risk evaluation is also discussed by Apostolakis (2004) and Creedy (2011), who question the probabilities and frequencies used in risk analysis. They affirm that these values are retrieved from outdated databases and may not fit the target of the study. Moreover, scarcity of data heavenly affects the probability calculation. Landucci et al. (2016a, 2016b), demonstrate how the impact of an unwanted event is influenced by a series of dynamic variables, which are not always considered for its prediction.

How do we know that we are not missing anything in our risk analysis? In order to be aware of this limitation, Aven and Krohn (2014) suggest including the knowledge (k) as new dimension in the definition of risk (R):

$$R = f(s, p, c, k) \tag{2}$$

This new dimension gives important insight on how we should treat risk assessment results and support the continuous improvement of the analysis. Calibration and correction based on new evidence would possibly allow the inclusion of missing details and evolving conditions (Paltrinieri, Comfort, & Reniers, 2019). For this reason, the focus of this contribution is the emerging topic of Dynamic Risk Analysis (DRA), as opposed to traditional risk analysis incapable of reflecting constantly evolving real-world risk.

Some risk analysis steps have the potential to be performed in a dynamic fashion. Inputs and models may be continuously improved, calibrated and corrected, based on new related evidence and lessons learned, in order to progressively approximate to the ever-changing reality. A wider range of data may be considered: not only technical, but also human and organizational information allowing for a proactive perspective. In addition, results may outline the economic impact of risk management in order to promote safety-oriented company strategies (Paltrinieri & Khan, 2016a, 2016b). However, the methodological evolution does not fully reflect the industrial application. DRA fundamentals may miss the chance to consolidate and provide the appropriate basis for a generation of risk analysis methodologies.

This work sheds a light on these fundamentals by addressing the main risk analysis concepts in terms of, respectively, risk, risk analysis and DRA. The latter is further discussed in its main components focusing on hazard identification, frequency analysis and consequence analysis. DRA methodological focus and its features are described by means of figurative examples. The framework in which DRA is encompassed is also described before showing some examples of further associated developments.

2. Dynamic risk analysis concepts

DRA has an evident focus on the concept of risk. However, we should not be misled into thinking that DRA aims at analyzing "dynamic risk." On the contrary, it is the analysis process that is designed to dynamically define risk. For this reason, the following sections provide relevant definitions of the concepts of "risk," "risk analysis" and "dynamic risk analysis" (Fig. 1).

2.1 Risk

Aven and Renn (2009, 2010), report a number of risk definitions from scientific literature, eventually refined by Aven (2012), demonstrating multiple perspectives on the understanding of the concept of risk. Table 1 lists these risk definitions.

As suggested by Villa et al. (2016a), the risk definitions from Table 1 may be grouped in four priority areas, based on the typology of risk considered as worst-case scenario. For instance, the risk definitions from priority area 1

Fig. 1 Components of Dynamic Risk Analysis.

Table 1 List of risk definitions from literature, their focus and priority areas on the consequence/probability diagram (Fig. 2).

Risk definition	Focus	Priority area
Risk equals the expected loss (Willis, 2007). Risk equals the expected disutility (Campbell, 2005).	Expected value (loss)	1
Risk is a situation or event where something of human value (including humans themselves) is at stake and where the outcome is uncertain (Rosa, 1998, 2003). Risk is an uncertain consequence of an event or an activity with respect to something that humans value (IRGC—International Risk Governance Council, 2009).	Event or consequence	
Risk means the likelihood of a specific effect originating from a certain hazard occurring within a specified period or in specified circumstances (Kirchsteiger, 2002).	Probability of an (undesirable) event	2
Risk is the potential for realization of unwanted, negative consequences of an event (Rowe, 1977).	Loss potential/ possibility	
Risk is a measure of the probability and severity of adverse effects (Lowrance, 1976). Risk is defined as a set of scenarios, each of which has a probability/frequency and a consequence—set of triplets. For a given scenario, risk is the product of estimated probability/frequency and event consequences (Kaplan, 1997; Kaplan & Garrick, 1981).	Event estimated frequency/ probability and event consequences	3–4
Risk is equal to the two-dimensional combination of events/consequences and associated uncertainties (will the events occur, what will be the consequences) (Aven, 2007).	Consequences/ damage/ severity and uncertainty	

associate risk mainly with consequence. For this reason, they may lead to identify worst-case scenarios in any point of the elongated priority area 1 in Fig. 1, without distinction on probability. Similarly, the priority area 2 shows where worst-case scenarios characterized by high probability can be identified by following the associated risk descriptions. The priority areas 3 and 4 rely on both consequence and probability to identify worst-case scenarios. However, while priority area 3 is described by

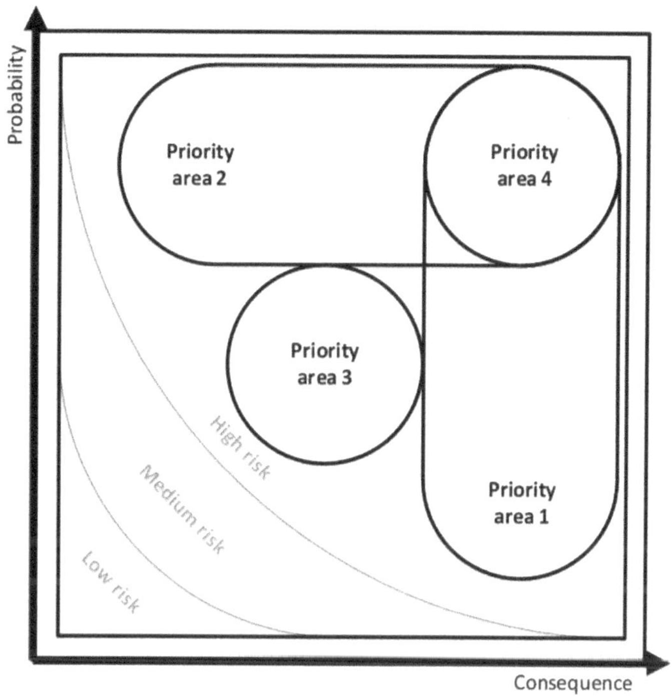

Fig. 2 Priority areas on a consequence/probability diagram.

Villa et al. (2016a) as less detailed, leading to uncalibrated assessment, priority area 4 may allow defining actual high-consequence and high-probability worst-case scenarios.

This indicates that different risk definitions may lead to different approaches for risk assessment and management. Three out of four priority areas may be potentially addressed by the risk strategy classes by Renn and Klinke (2004).

1. Priority on reducing the disaster potential, improving emergency preparedness and increasing resilience.
2. Due to the high probability of occurrence, priority on lessons learned from past events (not included within the original Renn and Klinke's classes).
3. Priority on implementing precautionary principle, improving knowledge and reduction/containment.
4. Priority on consciousness building and risk communication.

These different perspectives on risk definition and the potential associated strategies suggest that there may be several approaches leading to relatively different results, which may be all beneficial but intrinsically incomplete. In order to provide an overall comprehensive strategy, continuous enrichment and improvement of the risk picture are essential.

2.2 Risk analysis

Risk analysis is defined by various sectorial standards (ISO, 2018; NORSOK, 2010), which point at an established number of steps for its accomplishment (Fig. 3).

i. *Risk analysis planning*, which organizes the development of the study, such as objectives, responsibilities, methods, risk acceptance criteria and deliveries (throughout the process and the execution plan), in order to derive full value from the results obtained (Mannan, 2005).

ii. *System definition*, defining clear boundaries for the study and collecting the necessary system information for the risk estimation.

iii. *Hazard identification*, which outlines the potential scenario of what can go wrong—one of the main risk variables of Kaplan and Garrick's definition (1) (Kaplan & Garrick, 1981). The main aims of hazard

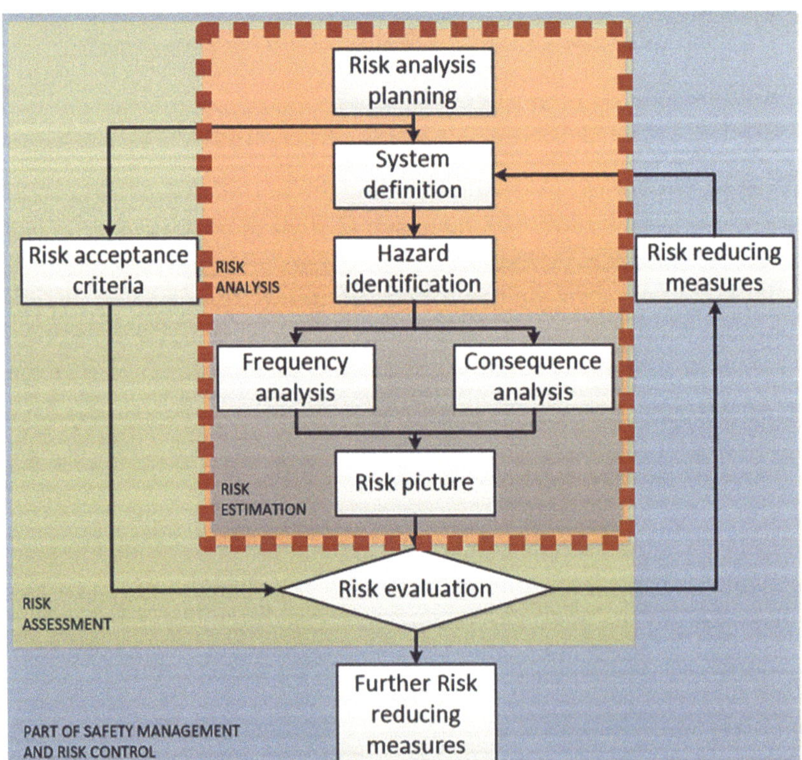

Fig. 3 Steps for risk analysis, estimation, assessment, control and safety management.

identification are highlighting possible malfunctions of the system, undesired situations (such as potentials of release of hazardous substances to the environment), and related consequences. Several approaches can be used for hazard identification (CCPS—Center for Chemical Process Safety, 2000): check lists, preliminary hazard analysis (PHA), failure modes and effects analysis (FMEA), fault tree analysis (FTA), bow-tie analysis, hazard and operability study (HAZOP), etc. Their applicability depends on the project lifecycle, as well as the amount of information required.

iv. *Frequency analysis*, which aims at the defining how likely the scenarios will be and, thus, addressing the second main risk variables of (1). Generic failure rates can be retrieved from specific databases, such as the ones reported by the Guidelines for Quantitative Risk Assessment "Purple Book" (TNO, 2005a). These failure data are largely based on the COVO study (COVO Commission, 1981), as reported by Beerens, Post, & Uijt De Haag (2006). Similarly, the Health & Safety Executive (HSE) provides sets of generic failure frequency data for process installations (HSE—Health & Safety Executive, 2009).

v. *Consequence analysis*, which will model the impact of the scenario and, thus, the last risk variable of (1). Consequence estimation is used to determine the potential for damage or injury from specific unwanted events. Quantification of consequences has been usually carried out in terms of production loss, human health loss, assets loss, and environmental loss (Khan & Haddara, 2004). The assessment of consequences can be performed using physical and empirical models (Arunraj & Maiti, 2009). For instance, consequences estimation for a system handling hazardous substances firstly includes source models, which allow assessing the loss of containment of such substance (e.g., the release rate, the degree of flashing and the evaporation rate) and the related physical effects, such as fires, explosions and toxic dispersions (TNO, 2005b). Damage models are then used to associate the modeled physical phenomena with the actual probability of damage on an asset, such as the probability of human death.

vi. Definition of the *risk picture*, which aims at recomposing frequency and consequence into an overall risk measure. The choice of risk metrics is critical as it directs what kind of information to obtain from the risk analysis and whether the results are considered legitimate and informative by decision-makers and stakeholders (Johansen & Rausand, 2014).

Until this point, the process can be defined as "risk analysis," although the last three steps can also be indicated as risk estimation (Fig. 3). As stated by Crowl & Louvar (2011), risk analysis basically involves the estimation of accident consequences and frequencies using engineering and mathematical techniques. Once the relevant risk metrics are obtained evaluation against risk acceptance criteria allows for risk assessment. The criteria are defined by both regulations and companies. Relevant examples of different acceptability criteria are the one developed by National Agencies (Pasman, 2011), as well as the Risk Matrix developed in the framework of European ARAMIS Project (Salvi & Debray, 2006). In case of unacceptable risk, adequate measures for its reduction are applied (Aven & Vinnem, 2005) throughout the life cycle of the plant. During the concept and engineering phase, risk reduction proposals should be focused on possible alternative process options, different layouts, equipment and location (NORSOK, 2010). During the operational phase, risk reduction proposals on operational nature and modification projects should be evaluated in order to demonstrate that the risk level during operations is the lowest, or provide a justification, if it is not (NORSOK, 2010). This last step concludes the safety management and risk control overall phase.

It must be underlined that risk analysis an iterative process (CCPS—Center for Chemical Process Safety, 2000) and should be updated maximum every 5 years or in case of major plant changes, as stated by the Seveso directive (European Parliament and Council, 2012). Communications and consultation activities, as well as monitoring, review and update activities should be performed throughout the process (NORSOK, 2010).

Being risk analysis only a part of the overall safety management, it should be clear that it does not include evaluation and definition of safety measures, although it may be also used to refer to risk assessment or risk management (Aven et al., 2018). Risks to human health and the environment, both built and natural, are the main interests within this domain (Society for Risk Analysis, 2019). Risks towards individuals may also entail risks to public and private sector organizations, and society at various geographic scales. For this reason, risk research also covers threats from physical, chemical, and biological agents and from a variety of human activities as well as natural events. Communication, consultation, monitoring and review activities should be performed prior to, during and after the analysis, in purpose to guarantee the achievement of its goals (Villa, Paltrinieri, & Cozzani, 2015).

Greenberg et al. (2012) describe some of the accomplishments that risk analysis has had addressing health, safety and the environment. Risk perception and behavior are increasingly affected by analytical evaluation, leading to less concern over technology introduction and more benefits perceived by the public (Siegrist, 2000; Siegrist & Cvetkovich, 2000; Slovic, 2000).

Risk analysis is the foundation for estimating the economic impact of major accidents (Kyaw & Paltrinieri, 2015, 2016a, 2016b; Paltrinieri, Khan, & Cozzani, 2015). Risk analysis provides reliable and useful information to the stakeholders and experts, preventing them to jump to conclusions (National Research Council, 1989). Finally, legal disputes may be resolved by risk-informed decisions, as demonstrated by the trials on the Buncefield explosion in 2005 (Paltrinieri, Dechy, Salzano, Wardman, & Cozzani, 2013).

On the other hand, a number of criticisms associated to risk analysis underline the need for improvements. Creedy (2011) states that the frequencies used by risk analysis are mostly based on outdated values and there is the need for both realistic failure probability values and improved event frequency to describe the actual conditions of the system. Bucelli, Paltrinieri, and Landucci (2018) reminds us that the classic risk analysis approach is static and decomposes the system by focusing on failure events of components. This approach is not sufficient to explain all what can go wrong, because it does not grasp the dynamics of unsafe interactions and fails to capture the variation of risks as deviations or changes in the process and plant (Kalantarnia, Khan, & Hawboldt, 2009; Pasman & Reniers, 2014). Moreover, Creedy (2011) affirms that the estimation of likelihood is typically treated in a cursory way and sometimes even with apparent reluctance, as if examining more deeply could call into question the validity of (or faith in) the risk assessment process. A specific mind-set based on the trust of this quantitative method should be built by involving decision-makers in the creation and refinement of the associated tools.

The accomplishments show the maturity of risk analysis. Its solidity is demonstrated by the wide application to several industrial sectors. However, risk analysis techniques generally the need for improvement, in order to overcome their dependence on outdated frequencies data and little capability of progressive update.

2.3 Dynamic risk analysis

DRA aims to overcome some of the limitations of the traditional risk analysis, in particular its difficulties at reflecting the evolving real-world risk. The adjective "dynamic" has a specific significance in relation to risk analysis: *(of a process or system) characterized by constant change, activity, or progress.*

The DRA concept assumes unique importance with the prevention of High Impact Low Probability (HILP) accidents. Pasman (2015) affirms that events in the process industry involving hazardous materials, such as the ones handled in Seveso sites, represent the most dreadful risk. Loss of control of such substances has the potential to cause HILP accidents (Pasman, 2015). High consequence means catastrophic losses, sometimes with huge loss of human life. On the other hand, such events have may have very low probability and may not happen in our lifetime.

Such events have been widely studied and discussed by several recent related works. For instance, Paltrinieri, Dechy, et al. (2013) and Paltrinieri et al. (2012) define the concept of "atypical accidents," which may address a specific aspect of Pasman's HILP events. Atypical accidents are event that are "not captured by hazard identification methodologies because deviating from normal expectations of unwanted events or worst-case reference scenarios." An atypical accident may occur when hazard identification does not produce a complete overview of system hazards (Paltrinieri, Dechy, et al., 2013; Paltrinieri et al., 2012). Taleb (2007) defines rare catastrophic events that have never been encountered before as Black Swans. According to him, such events can be explained only after the fact and cannot be anticipated. However, Paté-Cornell (2012) and Haugen & Vinnem (2015) warn against the misuse of the Black Swan concept. This should not be a reason for ignoring potential scenarios or waiting until a disaster happens to take safety measures and issue regulations against a predictable situation. On the contrary, it should represent an incentive to continuously learn and improve in a dynamic fashion.

Sornette (2009) uses metaphors to address HILP events by outlining the concept of "Dragon-Kings": "Kings" because such events are extreme and outliers (in analogy with the kings' wealth). "Dragons" because they are unlike anything else, out of the bestiary. One of the Sornette's earlier works addressed the prediction of HILP events such as the earthquakes (Sornette & Sornette, 1989). In these events, some degrees of organization and coordination can serve to amplify the fractures that are always present in the tectonic plates and turn them into large shaking of the Earth surface.

Major accidents are the result of a particular combination of single events, some of which are deviations from normal/optimal conditions. Paltrinieri, Tugnoli, and Cozzani (2016) associate such deviations with the concept of "Small Things." Small things may be recurring old issues in a plant or organization, needing the compliance with already present procedures to be prevented. Acting on Small Things would allow reducing the probability of unfortunate combinations of events leading to more serious and unexpected consequences.

In the last decade, increasing attention has been dedicated to evaluation and monitoring of early deviations through appropriate indicators, as a way to assess and control risk (Øien, Utne, & Herrera, 2011). However, the traditional risk analysis does not allow exploiting the growing amount of information from indicators and DRA is indicated as a possible solution (Khakzad, Khan, & Amyotte, 2013a, 2013b, 2012; Khakzad, Khan, & Paltrinieri, 2014; Villa, Paltrinieri, Khan, & Cozzani, 2016).

Following by Kaplan and Garrick's definition of risk (1) (Kaplan & Garrick, 1981), DRA can be performed on three levels by focusing on one of its three variables: *s*, *p* or *c*. Iteration of hazard identification, frequency analysis and consequence analysis (Fig. 3) would allow for continuous update of the risk picture, and, in turn, optimize the knowledge variable *k*, as expressed in Aven and Krohn's revised definition of risk (2) (Aven & Krohn, 2014).

Paltrinieri and Khan (2016a, 2016b) illustrate several methods that may iterate the identification accident scenarios, and the analysis of frequency and consequence (Fig. 3), in order to support the steps following risk analysis, i.e., evaluation and definition of safety measures. This contribution illustrates only a selection of these dynamic methods.

2.3.1 Dynamic Hazard Identification

As shown by Paltrinieri, Tugnoli, Buston, Wardman, and Cozzani (2013), common hazard identification methods do not always succeed at identifying HILP accident scenarios. Evidence of new hazards may be obtained from early deviations or warnings (Small Things). The method named Dynamic Procedure for Atypical Scenarios Identification (DyPASI) (Paltrinieri, Tugnoli, et al., 2013) was specifically developed to continuously improve risk analysis and prepare for atypical accident scenarios. Atypical accident scenarios may be well known to specialists, but not to the community of safety professionals, since their occurrence may be rare and/or limited to specific industrial activities (Paltrinieri, Tugnoli, & Cozzani, 2015). For this reason, dynamic hazard identification assumes special importance when dealing with new or emerging technologies with relative lack of risk experience.

The application of DyPASI entails a systematic screening process of early warnings and risk notions to update hazard identification. The Bow-Tie Analysis approach, such as the one suggested by the CCPS—Center for Chemical Process Safety (2000) or Delvosalle, Fievez, Pipart, and Debray (2006), is taken as a basis to develop the methodology. Table 2 reports

Table 2 DyPASI steps.

Step	Description
0.	DyPASI requires the application of the conventional bow-tie technique (CCPS—Center for Chemical Process Safety, 2000; Delvosalle et al., 2006) to identify the relevant critical events.
1.	A search for relevant information concerning undetected potential hazards is carried out. Search boundaries must be outlined and quoted in the formulation of the query. Algorithms can be applied to rank the relevance of the results obtained (Zhao, Cui, Zhao, Qiu, & Chen, 2009).
2.	Determination as to whether data are significant to trigger further action and proceed with the process of risk analysis. As a support of this process of prioritization, a register collecting the risk notions obtained from the retrieval process and showing their relative relevance and impact.
3.	Potential scenarios are isolated from the early warnings gathered and a cause-consequence chain consistent with the bow-tie diagram is developed. One or more suitable bow-tie diagrams from step 0 are identified for the process of integration.
4.	Experience on effectiveness and performance of safety barriers is encompassed in the analysis. Bow-tie diagrams are completed with safety barriers (technical, operational and organizational elements which are intended individually or collectively to reduce possibility/ for a specific error, hazard or accident to occur, or which limit its harm/disadvantages (Petroleum Safety Authority, 2013)).

Adapted from Paltrinieri, N., Tugnoli, A., & Cozzani, V., 2015. Hazard identification for innovative LNG regasification technologies. *Reliability Engineering and System Safety 137*, 18–28. https://doi.org/10.1016/j.ress.2014.12.006.

DyPASI steps. Further details can be found elsewhere (Paltrinieri, Landucci, Nelson, & Hauge, 2016; Paltrinieri, Tugnoli, & Cozzani, 2016; Paltrinieri, Tugnoli, et al., 2013).

2.3.2 Dynamic frequency analysis

An example of dynamic frequency analysis can be found in the method developed by Landucci and Paltrinieri (2016a), which uses a frequency modification methodology and is based on technical, operational, and organizational factors. For this reason, it is named TEC2O.

TEC2O is developed for the oil and gas sector and aims at updating the frequency of hydrocarbon release from the system equipment through the following the fundamental equation:

$$F(t) = F_0 \times TMF \times MMF \tag{3}$$

where $F(t)$ is the timely updated accident frequency, t is the time, F_0 is the baseline frequency value, *TMF* is the technical modification factor and *MMF* is the management modification factor. *TMF* and *MMF* are obtained by combining different scores, which are produced by monitoring the quantitative indicators. The TEC2O procedure is depicted in Fig. 4.

A set of indicators is measured and quantified on a standardized discrete scale. The impact of these indicators on the overall release frequency is calculated based on a weighted summation. Being TEC2O focused on the single equipment items included in the risk analysis, the set of indicators is outlined based on the specific equipment characteristics.

TMF aims to synthetically account for the lifecycle of equipment, penalizing "overused" units, which may be more prone to leaks and failure due to aging, erosion and/or corrosion phenomena. Moreover, external factors (environmental issues, seismic zone, and harsh weather areas) are considered. *TMF* only contributes as a worsening element since the failure likelihood of typical mechanical and electrical components or systems increases with time, with an increasing rate that approaches (or in some cases extends) the end of the design life. *TMF* is based on four subfactors, as indicated in Fig. 4A. Periodical monitoring of related indicators statuses enables an average score to be assigned to each of the four subfactors. The weighted combination of the scores enables the total technical score (ε) to be determined and converted to the *TMF*.

The evaluation of the *MMF* is based on the concept of resilience and follows the resilience-based early warning indicators (REWI) methodology (Øien, Massaiu, Tinmannsvik, & Størseth, 2010). Managerial aspects are related to the definition of the safety procedures, training, and competencies of the operators, safety culture, frequency of maintenance operations and communication at different levels of the organization. To introduce a quantitative evaluation of these factors, the REWI method proposes the use of indicators, which are quantitative parameters, so they can be monitored, modified and updated in time. According to Øien et al. (2010), the *MMF* is divided into two main subfactors (Fig. 4B): an operational subfactor and an organizational subfactor. Periodical monitoring of related indicators statuses enables an average score to be assigned to each subfactor. The weighted combination of the scores enables the total technical score (μ), which is converted into the *MMF*.

An extensive description of the method is reported elsewhere (Landucci & Paltrinieri, 2018, 2016a, 2016b; Lee et al. 2019; Paltrinieri, Landucci, & Salvo Rossi, 2017).

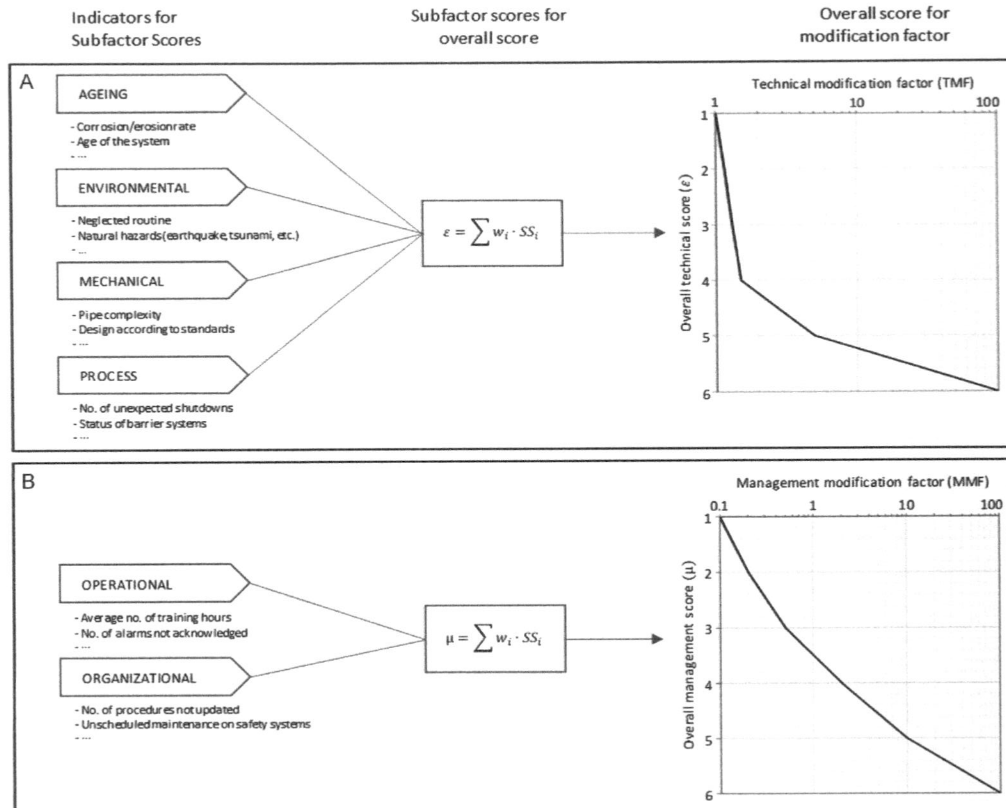

Fig. 4 Schematization of the procedure for the application of technical, operational, and organizational factors (TEC2O): (A) Evaluation of technical modification factor technical modification factor (TMF), (B) evaluation of management modification factor management modification factor (MMF). *Adapted from Lee, S., Landucci, G., Reniers, G., & Paltrinieri, N. (2019). Validation of dynamic risk analysis supporting integrated operations across systems.* Sustainability, 11, *6745. https://doi.org/10.3390/su11236745.*

2.3.3 Dynamic consequences analysis

A simplified DRA method considering runaway reaction potential consequence is proposed by Casson Moreno, Salzano, and Khan (2016). Runaway reactions are typically characterized by an exponential increase of the temperature inside the vessel, when the rate of heat generation becomes faster than the rate of heat removal/losses, with a consequent accumulation of heat and acceleration of the reaction rate (Varma, Morbidelli, & Wu, 2005). This event may be frequent for batch (and semi-batch) chemical processes, in which heat accumulation may occur (Copelli et al., 2014).

Batch processes are used for production of fine and speciality chemicals, which require the manufacturer to be flexible and meet variable market demands (Rippin, 1993). However, different raw materials and products have different behaviors with respect to the runaway reaction consequence. In addition, safety analysis addressing chemical reactions is mainly based on experimental activity due to the relative complexity of the problem—such phenomena might be yet to be fully understood (Stoessel, 2008).

This method considers the consequence of a runaway reaction as related to the heat that can be released from this uncontrolled event. Probability is defined based on reaction evolution, onset temperature of reaction and adiabatic time to maximum rate. A minimum set of "hazard parameters" (Table 3) allows estimating runaway reaction probability on a real-time basis (through reactor temperature) and updating the consequence with specific reaction features whenever a new batch process is started.

Table 4 reports the steps for dynamic analysis of runaway reaction risk. Further details can be found elsewhere (Casson Moreno et al., 2016).

Table 3 Definition of hazard parameters (Group, 2002).

Onset Temperature (T_{onset})	Lowest temperature at which exothermic reaction is observed.
Time to maximum rate (TMR)	Time to achieve the maximum self-heating rate under adiabatic conditions (due to side reactions or decomposition).
Heat evolved by the reaction (ΔH_R)	Heat produced by an exothermic reaction.

Table 4 Steps for dynamic analysis of runaway reaction risk.

Step	Description
1.	Monitor the temperature inside the reactor.
2.	Calculate the rate of temperature dT/dt and compare it with a threshold value (depending on the sensitivity of the equipment).
3.	Onset temperature is recorded for self-heating mode. The adiabatic time to maximum rate is calculated in real time.
4.	The ratio of probability is calculated based on a reference compound congruent with the type of reaction under analysis.
5.	The ratio of severity is calculated based on a reference compound congruent with the type of reaction under analysis.
6.	The thermal risk index is calculated by multiplying the ratios of probability and severity and compared with a predefined risk scale.

3. Methodological focus

Based on Fig. 1 and the related discussion of Section 2, DRA can be also defined as an approach aiming to convert the traditionally static process of risk analysis into a dynamic technique with the capacity to be regularly updated. The focus on the analysis, as depicted in Fig. 3, is intentional and reflects the domain priority of addressing the methodological core of the problem, by excluding the evaluation. The fundamentals of DRA cannot claim to be addressing risk assessment unless the research is related to specific applications for which there is common understanding of tolerability and acceptability of risk levels and a clear operational scope (Yang, Haugen, & Paltrinieri, 2017). At the same time, DRA cannot be treated as a synonym for real-time risk approaches, as the data collection and the frequency based on which the analysis is updated and/or iterated fall outside the purely methodological scope of the research.

An automotive metaphor can be used to describe the DRA fundamental characteristics. A car is a wheeled motor vehicle used for transportation. It is meant to carry the driver from point A to point B. Some cars may be faster than others (e.g., formula one cars), but they can all be considered dynamic as they have the possibility of easily moving and changing their position. Cars are designed to be dynamic and, whether they are used every day, or they are just parked in a garage, they will always keep this characteristic

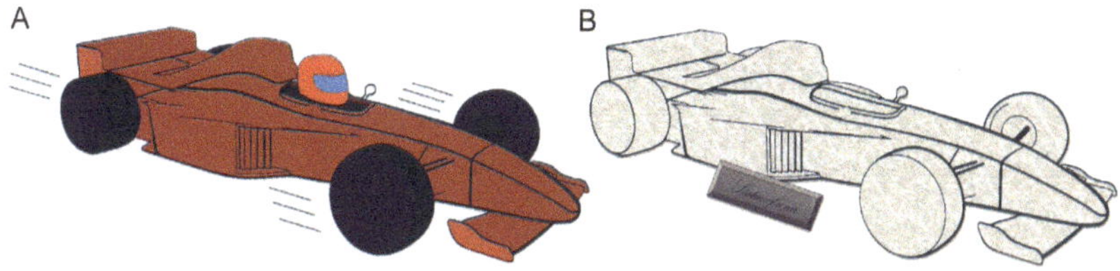

Fig. 5 (A) Formula one car representing the DRA approach; (B) Statue of a car representing the traditional risk analysis.

for which they are developed and produced. Similarly, DRA refers to the risk analysis methods designed to be dynamic, which have the ability to update and iterate when needed. Whether a DRA method provides instantaneous, yearly, or decennial updates, it will always keep its dynamic characteristic as it is independent of the actual use.

Fig. 5 should be considered to further discuss the metaphor. An untrained eye would see two similar cars. However, a better look (and the caption) allows the reader to understand that Fig. 5A depicts a running formula one car representing DRA, while Fig. 5B depicts the statue of a formula one car (such as the one that can be found at the Rond Point du Portier in Monaco) representing traditional static risk analysis.

Despite the resemblance with Fig. 5A and B is not designed to be dynamic, as the only way it has to go from point A to point B is with a lot of effort and the help of a crane. Analogously, a traditional risk analysis performed for a QRA (Quantitative Risk Assessment) has still the possibility to be updated, but this would require a considerable amount of time and energy as these studies are to be carried out by appointed teams of experts and they result in written reports of several hundred pages.

A similar example can be made by means of computer software coding. In fact, when a software code is meant to be updated based on the inputs provided while running, its structure and characteristics should be defined and designed accordingly. We can say that this software code is dynamic. When, instead, a software code does not accept any input for the sake of updates while running, the only option is modifying the lines of code that we want to be changed. We can define the latter code as static. Its update is still possible, but it requires increased effort and knowledge as it was not designed for it.

Thus, if we need to understand whether a method for risk analysis can be considered dynamic, we should ask ourselves whether it was intended and designed to be dynamic, as this characteristic must be taken into account in the definition of the technique scope, i.e., the very first phases of its development.

4. Dynamic Risk Management

DRA is encompassed by an overall management framework, as indicated for risk analysis in Fig. 3. For this reason, a specific Dynamic Risk Management Framework (DRMF) is presented by Grøtan and Paltrinieri (2016), focusing on continuous systematization of information from early signals of risk related to past events. This risk management approach is a development from previous pivotal frameworks (IRGC—International Risk Governance Council, 2009; ISO, 2018; NORSOK, 2010) and provides a support to the identification and assessment of potential atypical accident scenarios related to the substances, the equipment and the site considered, capturing available early warnings or risk notions.

DRMF, whose schematization is reported in Fig. 6, has a characteristic shape that is open to the outside. This shows the need to avoid vicious circles and self-sustained processes and open the process to external experience and early warnings. Following DRMF potentially allows

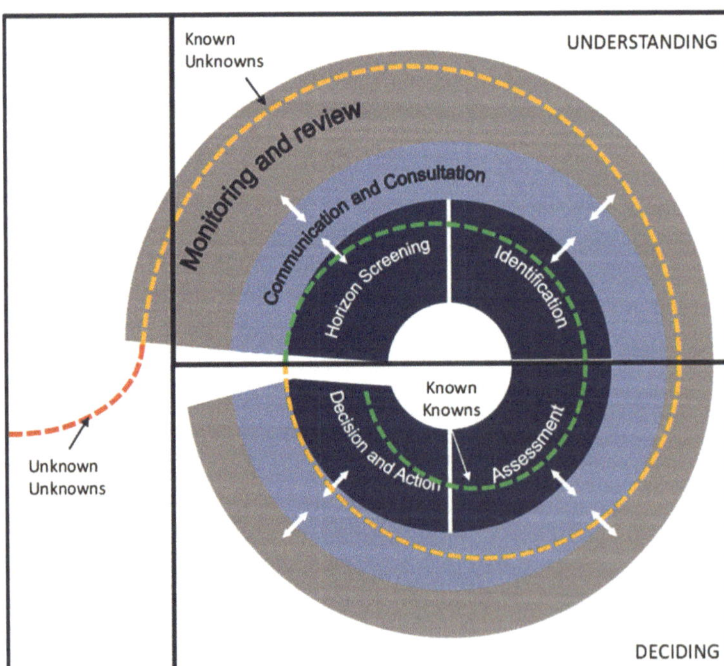

Fig. 6 Representation of the Dynamic Risk Management Framework—DRMF (Paltrinieri, Khan, Amyotte, & Cozzani, 2014).

integration of information on accident scenarios that we may be not aware we do not know (unknown unknowns—red line). Increased awareness of emerging risks, related to events we are now aware we do not know (known unknowns—yellow line), hypothetically leads to a learning and understanding phase (based on monitoring and review of accumulated information), supporting a following phase of decision on whether to assess such risks (based on communication and consultation). Horizon Screening, Hazard Identification, Assessment and final Decision/Action are the steps needed by DRMF to thoroughly evaluate the risks associated to potential accident scenarios, which we are eventually aware we know (known knowns—green line) and can manage. There is no end to the process, but continuous reiteration, in order to keep track of changes and process them for more effective and dynamic management of risk.

5. Further developments

The call for continuous risk analysis expressed by DRA has led to the definition of techniques to process relevant data, which are to be coupled with adequate capability to deal with unexpected events and provide the right support to enable risk management. For instance, Paltrinieri et al. (2019) suggest a risk assessment approach based on machine learning, as illustrated in Fig. 7. However, intrinsic model limitations should be considered, and appropriate model selection and customization should be carefully carried out to deliver appropriate support for safety-related decision-making.

For instance, indicators reporting the system performance on a regular basis represent an opportunity to consider changes and evolutions, and continuously update risk assessment. Considering operational and organizational factors (e.g., number of simulator hours carried out by the DPO in the last 3 months), in addition to technical ones (e.g., the number of thruster controls failures in the last 3 months), aims at producing proactive risk evaluation (Paltrinieri, Landucci, Nelson, & Hauge, 2016, Paltrinieri, Tugnoli, & Cozzani, 2016, Scarponi & Paltrinieri, 2016). Nevertheless, indicators may reflect different projections in time. A technical failure may be directly associated to the accident development, while early operational/organizational deviations have a lower degree of causality and may be disregarded and not registered. Moreover, operational and organizational indicators rely on personnel's feedback and may be collected less frequently than technical ones. For this reason, sparsity of data may be especially

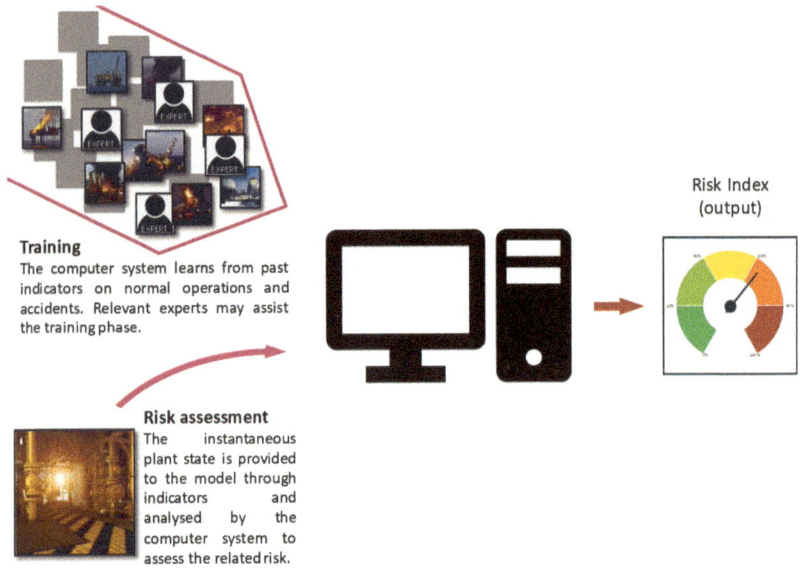

Fig. 7 Representation of a machine learning approach for risk assessment. *Adapted from Paltrinieri, N., Comfort, L., & Reniers, G., 2019. Learning about risk: Machine learning for risk assessment.* Safety Science 118, *475–486. https://doi.org/10.1016/j.ssci.2019.06.001.*

encountered for operational and organizational indicators, and this may undermine the dynamic capabilities of the model.

Moreover, it must be mentioned that some machine learning techniques, such as deep learning (Fig. 7), may have limitations concerning dynamicity. In fact, every time a new set of indicators is available, the model needs to be re-trained and updated through a computationally expensive process. To tackle this challenge, a warm-starting system may be implemented (Cheng et al., 2016), which avoids redesigning a solution from scratch and lay the groundwork for updates.

6. Conclusions

DRA is an emerging topic within the scientific risk literature as risk analysis is constantly affected by uncertainties on inputs, models and results. In fact, this confirms the need for continuous improvement, calibration and correction, based on new related evidence and lessons learned, with the purpose to progressively approximate to the ever-changing reality. This contribution discusses the DRA fundamentals in a systematic fashion.

The foundations of risk approaches are identified in the understanding of the concept of risk, as its very definition may affect the overall analysis. Having the knowledge variable entered the formula for risk definition, DRA assumes an even stronger role in ensuring appropriate risk evaluations. However, traditional techniques are not able to provide such a shift towards continuous methods. For this reason, DRA focus has been directed to the steps of risk analysis that reflect the main variables of the risk definition: hazard identification, frequency analysis and consequence analysis.

DRA has, thus, a primarily methodological focus, as it considers risk analysis and does not trespass into the assessment. The methodological focus is also reflected by the fact that DRA research is not concentrating on the use of dynamic methodologies. In other words, DRA main objective is not how often risk analysis is iterated and whether it is real-time, but it is rather its capability to be updated, considered as a basic feature of the method design.

DRA is encompassed in an overall framework for risk management. One example is provided by DRMF and described in this contribution in order to show how such dynamic an ever-improving attitude may be also adopted on a higher level. Within this framework, further techniques are emerging from the field of machine learning. These developments have the potential to lay the foundations for a second generation of DRA methods, but they are still characterized by intrinsic limitations and need dedicated research to become reliable risk tools.

References

Apostolakis, G. E. (2004). How useful is quantitative risk assessment? *Risk Analysis*, *24*, 515–520. https://doi.org/10.1111/j.0272-4332.2004.00455.x.

Arunraj, N. S., & Maiti, J. (2009). A methodology for overall consequence modeling in chemical industry. *Journal of Hazardous Materials*, *169*, 556–574. https://doi.org/10.1016/j.jhazmat.2009.03.133.

Aven, T. (2007). A unified framework for risk and vulnerability analysis covering both safety and security. *Reliability Engineering and System Safety*, *92*, 745–754. https://doi.org/10.1016/j.ress.2006.03.008.

Aven, T. (2012). The risk concept—Historical and recent development trends. *Reliability Engineering and System Safety*, *99*, 33–44. https://doi.org/10.1016/j.ress.2011.11.006.

Aven, T., Andersen, H. B., Cox, T., Droguett, E. L., Greenberg, M., Guikema, S., et al. (2018). *Core subjects of risk analysis*. Herndon, VA, USA: Society for Risk Analysis.

Aven, T., & Krohn, B. S. (2014). A new perspective on how to understand, assess and manage risk and the unforeseen. *Reliability Engineering and System Safety*, *121*, 1–10.

Aven, T., & Renn, O. (2009). On risk defined as an event where the outcome is uncertain. *Journal of Risk Research*, *12*, 1–11. https://doi.org/10.1080/13669870802488883.

Aven, T., & Renn, O. (2010). Response to Professor Eugene Rosa's viewpoint to our paper. *Journal of Risk Research*, *13*, 255–259. https://doi.org/10.1080/13669870903484369.

Aven, T., & Vinnem, J. E. (2005). On the use of risk acceptance criteria in the offshore oil and gas industry. *Reliability Engineering and System Safety*, *90*, 15–24. https://doi.org/10.1016/j.ress.2004.10.009.

Beerens, H. I., Post, J. G., & Uijt De Haag, P. A. M. (2006). The use of generic failure frequencies in QRA: The quality and use of failure frequencies and how to bring them up-to-date. *Journal of Hazardous Materials*, *130*, 265–270. https://doi.org/10.1016/j.jhazmat.2005.07.013.

Bucelli, M., Paltrinieri, N., & Landucci, G. (2018). Integrated risk assessment for oil and gas installations in sensitive areas. *Ocean Engineering*, *150*, 377–390. https://doi.org/10.1016/j.oceaneng.2017.12.035.

Campbell, S. (2005). Determining overall risk. *Journal of Risk Research*, *8*, 569–581. https://doi.org/10.1080/1366987050018329.

Casson Moreno, V., Salzano, E., & Khan, F. (2016). Chapter 11—assessing the severity of runaway reactions. In N. Paltrinieri & F. Khan (Eds.), *Dynamic risk analysis in the chemical and petroleum industry* (pp. 127–138). Butterworth-Heinemann. https://doi.org/10.1016/B978-0-12-803765-2.00011-1.

CCPS—Center for Chemical Process Safety. (2000). Guideline for chemical process quantitative risk analysis. New York, NY: American Institute of Chemical Engineers—Center of Chemical Process Safety.

Cheng, H.-T., Koc, L., Harmsen, J., Shaked, T., Chandra, T., Aradhye, H., et al. (2016). Wide & deep learning for recommender systems. In *Proceedings of the 1st workshop on deep learning for recommender systems* (pp. 7–10). ACM.

Copelli, S., Torretta, V., Pasturenzi, C., Derudi, M., Cattaneo, C. S., & Rota, R. (2014). On the divergence criterion for runaway detection: Application to complex controlled systems. *Journal of Loss Prevention in the Process Industries*, *28*, 92–100. https://doi.org/10.1016/j.jlp.2013.05.004.

COVO Commission. (1981). *Risk analysis of six potentially hazardous object in the Rijnmond area, a pilot study*. Schiedam, the Netherlands: Report to the Rijnmond Public Authority, Central Environmental Control Agency.

Creedy, G. D. (2011). Quantitative risk assessment: How realistic are those frequency assumptions? *Journal of Loss Prevention in the Process Industries*, *24*, 203–207. https://doi.org/10.1016/j.jlp.2010.08.013.

Crowl, D. A., & Louvar, J. F. (2011). *Chemical process safety: Fundamentals with applications* (3rd ed.). Boston, MA, United States of America: Prentice Hall.

Delvosalle, C., Fievez, C., Pipart, A., & Debray, B. (2006). ARAMIS project: A comprehensive methodology for the identification of reference accident scenarios in process industries. *Journal of Hazardous Materials*, *130*, 200–219.

European Parliament and Council. (2012). Directive 2012/18/EU of 4 July 2012 on the control of major-accident hazards involving dangerous substances, amending and subsequently repealing council directive 96/82/EC—Seveso III. *Official Journal of the European Union*, *197*, 1–37.

Greenberg, M., Haas, C., Cox, A., Lowrie, K., McComas, K., & North, W. (2012). Ten most important accomplishments in risk analysis, 1980-2010. *Risk Analysis*, *32*, 771–781. https://doi.org/10.1111/j.1539-6924.2012.01817.x.

Grøtan, T. O., & Paltrinieri, N. (2016). Dynamic risk management in the perspective of a resilient system. In N. Paltrinieri & F. Khan (Eds.), *Dynamic risk analysis in the chemical and petroleum industry: Evolution and interaction with parallel disciplines in the perspective of industrial application* (pp. 245–257). Butterworth-Heinemann. https://doi.org/10.1016/B978-0-12-803765-2.00020-2.

Group, H.W. (2002). *HarsBook: A technical guide for the assessment of highly reactive chemical systems, Safety engineering in practice*. DECHEMA.

Haugen, S., & Vinnem, J. E. (2015). Perspectives on risk and the unforeseen. *Reliability Engineering and System Safety*, *137*, 1–5.

HSE—Health & Safety Executive. (2009). *Failure rate and event data for use within land use planning risk assessments*. HSE Books.

IRGC—International Risk Governance Council. (2009). *Risk governance deficits. An analysis and illustration of the most common deficits in risk governance*. Geneva: Risk Governance Council.

ISO. (2009). *Risk management—Vocabulary. Guide 2009*. ISO.

ISO. (2018). *Risk management. ISO 31000:2018*. ISO.

Johansen, I. L., & Rausand, M. (2014). Foundations and choice of risk metrics. *Safety Science*, *62*, 386–399. https://doi.org/10.1016/j.ssci.2013.09.011.

Kalantarnia, M., Khan, F., & Hawboldt, K. (2009). Dynamic risk assessment using failure assessment and Bayesian theory. *Journal of Loss Prevention in the Process Industries* https://doi.org/10.1016/j.jlp.2009.04.006.

Kaplan, S. (1997). The words of risk analysis. *Risk Analysis*, *17*, 407–417. https://doi.org/10.1111/j.1539-6924.1997.tb00881.x.

Kaplan, S., & Garrick, B. J. (1981). On the quantitative definition of risk. *Risk Analysis*, *1*, 11–27. https://doi.org/10.1111/j.1539-6924.1981.tb01350.x.

Khakzad, N., Khan, F., & Amyotte, P. (2012). Dynamic risk analysis using bow-tie approach. *Reliability Engineering and System Safety*, *104*, 36–44. https://doi.org/10.1016/j.ress.2012.04.003.

Khakzad, N., Khan, F., & Amyotte, P. (2013a). Quantitative risk analysis of offshore drilling operations: A Bayesian approach. *Safety Science*, *57*, 108–117. https://doi.org/10.1016/J.SSCI.2013.01.022.

Khakzad, N., Khan, F., & Amyotte, P. (2013b). Dynamic safety analysis of process systems by mapping bow-tie into Bayesian network. *Process Safety and Environment Protection*, *91*, 46–53. https://doi.org/10.1016/j.psep.2012.01.005.

Khakzad, N., Khan, F., & Paltrinieri, N. (2014). On the application of near accident data to risk analysis of major accidents. *Reliability Engineering and System Safety*, *126*. https://doi.org/10.1016/j.ress.2014.01.015.

Khan, F., & Haddara, M. R. (2004). Risk-based maintenance (RBM): A new approach for process plant inspection and maintenance. *Process Safety Progress*, *23*, 252–265. https://doi.org/10.1002/prs.10010.

Kirchsteiger, C. (2002). Preface. International workshop on promotion of technical harmonisation on risk-based decision-making. *Safety Science*, *40*, 1–15.

Knight, F. H. (1921). *Risk, uncertainty, and profit*. Boston: Houghton Mifflin Co.

Kyaw, K., & Paltrinieri, N. (2015). The cost of reputational damage when a major accident occurs. In *Safety and reliability of complex engineered systems—Proceedings of the 25th European safety and reliability conference, ESREL 2015*.

Kyaw, K., & Paltrinieri, N. (2016a). Reputational damage after major accidents. In N. Paltrinieri & F. Khan (Eds.), *Dynamic risk analysis in the chemical and petroleum industry: Evolution and interaction with parallel disciplines in the perspective of industrial application* (pp. 221–229). Butterworth-Heinemann. https://doi.org/10.1016/B978-0-12-803765-2.00018-4.

Kyaw, K., & Paltrinieri, N. (2016b). Estimation of reputational damage from major accidents: Tutorial and examples. In N. Paltrinieri & F. Khan (Eds.), *Dynamic risk analysis in the chemical and petroleum industry: Evolution and interaction with parallel disciplines in the perspective of industrial application* (pp. 231–241). Butterworth-Heinemann. https://doi.org/10.1016/B978-0-12-803765-2.00019-6.

Landucci, G., & Paltrinieri, N. (2016a). A methodology for frequency tailorization dedicated to the oil & gas sector. *Process Safety and Environment Protection*, *104*, 123–141. https://doi.org/10.1016/j.psep.2016.08.012.

Landucci, G., & Paltrinieri, N. (2016b). Dynamic evaluation of risk: From safety indicators to proactive techniques. *Chemical Engineering Transactions*, *53*, 169–174. https://doi.org/10.3303/CET1653029.

Landucci, G., & Paltrinieri, N. (2018). Proactive monitoring of risk-based indicators: Example of application in the oil & gas integrated operations. In *Institution of chemical engineers symposium series*.

Landucci, G., Pontiggia, M., Paltrinieri, N., & Cozzani, V. (2016a). Dynamic consequence analysis through computational fluid dynamics modeling. In N. Paltrinieri & F. Khan (Eds.), *Dynamic risk analysis in the chemical and petroleum industry: Evolution and interaction with parallel disciplines in the perspective of industrial application* (pp. 105–114). Butterworth-Heinemann. https://doi.org/10.1016/B978-0-12-803765-2.00009-3.

Landucci, G., Pontiggia, M., Paltrinieri, N., & Cozzani, V. (2016b). Computational fluid dynamics modeling: Tutorial and examples. In N. Paltrinieri & F. Khan (Eds.), *Dynamic risk analysis in the chemical and petroleum industry: Evolution and interaction with parallel disciplines in the perspective of industrial application* (pp. 115–125). Butterworth-Heinemann. https://doi.org/10.1016/B978-0-12-803765-2.00010-X.

Lee, S., Landucci, G., Reniers, G., & Paltrinieri, N. (2019). Validation of dynamic risk analysis supporting integrated operations across systems. *Sustainability*, *11*, 6745. https://doi.org/10.3390/su11236745.

Lowrance, W.W., 1976. Of acceptable risk—Science and the determination of safety., William Ka. Ed. ERIC, Los Altos, CA.

Mannan, M. S. (2005). Lees's loss prevention in the process industries (3rd ed.). Burlington, MA: Elsevier Butterworth-Heinemann.

National Research Council. (1989). Improving risk communication. Washington, DC, USA: The National Academies Press. https://doi.org/10.17226/1189.

NORSOK. (2010). Standard Z-013, risk and emergency preparedness analysis (3rd ed.). Standards Norway, Lysaker, Norway.

Øien, K., Massaiu, S., Tinmannsvik, R. K., & Størseth, F. (2010). Development of early warning indicators based on resilience engineering. In *10th international conference on probabilistic safety assessment and management 2010, PSAM 2010* (pp. 1762–1771). .

Øien, K., Utne, I. B., & Herrera, I. A. (2011). Building safety indicators: Part 1—Theoretical foundation. *Safety Science*, *49*, 148–161. https://doi.org/10.1016/j.ssci.2010.05.012.

Paltrinieri, N., Comfort, L., & Reniers, G. (2019). Learning about risk: Machine learning for risk assessment. *Safety Science*, *118*, 475–486. https://doi.org/10.1016/j.ssci.2019.06.001.

Paltrinieri, N., Dechy, N., Salzano, E., Wardman, M., & Cozzani, V. (2012). Lessons learned from Toulouse and Buncefield disasters: From risk analysis failures to the identification of atypical scenarios through a better knowledge management. *Risk Analysis*, *32*, 1404–1419. https://doi.org/10.1111/j.1539-6924.2011.01749.x.

Paltrinieri, N., Dechy, N., Salzano, E., Wardman, M., & Cozzani, V. (2013). Towards a new approach for the identification of atypical accident scenarios. *Journal Risk Research*, *16*, 337–354. https://doi.org/10.1080/13669877.2012.729518.

Paltrinieri, N., & Khan, F. (2016a). Dynamic risk analysis in the chemical and petroleum industry: Evolution and interaction with parallel disciplines in the perspective of industrial application (1st ed.). Butterworth-Heinemann. https://doi.org/10.1016/B978-0-12-803765-2.01001-5.

Paltrinieri, N., & Khan, F. (2016b). New definitions of old issues and need for continuous improvement. In N. Paltrinieri & F. Khan (Eds.), *Dynamic risk analysis in the chemical and petroleum industry: Evolution and interaction with parallel disciplines in the perspective of industrial application* (pp. 13–21). Butterworth-Heinemann. https://doi.org/10.1016/B978-0-12-803765-2.00002-0.

Paltrinieri, N., Khan, F., Amyotte, P., & Cozzani, V. (2014). Dynamic approach to risk management: Application to the Hoeganaes metal dust accidents. *Process Safety and Environment Protection*, *92*, 669–679. https://doi.org/10.1016/j.psep.2013.11.008.

Paltrinieri, N., Khan, F., & Cozzani, V. (2015). Coupling of advanced techniques for dynamic risk management. *Journal Risk Research, 18*, 910–930. https://doi.org/10.1080/13669877.2014.919515.

Paltrinieri, N., Landucci, G., Nelson, W. R., & Hauge, S. (2016). Proactive approaches of dynamic risk assessment based on indicators. In N. Paltrinieri & F. Khan (Eds.), *Dynamic risk analysis in the chemical and petroleum industry: Evolution and interaction with parallel disciplines in the perspective of industrial application* (pp. 63–73). Butterworth-Heinemann. *https://doi.org/10.1016/B978-0-12-803765-2.00006-8*.

Paltrinieri, N., Landucci, G., & Salvo Rossi, P. (2017). Real-time data for risk assessment in the offshore oil&gas industry. In *Proceedings of the international conference on offshore mechanics and Arctic engineering—OMAE*. https://doi.org/10.1115/OMAE201761486.

Paltrinieri, N., Tugnoli, A., Buston, J., Wardman, M., & Cozzani, V. (2013). Dynamic procedure for atypical scenarios identification (DyPASI): A new systematic HAZID tool. *Journal of Loss Prevention in the Process Industries, 26*, 683–695. https://doi.org/10.1016/j.jlp.2013.01.006.

Paltrinieri, N., Tugnoli, A., & Cozzani, V. (2015). Hazard identification for innovative LNG regasification technologies. *Reliability Engineering and System Safety, 137*, 18–28. https://doi.org/10.1016/j.ress.2014.12.006.

Paltrinieri, N., Tugnoli, A., & Cozzani, V. (2016). Chapter 4—Dynamic hazard identification: Tutorial and examples. In N. Paltrinieri & F. Khan (Eds.), *Dynamic risk analysis in the chemical and petroleum industry* (pp. 37–48). Butterworth-Heinemann. https://doi.org/10.1016/B978-0-12-803765-2.00004-4.

Pasman, H. J. (2011). History of Dutch process equipment failure frequencies and the purple book. *Journal of Loss Prevention in the Process Industries, 24*, 208–213. https://doi.org/10.1016/j.jlp.2010.08.012.

Pasman, H. J. (2015). Risk analysis and control for industrial processes—Gas, oil and chemicals: A system perspective for assessing and avoiding low-probability, high-consequence events. Elsevier Science.

Pasman, H., & Reniers, G. (2014). Past, present and future of quantitative risk assessment (QRA) and the incentive it obtained from land-use planning (LUP). *Journal of Loss Prevention in the Process Industries, 28*, 2–9.

Paté-Cornell, E. (2012). On "black swans" and "perfect storms": Risk analysis and management when statistics are not enough. *Risk Analysis, 32*, 1823–1833. https://doi.org/10.1111/j.1539-6924.2011.01787.x.

Petroleum Safety Authority. (2013). Principles for barrier management in the petroleum industry. Stavanger, Norway: PSA.

Renn, O., & Klinke, A. (2004). Systemic risks: A new challenge for risk management. *EMBO Reports, 5*, S41–S46. Spec No.

Rippin, D. W. T. (1993). Batch process systems engineering: A retrospective and prospective review. *Computers and Chemical Engineering, 17*, S1–S13. https://doi.org/10.1016/0098-1354(93)80201-W.

Rosa, E. A. (1998). Metatheoretical foundations for post-normal risk. *Journal Risk Research, 1*, 15–44. https://doi.org/10.1080/136698798377303.

Rosa, E. A. (2003). The logical structure of the social amplification of risk framework (SARF): Metatheoretical foundation and policy implications. In N. Pidgeon, R. E. Kaspersen, & P. Slovic (Eds.), *The social amplification of risk* (pp. 47–79). Cambridge, UK: Cambridge University Press.

Rowe, W. D. (1977). *An anatomy of risk*. New York: John Wiley & Sons.

Salvi, O., & Debray, B. (2006). A global view on ARAMIS, a risk assessment methodology for industries in the framework of the SEVESO II directive. *Journal of Hazardous Materials, 130*, 187–199. https://doi.org/10.1016/j.jhazmat.2005.07.034.

Scarponi, G. E., & Paltrinieri, N. (2016). Comparison and complementarity between reactive and proactive approaches. In N. Paltrinieri & F. Khan (Eds.), *Dynamic risk analysis in the chemical and petroleum industry: Evolution and interaction with parallel disciplines in the perspective of industrial application* (pp. 93–101). Butterworth-Heinemann. https://doi.org/10.1016/B978-0-12-803765-2.00008-1.

Siegrist, M. (2000). The influence of trust and perceptions of risks and benefits on the acceptance of Gene Technology. *Risk Analysis*, *20*, 195–204. https://doi.org/10.1111/0272-4332.202020.

Siegrist, M., & Cvetkovich, G. (2000). Perception of hazards: The role of social trust and knowledge. *Risk Analysis*, *20*, 713–719. https://doi.org/10.1111/0272-4332.205064.

Slovic, P. (2000). Trust, emotion, sex, politics, and science: Surveying the risk-assessment battlefield. *Risk Analysis*, *19*, 390–412. https://doi.org/10.1111/j.1539-6924.1999.tb00439.x.

Society for Risk Analysis. (2019). *SRA website*. Herndon, VA, USA: Society for Risk Analysis.

Sornette, D. (2009). Dragon-kings, black swans, and the prediction of crises. *International Journal of Terraspace Science and Engineering*, *2*, 1–18.

Sornette, A., & Sornette, D. (1989). Self-organized criticality and earthquakes. *Europhysics Letters*, *9*, 197–202. https://doi.org/10.1209/0295-5075/9/3/002.

Stoessel, F. (2008). Thermal Safety of chemical processes. Weinheim.

Taleb, N. N. (2007). The black swan : The impact of the highly improbable. Penguin Books Ltd. https://doi.org/10.5465/AMP.2011.61020810.

TNO. (2005a). The "Purple book"—Guidelines for quantitative risk assessment, CPR 18 E. In *Publication series on dangerous substances (PGS 3)*.

TNO. (2005b). The "Yellow Book"—Methods for the calculation of physical effects—CPR 14E. In *Publication series on dangerous substances (PGS 2)*.

Varma, A., Morbidelli, M., & Wu, H. (2005). *Parametric sensitivity in chemical systems*. Cambridge University Press.

Villa, V., Paltrinieri, N., & Cozzani, V. (2015). Overview on dynamic approaches to risk management in process facilities. *Chemical Engineering Transactions*, *43*, 2497–2502. https://doi.org/10.3303/CET1543417.

Villa, V., Paltrinieri, N., Khan, F., & Cozzani, V. (2016). Towards dynamic risk analysis: A review of the risk assessment approach and its limitations in the chemical process industry. *Safety Science*, *89*, 77–93. https://doi.org/10.1016/j.ssci.2016.06.002.

Villa, V., Paltrinieri, N., Khan, F., & Cozzani, V. (2016a). Towards dynamic risk analysis: A review of the risk assessment approach and its limitations in the chemical process industry. *Safety Science*, *89*, 77–93. https://doi.org/10.1016/j.ssci.2016.06.002.

Villa, V., Paltrinieri, N., Khan, F., & Cozzani, V. (2016b). A short overview of risk analysis background and recent developments. In N. Paltrinieri & F. Khan (Eds.), *Dynamic risk analysis in the chemical and petroleum industry: Evolution and interaction with parallel disciplines in the perspective of industrial application* (pp. 3–12). Butterworth-Heinemann. https://doi.org/10.1016/B978-0-12-803765-2.00001-9.

Willis, H. H. (2007). Guiding resource allocations based on terrorism risk. *Risk Analysis*, *27*, 597–606. https://doi.org/10.1111/j.1539-6924.2007.00909.x.

Yang, X., Haugen, S., & Paltrinieri, N. (2017). Clarifying the concept of operational risk assessment in the oil and gas industry. *Safety Science*, *108*, 259–268. https://doi.org/10.1016/j.ssci.2017.12.019.

Zhao, J., Cui, L., Zhao, L., Qiu, T., & Chen, B. (2009). Learning HAZOP expert system by case-based reasoning and ontology. *Computers and Chemical Engineering*, *33*, 371–378. https://doi.org/10.1016/j.compchemeng.2008.10.006.

CHAPTER THREE

Logic based methods for dynamic risk assessment

Hans J. Pasman[a,]*, William J. Rogers[b]
[a]TEES Mary Kay O'Connor Process Safety Center, Texas A&M University System, College Station, TX, United States
[b]Artie McFerrin Department of Chemical Engineering, Texas A&M University, College Station, TX, United States
*Corresponding author: e-mail address: hjpasman@gmail.com.

Contents

Methods in Chemical Process Safety, Volume 4
ISSN 2468-6514
https://doi.org/10.1016/bs.mcps.2020.02.001

1. Introduction

The rise of process industry has been very beneficial but due to mishaps with disastrous effects, demand to assess process and plant risks grew in the early 1970s. Methods were applied originally developed fragmentedly in science communities, such as the nuclear power one, the military and aviation, while only in due time also methods came forward specifically for the process industry. Names of methods are established sometimes only over time, so the earliest forms of a method have often been indicated differently from how we know this method now. Until recently, risk situations have been considered without dynamic elements and largely focused on a design or a change in an existing plant.

As mentioned in previous chapters, because of accidents and rapid expansion of the process industry, often in densely populated areas, governmental authorities started to develop regulation requiring risk-based assessment with the goal to protect workers and residents. Interests of the industry and the population are rarely fully parallel, so a need arose for clear acceptance criteria for both groups. In some countries this became for residents a safety distance based on maximum possible spread of injuring effects of a hazardous material mishap present and an accepted toxic concentration, blast overpressure or radiant heat. In other countries the calculation became required of an accepted fatality risk contour around a risk source (known as individual risk), and given the local population density that of the exceedance frequency of a number of fatalities versus the number of fatalities (societal or group risk, or *F–N* curve) with an accepted ceiling.

The basic idea of risk assessment is simple and appealed to many. As repeated many times it is about the questions what can go wrong, how often and with what consequence, and risk was considered simply as the product of mishap frequency or probability and the severity of its consequence. In the late 1970s and, early 1980s many research groups started developing risk assessment methods, with the first 20 years in particular the physical models of consequence analysis methods forming the main focus. Many problems appeared describing the effect area when a hazardous material spill occurs. In the mid-1980s the European Union (EU) embarked on a program, also in the framework of the Seveso Directive, to support development work and by funding field tests. It became apparent though, that calculations by different groups predicting a certain risk level at fixed conditions produced large differences in results, while also acceptance criteria differed from one country to the other. In an attempt to obtain a better comparative basis and unity, the EU initiated benchmarking exercises and consultation lasting the whole of the 1990s and even beyond.

Most needed appeared to be improvement of hazard identification and scenario definition followed by failure and other data for realistic estimations of uncertainties. Just primary possible accident scenario definition is already problematic, let alone domino effects and effects by earthquake, flooding and extreme weather (NaTech). Human and organizational factors are not routinely included yet in assessments, although because of the growing interest in dynamic and operational risk assessment these factors become critically important (for an interesting review and suggested way forward, see Taylor, 2020). Security risk assessment comes with its own problems. Uncertainties, persistent problems and weaknesses in the assessments generate lack of credibility. Risk assessment can be called a success if not only risks can be determined sufficiently accurate at the design stage and with regard to modifications during the life cycle, but also can be continuously monitored during operations with all the dynamic elements that changes its value. This will not ban the occurrence of mishap events but it may provide the best warning tool for management.

Emphasis in this chapter is on the logic based methods with special attention on dynamic risk assessment applications. This concerns mostly hazard identification and failure or other event probability. However, in risk assessment consequence analysis is playing a central role as well and this is for the most important part based on physical models, largely covered by Casal (2017), although probabilistics appear too, in particular in damage models, hence asset vulnerability.

Section 2 will be mainly on the development of methods and data, up till the benchmarking episode in the 1990s; Section 3 will describe the various EU and other efforts to compare results of various analyst teams and to seek the causes of the differences observed. In Section 4, improvements and new approaches will be touched upon. Recently, *Cyber-security* obtained much attention, as appears from standard IEC 62443-4-2:2019

(2019) and its predecessors, all having a risk assessment basis. For a major part, risk assessment on cyber-attack effects makes use of the regular risk assessment methods, including vulnerability assessment, but in Section 4.8 "Game Theory" will be touched upon as being a tool specifically suited for terrorist and cyber-attack analysis.

2. Review of risk assessment methods

2.1 Pre-1970: Earliest logic based risk assessment concepts and methods

At the time many minds were focused on accident investigation, how could it happen? Exothermic decomposition, fires, gas and dust explosions, lack of process control, and unexplained leakages occurred, all caused or co-caused by erroneous decisions and lack of knowledge. Only few system thinking individuals developed logical analytical concepts; and it took time. One of those individuals was Houston at Imperial Chemical Industries (ICI) who presented a paper on the first Loss Prevention Symposium in the United Kingdom in 1971, entitled "New Approaches to the Safety Problem," in which he laid down several risk assessment method concepts. After recognizing that the need to tackle the safety problem was increasing due to on one hand a deteriorating record of fire accidents, and on the other, the larger scale of units and higher public sensitivity, Houston stressed the imperative to remove or to reduce risks at the design stage. Due to growing process and plant complexity, in his eyes the need for a system approach was clear. Houston mentions two methods, the qualitative "Operability Study" and the quantitative "Hazard Analysis," the latter applied in case of more serious risk.

Houston starts out to describe the team effort of performing an operability study. A practice that developed in ICI since 1963. The team should consist of a mixture of as he called it "skills" and the study would have the character of a "critique." The assumption is that due to the complexity of the design, problems are missed but not that there are knowledge gaps. This assumption should be verified at the outset. So, the "skills" should cover all knowledge required, but the team should still be small enough to be able to act efficiently and the members should have the power to follow up on their decisions. Information on the process will be given in the form of a flowsheet. The procedure is to go systematically, line by line, and imagine a process deviation, a possible cause and consequence given the conditions. Houston indicates here directions, namely backwards

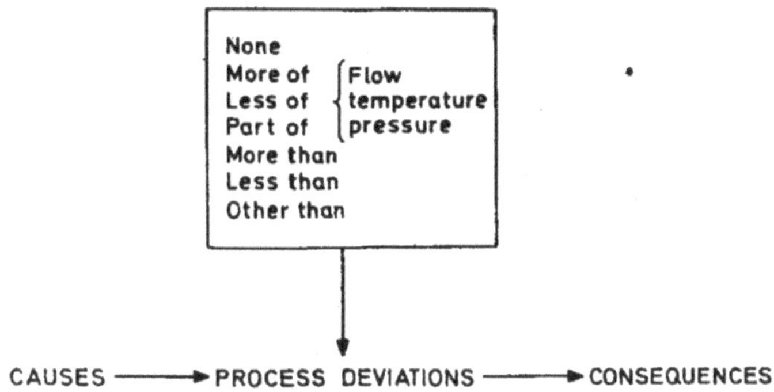

Fig. 1 Checklist of "Operability study" deviations by Houston (1971).

to the cause and forward to the consequence. If needed corrective action is proposed. Deviations are a limited set, and are given as a checklist, see Fig. 1, which boils down to "non-occurrence, change in degree and change in kind." Houston further describes enthusiastically advantages of the method, e.g., its flexibility with respect to the level of detail. The method development has been throughout ICI, so that there is no one person that can be merited for it. Additionally, he mentions that it sometimes is useful to draw-up a logic diagram as presented in Fig. 2 to show the various events that can occur, while the diagram also helps to find the optimum location of protective systems.

The reader with some background in process safety will have recognized from Houston's description the basic form of what we now call HAZOP, Hazard & Operability Study, that conquered the world and which we shall describe in more detail later in this chapter. In the logic diagram one can recognize a bowtie diagram. Unlike present convention, it consists at the right hand of a causal structure of possible faults (a concept broader and including failure[a]), building up to a critical event, "the explosion, " and at the left-hand side consequence events. Today one would call the right side a fault tree and the left side an event tree, whereas convention is now to direct all the arrows from left to right.

Houston continues about quantitative methods by first citing Heinrich's pyramid of few disastrous accidents at the top, and down with increasing numbers to near misses, less serious events and dangerous situations. Loss control should be based on the whole pyramid and not only

[a] According to current standard IEC 60812 *failure* means "termination of the ability of an item to perform a required function," while *fault* is "state of an item characterized by the inability to perform a required function," with some exceptions, such as preventive maintenance or lack of external resources.

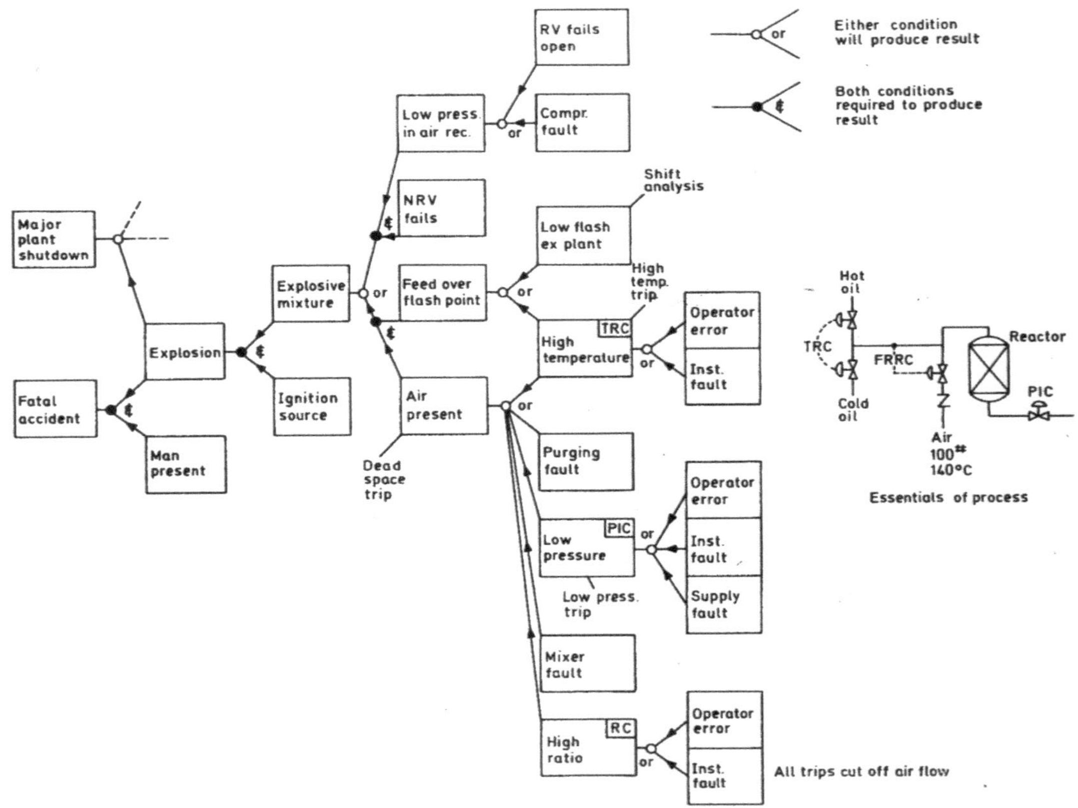

Fig. 2 Logic diagram presented by Houston (1971) as example of how an “Operability Study” can assist in unraveling complex cases; here possibly leading to fatal accident and plant shutdown.

on the major accidents. He further mentions that plant managers are interested in the severity of a potential event and its probability. Houston is of the opinion that the calculation of severity is relatively easy, which is a very optimistic statement in view of all the later efforts worldwide over many years to do it sufficiently realistic. On the other hand, "relatively" it is true, because also still today the problem to find data to derive credible probability values is large, although that may change in the future with the digitalization. Anyhow, referring to work in nuclear and aerospace industry for calculating failure probability rates, Houston turns for a detailed example of a failing vessel to reliability engineering and refers to the logic diagram of Fig. 2, in which he indicates OR junctions (probabilities added) and AND junctions (probabilities multiplied). This is Boolean logic applied to a Fault tree but both names are not mentioned. A New Data Bank should facilitate the availability of data in a ready-to-use form. Despite lack of reliability data development of protective devices has been performed for several ICI plants. In his chapter he further mentions the need to have an injury tolerability limit for industry and a standard for risk acceptance, while he makes a difference between voluntary and involuntary risk, risk to employees and the general public, financial losses and insurance. Finally, he describes the generic accident process and distinguishes three phases (induction, development, and resolution), presents these in a logic diagram, a model accident report with much detail, and acknowledges his ICI colleagues naming Trevor Kletz, who in the same conference presented a paper on hazard analysis. In conclusion, Houston's paper laid down a safety assessment fundament for many years to come.

Kletz's (1971) paper on hazard analysis was on quantification of the Fatal Accident Frequency Rate (FAFR, now FAR) per 10^8 h at work, concerning more personal safety than process safety. The metric was used for safety measure priority setting. Kletz made comparisons and gave examples.

2.2 1970–90: Development of logic based failure analysis methods

We shall consider in this section the systematic and logic based methods of HAZOP, FMECA, Fault tree and Event tree. The first two are qualitative, the latter two quantitative. They all serve hazard identification and scenario definition.

In fact, the logic concept of *Fault Tree* has been developed by Watson at Bell Laboratories in 1961 (Watson, 1961) for safe missile launch and found its way around. *Failure Mode Effect and Critical Analysis* (FMECA) or the shorter FMEA was developed by the military (DoD, 1949) to predict effects in the functioning of equipment if components or parts fail. In the 1970s its use spread via NASA to other engineering disciplines, such as nuclear, automotive and aviation.

Besides the methods mentioned there are many other less structured and logic ones applied for hazard identification. In fact, it all starts with knowledge about the properties of the hazardous materials one must deal with, while further methods are applied, such as audits, checklists, What-If, human error analysis, Master-Logic Diagram, and quite a few more, see Mannan (2015, chapter 8).

Details and references of all of the early and basic logic methods described in this chapter can be found in Mannan (2015).

2.2.1 HAZOP, hazard and operability study

HAZOP is a systematic approach for hazard identification and therefore used as an important tool to enable formulating top events or otherwise for the logic based tree methods to be explained later. It took years before HAZOP was recognized as a technique or tool of its own in the hazard identification step of a risk assessment. Lawley (1974) published a paper in line with that of Houston (1971), in which hazard identification was still called operability study and the quantification of risk by means of a logic tree hazard analysis. After the Chemical Industries Association in the United Kingdom recommended the use of Hazard and Operability studies to prevent accident (CIA, 1977), it got its name and its conquest tour around the world started. It also took a while till the short name HAZOP as it is known now evolved, not being the acronym HAZOP but a contraction of Haz & Op, so it is in older articles also indicated as hazop, Hazop and HazOp.

Much has been published on HAZOP, numerous descriptions of the procedure to follow are given and it is part of the standard IEC 61511: 2016 (E) (2016). An overview of literature on HAZOP and worked out examples can be found in Mannan (2015, chapter 8). For conducting a HAZOP data must be collected. Important sources are a Process Flow Sheet, a Piping & Instrumentation Diagram (P&ID) and other supporting documentation. A brief description of the procedure is taken from Pasman (2015), based on a guide by Crawley, Preston, and Tyler (2000).

HAZOP is essentially a qualitative procedure in which a small team consisting, e.g., of a process engineer, a control instrument expert, a risk analyst, and an operator, guided by an experienced chairman, examines a proposed design by generating questions about it in a systematic manner. For this purpose, a number of "guide-words" are used indicating potential deviations of a process parameter from the design intention. These deviation questions may reveal (often hidden) possible causes and the effects on process behavior, hence causes and effects of a possible fault. Commonly used guide-words are given in Table 1. The HAZOP study is function driven. Thus, potential safety and operability problems are identified and appropriate actions can be taken. A poorly operable process is usually also unsafe.

Process parameters can be chosen from a wide variety, but the first five of the set mentioned hereafter are the most frequently used: flow, pressure, temperature, mixing, stirring, control, pH, level, sequence, viscosity, signal, reaction, start/stop, composition addition, separation, services, operate, maintain, time, and communication. Meaningful combinations shall be selected, as suggested in Table 2.

The questions arise creatively through interaction (brainstorming) among the team members. Sometimes a guide-word first approach is chosen, other times preference is given to calling the parameter first. The team secretary or scribe keeps track of the findings in a worksheet format shown in Table 3.

Table 1 Standard HAZOP guide-words and their generic meanings.

Guide-word	Meaning
No (not, none)	None of the design intent is achieved
More (more of, higher)	Quantitative increase in a parameter
Less (less of, lower)	Quantitative decrease in a parameter
As well as (more than)	An additional activity occurs
Part of	Only some of the design intention is achieved
Reverse	Logical opposite of the design intention occurs
Other than (other)	Complete substitution—another activity takes place
Other useful guide-words include:	
Where else	Applicable for flows, transfers, sources, and destinations
Before/after	The step (or some part of it) is effected out of sequence
Early/late sequence	The timing is different from the intention
Faster/slower	The step is done/not done with the right timing

Table 2 Examples of meaningful combinations of parameters and guide-words.

Parameter	Guide-words that provide a meaningful combination
Flow	None; more of; less of; reverse; elsewhere; as well as
Temperature	Higher; lower
Pressure	Higher; lower, reverse
Level	Higher; lower; none
Mixing	Less; more; none
Reaction	Higher (rate of); lower (rate of); none; reverse; as well as/other than; part of
Phase	Other; reverse; as well as
Composition	Part of; as well as
Communication	None; part of; more of; less of; other, as well as

Table 3 Simplified hazard and operability study worksheet format.

Number	Parameter	Deviation	Cause	Consequence	Safeguard	Action	On person

For the exercise the Piping and Instrumentation Diagram (P&ID) is sectioned in analysis nodes to enable paying attention to required details. Optimum order of discussing the sections can be an issue. A consequence in one node may by its impact be a cause for a more severe consequence in a next node. To assist in the identification of hazardous deviations, the team will usually find it helpful during the exercise to compare the proposed design with relevant engineering standards. In using the HAZOP method, the need for action is decided semi-quantitatively based on the team's experience and judgment of the seriousness of the consequences, together with the expected probability (frequency) of the occurrence. In a situation in which uncertainty remains, analysis using quantitative techniques may be helpful, such as prioritization of actions to reduce risk. Although identification is carried out vigorously and to a simple extent fault tree paths are probed, very detailed fault analysis is not normally a systematic part of the HAZOP procedure. Its main purpose is to identify the main hazards and operability problems. In depth analysis of high hazard/low-probability events is the domain of the next steps in risk analysis. Because in principle HAZOP can be performed in any stage a process is in, as a technique it is suited for dynamic analysis.

In addition to its open-ended approach, a fundamental strength of HAZOP is the brainstorming and thinking together among members of the study team. Its success depends on the degree of co-operation among individuals, their experience, competence, and the commitment of the team as a whole. A guarantee cannot be given that during a study all possible potential problems have been unearthed. Human factors—e.g., competence, stress, emergency preparedness, procedure weaknesses, maintenance issues and many other are easily overlooked. To maintain focus of the team (it is often not a highly inspiring activity!) and not to interfere too much with other work of the members, time devoted is often limited to half-a-day per day. Despite various trials to automate at least part of the procedure, IT support is mainly constrained to administrative aspects. The main complaint is therefore the time involved performing a HAZOP and therefore the costs (e.g., 5 × 20 h for an average P&ID, hence for a plant 1–8 weeks). Since a HAZOP should be repeated after a major change in the process or once every 5 years, the search for more efficient methods is continuing. It also turned out that the completeness of many HAZOPs is questionable. On average perhaps, just more than half of the possible major events are identified (Suokas & Rouhiainen, 1989), while Baybutt (2015) and a score of other papers gives many reasons why. Jarvis and Goddard (2016) analyzing causes of the 100 major losses in the in the onshore oil, gas and petrochemical industries over the period 1996–2015, concluded that a major contributor (48% of integrity failures) is inadequate process hazard analysis, in other words failing hazard identification, hence, a.o., failing HAZOP, in part due to lack of quality control of it. In the section on EU Benchmarking of QRA it will be shown that the largest uncertainty appeared to be scenario definition, hence again hazard identification.

2.2.2 Failure mode and effect analysis

Where HAZOP is process function-driven analysis, Failure Mode and Effect Analysis (FMEA), is fully focused on system/equipment/components failure. The technique has been described in detail in standard IEC 60812:2018 (2018) and can be applied several times during the lifetime of an equipment. The procedure for a system, of which boundaries have been defined, is to break first its hierarchy down to basic components followed by a bottom-up approach.

Adapting the description by Pasman (2015) the further procedure is as follows: FMEA is a qualitative method used to analyze systematically and inductively an equipment unit or part of it for possible failures and to identify the effect the failures will have on its function and on the function of the system it is part of. FMEA is by preference a team activity performed by, e.g., an operator, a maintenance specialist, a process engineer, and a product expert, and is under the control of an experienced team leader. General failure modes are failure during operation, failure to operate or to cease operation at a prescribed time, or to operate prematurely. The specific failure modes for a component can be found by checking potential errors in Man, Material, Machine, Measurement, Method, and environment (in Dutch "Milieu")—the "6 M"s. FMEA is organized suspicion. It includes considering common cause failures, software errors, and human factor effects; this will however be only to a limited extent, because FMEA cannot deal with dependent faults. It also is of limited use in complex equipment with, e.g., software-hardware interactions.

FMEA can be extended to Failure Mode Effects and Criticality Analysis (FMECA), which considers the severity and probability of failure qualitatively and expresses it as a Risk Priority Number (RPN). This metric is the product of the rank parameters severity (S), occurrence (O), and likelihood of detection (D), all on a linguistically graded 10-point scale. Results are described in a format as in Tables 3 and 4. If desired severity can be plotted versus occurrence likelihood in a 4×5 criticality matrix. Personnel other than team members shall be asked to comment on the first result of a FMEA or FMECA analysis. If a quantitative analysis is required then a FMEA shall be followed by a Fault Tree Analysis.

2.2.3 Fault tree analysis

Fault tree analysis (FTA) is a deductive tool analyzing in a graphical representation the underlying possible causes of a top event, which can be a failure or another undesired event, such as the unintended release of

Table 4 Sample table for failure mode effect and criticality analysis (FMECA), abridged.

Component #	Failure		Effects	Criticality		
ID; function	Error mode	Causes	Local system	Severity frequency	Detection method	Remarks

hazardous material. After its inception by Watson (1961), the technique was taken up by the US military, the aviation industry, the US Nuclear Regulatory Commission (NRC), and has been described in detail in the NASA Fault Tree Handbook (Vesely et al., 2002). Standard IEC 61025:2006 (2006) is dedicated to the FTA technique.

In the analysis, causes of the top event of a system will be traced down to faults, which occur either in parallel or in series, all the way down via faults of intermediate components to basic component faults for which there are (possibly) observed failure data. These faults can be equipment failure events, human failure, or software errors. The qualitative FT graph shows the logical relationships between the causal events leading up to the top event. Fault trees are modeled with Boolean AND- and OR-gates or operators in a Boolean logic tree with binary component or subsystem function states: on = true, and off = false. Assuming fault independency, an AND-gate is a conjunction or intersection, meaning that if one of the components is false the gate outcome becomes false; the OR-gate is a disjunction or union, meaning that the gate will be false, if all components are false. The subsystems are either core processing functions or controls. The deductive defect tracking analysis proceeds downwards to ever more subsystems and components that together form the system. Finally, failing base components will be reached that can be regarded as non-decomposable units. If failure rates of components are known or initially can be estimated from generic data for similar units, the FT can be evaluated quantitatively bottom-up. This will enable calculation of the probability per unit of time, e.g., a year, or frequency of the top event. An example of a FT is given in Fig. 3.

An important qualitative result of the analysis is the identification of minimum cut sets (MCSs). A MCS is the minimal combination of basic and intermediate fault events leading up to the top event, hence it represents a failure path scenario. Each cut set can be evaluated quantitatively using the failure rates p_i of i basic events, an AND gate yielding the product of the rates of n components involved $\prod_{i=1}^{n} p_i$, and an OR-gate the sum minus the product. The latter is the overlapping fraction of both components failing if the combination is by two components. In case of n basic faults the OR-gate output is calculated easiest as $p = 1 - \prod_{i=1}^{n}(1 - p_i)$. This is called a Noisy-OR, a name borrowed from signal processing. Besides this type of most common OR-gate, there is an Exclusive-OR which fails if just one of the components fails, while a Priority-AND gate only fails if upon a condition a prescribed sequence

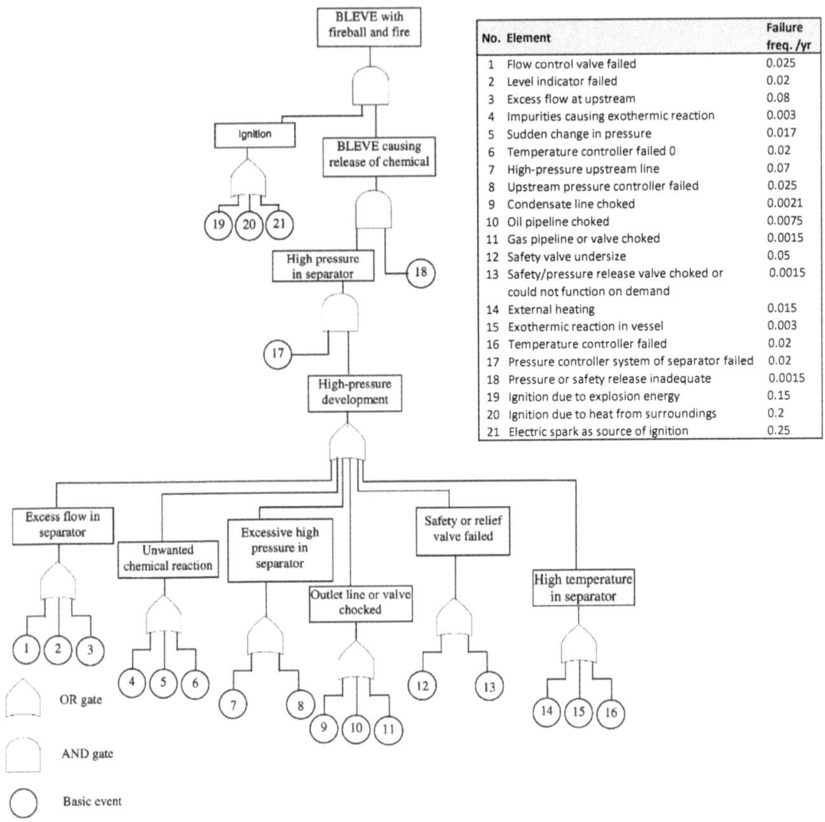

No.	Element	Failure freq. /yr
1	Flow control valve failed	0.025
2	Level indicator failed	0.02
3	Excess flow at upstream	0.08
4	Impurities causing exothermic reaction	0.003
5	Sudden change in pressure	0.017
6	Temperature controller failed 0	0.02
7	High-pressure upstream line	0.07
8	Upstream pressure controller failed	0.025
9	Condensate line choked	0.0021
10	Oil pipeline choked	0.0075
11	Gas pipeline or valve choked	0.0015
12	Safety valve undersize	0.05
13	Safety/pressure release valve choked or could not function on demand	0.0015
14	External heating	0.015
15	Exothermic reaction in vessel	0.003
16	Temperature controller failed	0.02
17	Pressure controller system of separator failed	0.02
18	Pressure or safety release inadequate	0.0015
19	Ignition due to explosion energy	0.15
20	Ignition due to heat from surroundings	0.2
21	Electric spark as source of ignition	0.25

Fig. 3 Example fault tree for an offshore platform separator rupture event resulting in a BLEVE, according to Khan, Sadiq, and Husain (2002). Given the failure rates of components 1–21 the expected frequency of the top event is calculated, as explained in the text.

of failing components is followed. For other types and symbols, more complex situations, e.g., with dependency, time dependence, and examples, see Vesely et al. (2002). For time dependence effects, hence dynamic fault tree analysis, Markov modeling is applied.

2.2.4 Markov modeling

A (first order) Markov model represents a chain of stochastic events, in which the probability of each event transition depends only on the state reached of the previous event. So, there is no "memory" beyond the previous event. The chain of successive events is called a Markov process, which is continuous, if transitions can occur any time, or discrete when this is only possible at fixed times. Continuous Markov process is well applicable in case failure rate λ_i of a component i is constant and probability of functioning reliably at time t, $P_i(t)$, decreases with time as $P_i(t) = e^{-\lambda_i t}$. This is because the exponential distribution is memoryless and ensures that the time to the next transition is not influenced by the time being in the current state. For a small time increment Δt it means that the probability change of i being still reliable is: $P_i(t+\Delta t) = P_i(t)e^{-\lambda_i \Delta t} \approx P_i(t)(1-\lambda_i \Delta t)$. If all λ_i are the same the process is called *homogeneous*, but usually for a system this will not be the case and is the process called *semi-Markov*. The equation describing the probability of system states P of a homogeneous process as a function of time t is:

$$P_j(t+\Delta t) = \sum_{k \neq j} P_k(t)\lambda_{kj}(\Delta t) + P_j(t)\left(1 - \sum_{k \neq j} \lambda_{jk}\Delta t\right),$$

in which λ_{kj} is the rate of transition of the system from state k to state j and λ_{jk} back to k. In case of a semi-Markov model solving a system of n components, it requires solving n simultaneous differential equations to derive all possible states of the system. For a sizable system this can be done only by Markov Chain Monte Carlo (MCMC) simulation.[b] In Fig. 4 the simplest case of a two-component system is shown, see also Ebeling (2019).

Markov modeling is thus used much in reliability analysis and it can model redundancy and repair time, so it is suitable as a tool in determining expected reliability and availability over time of a system. The same way as

4 Possible states of the system

Component 1	Component 2	Probability
Not failed	Not failed	$P_0(t)$
Failed	Not failed	$P_1(t)$
Not failed	Failed	$P_2(t)$
Failed	Failed	$P_3(t)$

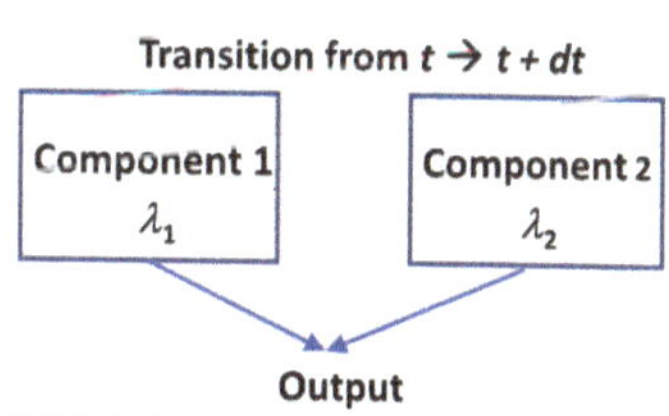

System reliability states with $P(t) = e^{-\lambda t}$	Or at infinitesimal change
$P_0(t+\Delta t) = P_0(t)(1-(\lambda_1+\lambda_2)\Delta t)$	$dP_0(t)/dt = -P_0(t)(\lambda_1+\lambda_2)$
$P_1(t+\Delta t) = P_0(t)\lambda_1\Delta t + P_1(t)(1-\lambda_2\Delta t)$	$dP_1(t)/dt = P_0(t)\lambda_1 - P_1(t)\lambda_2$
$P_2(t+\Delta t) = P_0(t)\lambda_2\Delta t + P_2(t)(1-\lambda_1\Delta t)$	$dP_2(t)/dt = P_0(t)\lambda_2 - P_2(t)\lambda_1$
$P_3(t+\Delta t) = P_1(t)\lambda_1\Delta t + P_2(t)\lambda_2\Delta t + P_3(t)$	$dP_3(t)/dt = P_1(t)\lambda_1 + P_2(t)\lambda_2$

AND gate	OR gate
$P_0(t) = e^{-(\lambda 1+\lambda 2)t}$	$P_0(t) = e^{-\lambda 1t} + e^{-\lambda 2t} - e^{-(\lambda 1+\lambda 2)t}$
$dP_0(t)/dt = -P_0(t)(\lambda_1+\lambda_2)$	$dP_0(t)/dt = -\lambda_1 e^{-\lambda 1t} - \lambda_2 e^{-\lambda 2t} + (\lambda_1+\lambda_2)e^{-(\lambda 1+\lambda 2)t}$

Fig. 4 Markov model of a simple two-component system: *left*: individual states; *right*: overall output reliability.

[b] The acronym MCMC or Markov Chain Monte Carlo is the name of a technique to solve not analytically feasible problems, e.g., by successive random sampling from a multidimensional integral building a Markov chain a numerical approximation can be generated.

failure rates, also repair rates can be introduced restoring functioning of the component. However, the method is limited to the use of the exponential distribution, while it is not suited for complex systems due to what is called state-space explosion. Instead, either dynamic Bayesian net, or even better, the more flexible stochastic and colored Petri net will be needed, but that are later developments and shall be briefly treated in the third subsection.

2.2.5 Event tree analysis

Event tree was first presented in 1975 in the Wash-1400 Reactor safety Study (Rasmussen, 1975). Event tree analysis (ETA) is an inductive logic tree branching consequences out from an initiating critical event, e.g., a hazardous material release, to various possible intermediate consequences and then on to end-points of final major hazard phenomena, such as explosion or fire. The event occurrences are quantifiable in terms of probability of the initiating event and of the probability pairs at each branch node. An example is shown in Fig. 5. If independence of the mechanisms at the branch points are assumed the final event probability is just the product of initiating event and subsequent branch probabilities. If the

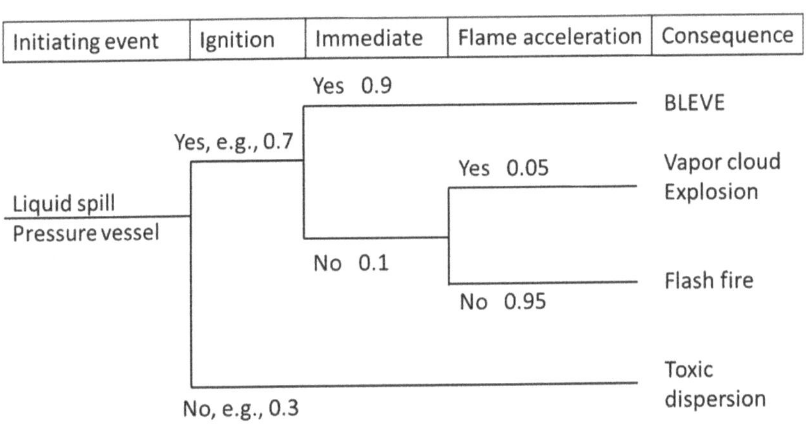

Fig. 5 Example of a simple event tree for a pressure vessel rupture causing as initiating event a flammable toxic boiling liquid spill, which may result in a BLEVE fireball, a vapor cloud explosion, a cloud flash fire, or a toxic cloud. At each node a probability point value can be assigned, as shown in the example. *After Pasman, H.J. (2015).* Risk analysis and control for industrial processes—gas, oil and chemicals. A system perspective for assessing and avoiding low-probability, high consequence events. *Butterworth Heinemann, Copyright © 2015 Elsevier Inc., ISBN:978-0-12-800057-1.*

events concern actions or active failures occurring in a certain sequence producing a range of results, it is called an event sequence diagram (adapted after Pasman, 2015).

Besides its use in the consequence analysis part of risk assessment, event tree analysis became also much used also as tool in the most useful Layer of Protection Analysis (LOPA), which we shall further consider in the next section. As we also shall see in the next section, in the 1990s as a "logic" further step the critical top event of the logic fault tree was taken at the same time as the initiating event of the logic event tree, the two forming a "bowtie." This concept became an important tool providing overview of the various scenarios that shared the same initiating top event.

2.3 1980–2000: Development of full-fledged risk assessment methods and tools

2.3.1 The first studies

As a result of various explosions, fires and toxic releases in the 1960s and '70s, unrest arose among population and employees about major hazards the developing industry posed. As a result, both in the United Kingdom (Canvey, 1978/1981) and in the Netherlands (COVO, 1982) in the late 1970s the first risk assessments were initiated. Canvey risk analysis has been carried out by SRD, then a systems research department of UK Atomic Energy Authority, meanwhile absorbed by Health and Safety Laboratory, and the Rijnmond, so-called COVO study, by Cremer & Warner, a consultant firm, which had gained experience in risk analysis for the nuclear power industry in the United Kingdom. A rather detailed description of both assessments was given by Lees (1996).

Both studies were already rather detailed quantitative risk assessments (QRA). Given that it has been first analyses, it is surprising to see the large number and variety of possible failures accounted for, including human failure. The latter appeared often in the fault trees used to estimate the frequencies of top events. Historical failure frequency data were available only in limited quantity; much had to be elicited from experts. In all, the studies answered the renowned three questions (the triplet) posed by Kaplan and Garrick (1981):

(i). What can happen? (i.e., What can go wrong)
(ii). How likely is it to happen?
(iii). If it does happen, what are the consequences?

2.3.2 Development of QRA software packages

After these initial studies, governments together with major companies and technology institutes embarked on research and development of more dedicated models and software. A logic sequence of risk analysis steps in the early days presented as an Ishikawa fish bone diagram is shown in Fig. 6. The "meat" at the bones are the tools to be used, while later measures were added to reduce risk, such as bowtie and layer of protection analysis, and a safety management system (first introduced in the United States in 1992 with OSHA's 29 CFR $ 1910.119). In flow diagrams the steps of Consequence analysis and Probability quantification are often shown in parallel blocks. But following the emphasis during the early days of modeling source terms, dispersion, vapor cloud explosion and ensuing damage, consequence analysis got priority. Probability quantification relied much on reliability engineering and is dependent on historical data under the conditions where the actual vessel or other equipment for which rupture or leak frequency must be estimated, is subjected too. As we shall see later this is still one of the weaknesses of QRA. Later, it was also realized that prior to an analysis attention must be given to defining scope and objectives, stakeholders, and criteria, during the assessment communication, deliberation, and consulting among team members encouraged, and the process monitored, while after the assessment reviews are required.

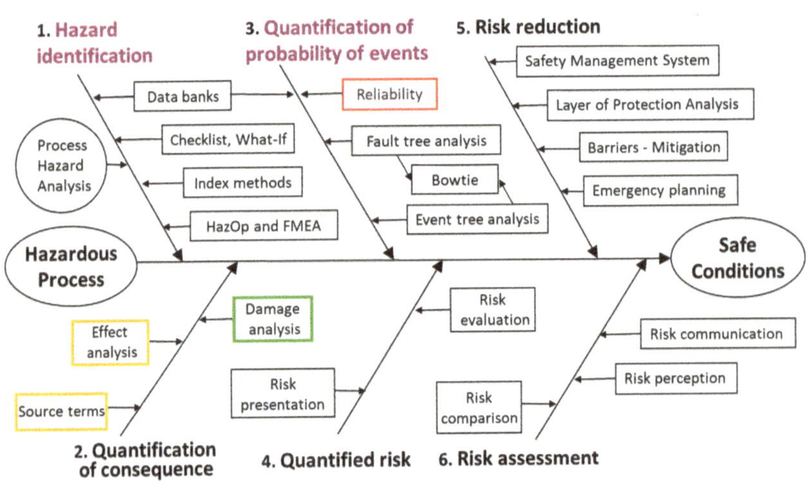

Fig. 6 Ishikawa or fish bone scheme of the six steps of a Quantitative Risk Assessment, QRA, showing various tools to perform the process, after Pasman, 2015. (Colors refer to the Dutch "colored books" developed in the 1980s and '90s: Step 1 and 3 purple book, Reliability red; Step 2 left yellow, right green; https://www.publicatiereeksgevaarlijkestoffen.nl/).

The problems identified when performing the first risk analyses are lack of insight, data and models to predict risk. That lack hampered conducting calculation of source terms (spill rates, spill pool extensions, evaporation, pool fire), gas/vapor dispersion in the atmosphere, vapor cloud explosion blast as sometimes observed in, e.g., refineries, the type of boiling liquid expanding vapor explosions (BLEVE) that had caused the Feyzin disaster in 1966, toxic concentration levels, and dangerous blast and radiant heat levels for humans and structures. Regarding fires, models of pool fires, jet fires, and flash fires provided fewer problems than dispersion and explosion. There was some awareness of the influence of human failure but that did not play a role in early QRA. Reactor runaway was a serious problem but was handled mainly in a separate community, as were gas and dust explosion research. Connection with risk analysis was made only much later.

So, around 1980 in Europe and the USA sponsorships by national governments research started to fill the empty spots. Also, in European countries with respect of the process industry, and in the United States in case of nuclear power industry and the waterways under vigilance of the US Coast Guard, a heated debate arose on what risk level could the population be exposed too and be considered acceptable. Sociologists and psychologists performed analyses. Solutions were found mainly according to two types of acceptance criteria: a safety distance to risk sources determined by the intensity of the threat to people that was considered not really harmful anymore, and the distance at which the overall risk of being killed by the risk source, the so-called individual risk was lower than a pre-set probability level of, e.g., 10^{-6} per year, also expressed as safety index SI 6. To the latter, a group or societal risk level was added to account for population density near the plant location resulting in a curve of cumulative annual frequency of events, F causing a number of fatalities of N or more, hence the probability of exceeding N, versus N, the $F-N$ (log-log) plot. This criterion was less easy to understand and became secondary.

The development of consequence analysis methods required also large-scale field tests both in the United States and Europe of dispersion of gases, in particular, gases heavier than air, such as chlorine and many organic chemicals. Cryogenic liquefied gases, such as LNG, in molecular mass lighter than air, are also heavy just after escape and not warmed up yet. Mixing of heavy gas with air is slow depending on atmospheric conditions and heavy clouds can drift over a large range. Vapor cloud explosion (VCE) has initially been called unconfined VCE (UVCE) to distinguish this open-air explosion from the ones inside equipment. Not only in dispersion but also in vapor cloud explosion obstacles to the flow play a role. Tests of explosive mixtures in balloons did not generate significant blast, unless initiated by a high explosive charge. Only, after the flame acceleration effect of obstacles was discovered, blast was measured (Lees, 1996, chapter 17). Many tests were also dedicated to evaporation, dispersion, and pool fires of liquefied natural gas (LNG). At the same time toxicologists in the United States and Europe have been engaged in defining harmful toxic concentration and dose, and because of the probabilistic effect on people expressed in normal distributions converted to the linear probit relations. Failure rate data were (and still are) another problem. Initially, much was taken from nuclear power databases, although additionally over the years governments also agreed to (sanitized) data brought in by process companies. All these activities are still continuing, albeit on a slow pace.

The experiments have been essential for developing the consequence models. One of the most known with respect to gas dispersion is the Thorney Island trials conducted during the first half of the 1980s. A concise summary of test arrangements and results is presented by Lees (1996, chapter 15). Models developed have been more extensively described in Mannan (2015, chapter 15). The trial outcomes and the evaluation of models resulted in the known Britter and McQuaid Workbook (Britter & McQuaid, 1988). For neutral gases the Gaussian model is available for which in field trials coefficients had been determined already long before in view of chemical warfare defense. Initially, for heavy gases, integral models, which are basically 2D and consist of ordinary differential equations, have been developed and tweaked to fit field trial outcomes. In the late 1980s and 1990s a number of Computational Fluid Dynamics (CFD) models based on the partial differential equations of Navier-Stokes have been launched enabling 3D simulation of hilly terrain, buildings and other obstacles to the flow. Optimizing these models continues till today.

In the same period good progress was made with respect to vapor cloud explosion research. A summary is given in Mannan (2015, chapter 17). Several simplified blast models, such as the TNO Multi-Energy Method and the Tang and Baker model have seen the light of day. However, also here CFD has proven a much better solution. The most known became FLACS, Flame Acceleration Simulation. It was much "calibrated" against outcomes of field trials, such as the EU sponsored MERGE and EMERGE and later ones of the UK Steel Construction Institute. FLACS also became suitable for dispersion simulation.

In parallel to the above-described activities, software packages comprehending an entire risk analysis have been developed, such as PHAST Process Hazard Analysis Software marketed now by DNV-GL, and the TNO developed EFFECTS and RiskCurves software, now owned by Gexcon. These developments have been a process of many years and are still not finished. User groups help strongly to make further advances. However, all of this has been for static situations, although some modules of the packages would be suitable to make the analysis dynamic.

2.3.3 Layers of protection analysis

Toward the end of the 1990s a few new methods emerged based on logic: Layers of Protection Analysis (LOPA) and Bowtie. LOPA is an event tree-based barrier effectiveness analysis method. Barrier thinking stems from Haddon (1973). The development of LOPA comes together with a strong wave of automation in the process industry and the development of highly reliable electronic programmable logic controllers (PLCs). This led to the definition of Safety Instrumented Functions (SIFs) to be applied as protective measures of last resort. Realization of these functions is by Safety Instrumented Systems (SISs). The required Safety Integrity Levels (SILs) were for the first time specified for the industry in general in standard IEC 61508 in 2001, in 2003 followed by IEC 611511 special for the process industry, which was Americanized in 2004 with ANSI/ISA-84.00.01-2004 (n.d.). Meanwhile, IEC 61508 2nd edition is from, 2010 and the IEC 61511 one from, 2016. The SIL levels are presented in Table 5. The idea is that SIS devices are offered for a certain SIL, however,

Table 5 SIL levels according to IEC 61511: 2016 (E) (2016).

	Low demand SIS	**High demand SIS**
SIL	Probability of dangerous failure on demand, *PFD*	Probability of dangerous failure λ_{AVG} (h^{-1}), *PFH*
3	$\geq 10^{-4}$ to $<10^{-3}$	$\geq 10^{-8}$ to $<10^{-7}$
2	$\geq 10^{-3}$ to $<10^{-2}$	$\geq 10^{-7}$ to $<10^{-6}$
1	$\geq 10^{-2}$ to $<10^{-1}$	$\geq 10^{-6}$ to $<10^{-5}$

unfavorable equipment design, lack of SIS inspection, testing and maintenance can lower the level. When going into detail it does not remain that simple (redundancy, voting systems, undetected dangerous failure, common cause failure) to determine the actual SIL.

Now, SISs are not the only independent protective devices that can be used. SISs consist of sensor(s), PLC and actuator circuit, they are expensive and there are additional cheaper solutions. If as far as possible inherent safer design has been applied and if the process still becomes uncontrollable because the safety provisions in the basic process control are exhausted, there are various options to bring the process back to normal conditions, or to apply an emergency shutdown. In the first place, after critical alarm is activated, there is active control either by intervention of operator/supervisor or shutdown (trip); but if the situation develops faster and more serious the safety layer of emergency shutdown can be activated (manually or automated), while in addition active safety systems shall protect for larger damage (safety instrumented systems (SIS) and actively effect reducing devices, such as steam curtains and water deluge); in case the latter are breached, permanently present passive layers (zoning, bunds, walls) shall mitigate effects, and ultimately emergency response force of the company or finally of the city fire brigade shall act. All these barriers form layers, which in first instance are considered independent. Given a HAZOP that concludes that protective layers must be installed a LOPA is to be performed, which can be considered as a mini-QRA (see Fig. 7).

Depending on the severity of the severest possible consequence, presence of employees and other conditions the standard IEC 61511 prescribes through a Risk Graph what SIL is required.

2.3.4 Bowtie

Bowtie is the very useful combination of fault and event tree with the top/initiating event in the center. Visser (1998) was the first to publish the concept which was in use at Shell E&P. It was picked up by the team working on the EU risk assessment improvement ARAMIS project that we shall describe near the end of the next section (de Dianous & Fiévez, 2006). Their picture of the concept is quite clear and is reproduced here as Fig. 8.

Any branch from basic event to any major event forms a possible scenario. Bowtie is used both in a qualitative sense in design for overview of all scenarios given a critical event and in operation for checking the health

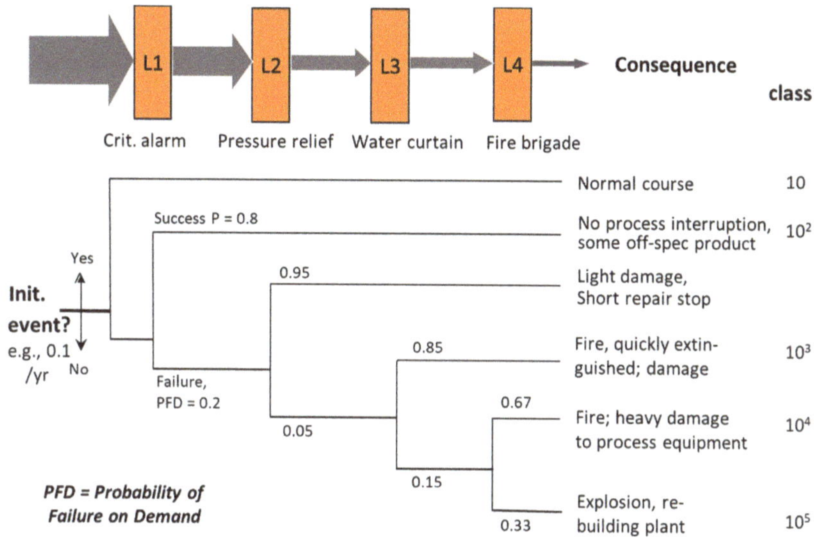

Fig. 7 Example of a LOPA event tree of four successively activated layers with at the second layer a SIL 1 device. As long as the layers can be considered acceptably independent (no significant disturbance of a later layer by a previous one failing) the probabilities of failure on demand, PFDs, can be multiplied, hence here given an initiating event, IE, resulting in an explosion probability of $5 \cdot 10^{-4}$ or fulfilling SIL 3 requirement. So, with IE frequency of 0.1/year, the expected frequency value of explosion is $5 \cdot 10^{-5}$/year. *After Pasman, H.J. (2015).* Risk analysis and control for industrial processes—gas, oil and chemicals. A system perspective for assessing and avoiding low-probability, high consequence events. *Butterworth Heinemann, Copyright © 2015 Elsevier Inc., ISBN:978-0-12-800057-1.*

of preventative, protective or mitigative barriers, but it can be used too in a quantitative risk analysis. If all failure frequency data, including those of the barriers, are known the bowtie can be evaluated most conveniently by applying Bayesian network (BN). This can also in a dynamic fashion simulating slow degradation or sudden change in failure. Bayesian network consists of a causal structure in which each node represents an event and the connecting arcs or edges the dependencies between node variables. Thanks to software developed in the last 20 years, given inputs at the basic nodes (parents or roots) the probability value of intermediate nodes (children or leaves) and output nodes are calculated. There are many aspects to BNs; these will be explained in Section 4.

A full treatment of the (static) risk assessment methodology up to about 2000 can be found in CCPS (2000) and Cameron and Raman (2005).

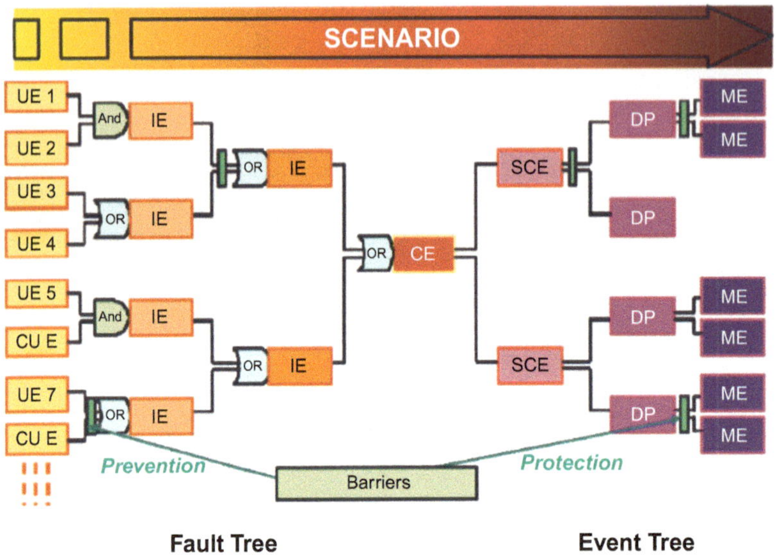

Fig. 8 Bowtie diagram (de Dianous & Fiévez, 2006); Basic causes: UE = Unwanted Event, e.g., human act, and CU E = Current Event condition, direct cause; IE = Intermediate Event, e.g., pump fails, CE = Critical Event, 12 types: leak, start of fire, etc., SCE = Secondary CE, escalation, DP = Dangerous Phenomena, 13 types VCE, pool fire, jet fire etc., ME = Major Event, four types: overpressure, heat radiation, toxic load, pollution.

3. Benchmarking of risk assessment methods

3.1 First benchmark, 1990

Because risk assessment had obtained such an essential role in the process industry safety regulation with regard to land use planning and licensing plant and the number of risk analysis models had been growing, the European Community (EC, not EU yet) sponsored a benchmark project in the late 1980s, early 1990s, managed by the EC Joint Research Centre, JRC in Ispra, Italy. A common element in all risk analysis methods in the various countries in Europe was the dispersion of heavy gas. One was aware of differences in model approaches and uncertainties in outcomes since the Thorney Island trials. Amendola, Contini, and Ziomas (1992) reported about the exercise in which 11 teams calculated the dispersion of spills from an ammonia storage plant of 15,000 ton refrigerated NH_3 supplied by ship tanker via subsea pipelines, and in turn after gasification supplying via 6 km long pipelines pressurized tanks at the site of a fertilizer plant.

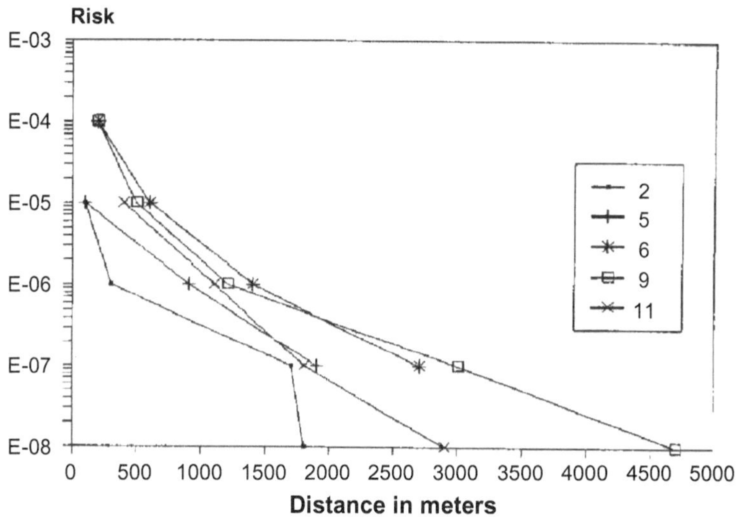

Fig. 9 Result of EC Benchmark Phase I: risk analysis of pressurized tank section of the NH_3 storage plant; risk is individual risk, hence probability to be killed when exposed for a year. *After Amendola, A., Contini, S., & Ziomas, I. (1992). Uncertainties in chemical risk assessment: Results of a European benchmark exercise.* Journal of Hazardous Materials, *29, 347–363.*

In Phase I a risk analysis of a plant section was asked to perform and in Phase II the teams did a system reliability analysis for an overpressurized scenario of the refrigerated tank, a human action analysis to isolate a pipe break, and four spill scenarios among which guillotine pipe fractures. The results showed order of magnitude differences. An example of results of Phase I is presented in Fig. 9. No uncertainty estimates were made.

A spill of refrigerated liquid ammonia generates at first a plume heavier than air and rainout, while after warming it is much lighter than air. So, it is a challenge for the dispersion models. However, prior to the dispersion calculation the teams' insights about the source term (spill rate, pool, evaporation) differed significantly contributing to the spread. The Phase II results of one of the three weather class dispersion predictions are presented in Fig. 10.

This was followed by a prediction of the individual risk contours following a liquid spill (see Figs. 3 and 11. For competent authority decision making the uncertainty in outcome seems unacceptable.

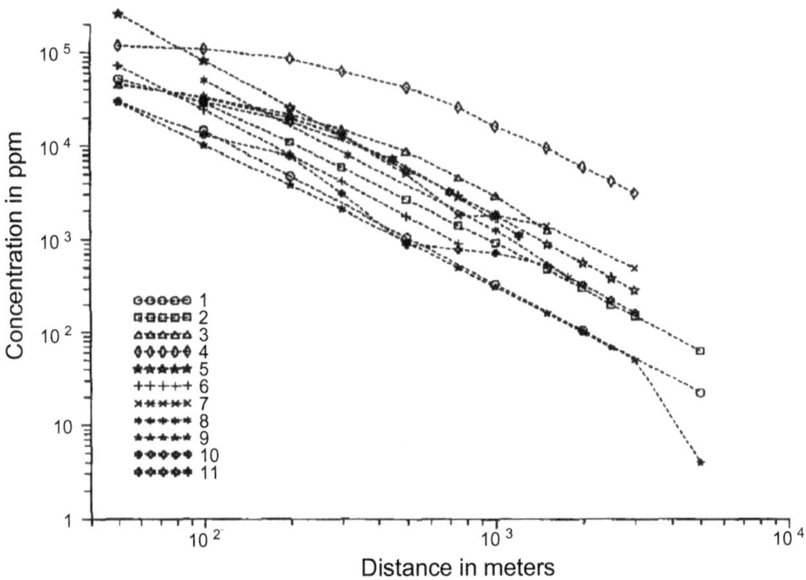

Fig. 10 Phase II weather class 2 NH_3 concentration predictions of 11 teams. *After Amendola, A., Contini, S., & Ziomas, I. (1992). Uncertainties in chemical risk assessment: Results of a European benchmark exercise.* Journal of Hazardous Materials, 29, *347–363.*

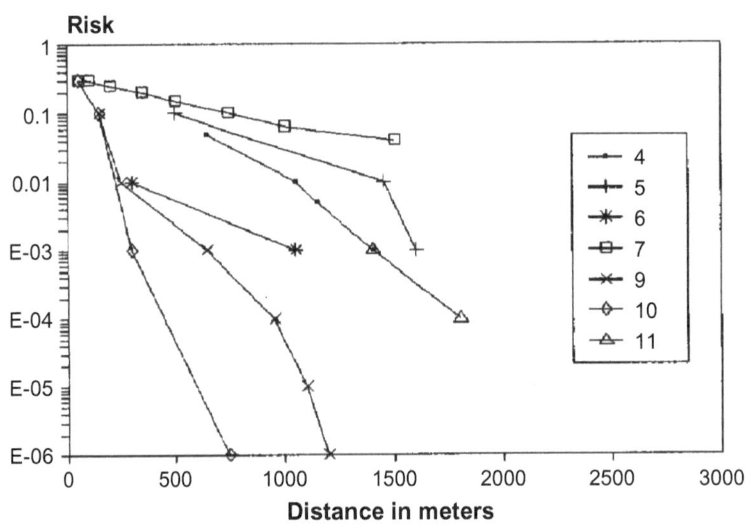

Fig. 11 Phase II result of individual risk as function of distance. Already at a risk level of 10^{-3} the spread in predicted distances ranges over a few kilometers. *After Amendola, A., Contini, S., & Ziomas, I. (1992). Uncertainties in chemical risk assessment: Results of a European benchmark exercise.* Journal of Hazardous Materials, 29, *347–363.*

3.2 Model evaluation group and EU SMEDIS, 1995

In the years of the benchmark and after, EC Directorate General XII, Science, Research and Development took initiatives attempting to ameliorate the QRA accuracy, stimulating coordinative work on dense gas dispersion models and sponsoring research on vapor cloud explosion. In all of this, UK HSE has been very active. With respect to gas dispersion the evaluation work in the United States by Hanna, Chang, and Strimaitis (1993) has been stimulating. In 1992 the Commission of the EC (CEC) founded the Model Evaluation Group (MEG) to develop for model quality assurance a protocol and guidelines (Petersen, 1994). A broad European discussion with ties to the United States took place and a comparison was made of heavy gas dispersion models for instantaneous releases identifying weak representation, such as low-wind speed and Pasquill F atmospheric stability (Brighton et al., 1994), which by the way from accidents' experience appear to be the most hazardous conditions (e.g., Chamberlain, Oran, & Pekalski, 2019; Havens, Walker, & Spicer, 2012). Following the preparatory work of the MEG, the CEC funded an exercise called Scientific Model Evaluation of Dense Gas Dispersion Models (SMEDIS) coordinated by UK HSE and running between 1996 and 1999 (Duijm, Carissimo, Mercer, Bartholome, & Giesbrecht, 1997). A model evaluation should be in six steps:

1. description of the model (origin, type, documentation, etc.);
2. description of the database used for the validation (references, type, release conditions, etc.);
3. scientific assessment of the model (completeness of the description of the physical and chemical phenomena, assumptions made, use of model constants, solution techniques);
4. user-oriented assessment of the model ("user-friendliness," guidance and assistance of the user, quality of user documentation, computer requirements);
5. verification of the code (software errors);
6. validation of the model by comparing model predictions with (experimental) observations.

A model that has been assessed according to the SMEDIS protocol by an independent expert in 2000 has been UDM 6.0, the gas dispersion model in DNV-GL's PHAST QRA software package. Lately, in Europe a new partly EU sponsored consequence model evaluation has been initiated, called SAPHEDRA, a EU collaborative platform. Objectives, history, participants, protocol and way of working are described in a UK HSE (2017) report.

3.3 ASSURANCE and later benchmarks, 2000

Next, EU sponsored benchmark project, named ASSURANCE (ASSessment of Uncertainties in Risk Analysis of Chemical Establishments). It has been running from 1998 to 2001 (Lauridsen et al., 2002). It concerned again an ammonia storage plant, but this time each of the seven experience participating teams conducted a QRA of the complete plant independently. The results have been devastating again in the sense that differences in outcomes from a decision maker point of view have to be qualified as unacceptably large. In Fig. 12A the largest and smallest 10^{-5}/year individual risk contour of the seven calculated are plotted; the largest cutting through a residential area. In Fig. 12B the *F–N* curves are depicted showing at least two orders of magnitude difference.

After results had been produced, a detailed investigation was made of the data used by the different teams, scenarios defined, etc. This led to a ranking of uncertainties as summarized in Table 6. It is clear that scenario definition is the most important factor followed by failure rate data and choice of dispersion model. Application of QRA for licensing became therefore either fully standardized as in the Netherlands or performed by the same team with fixed models and data as by HSE in the UK.

After this benchmark exercise, the EU sponsored from 2002 to 2004 the development project of an improved risk assessment model ARAMIS (Accident Risk Assessment Methodology for IndustrieS). The various project parts have all been described in a special issue of the *Journal of Hazardous Materials* starting with an overview paper of Salvi and Debray (2006), while some prominent aspects will be treated in the next section.

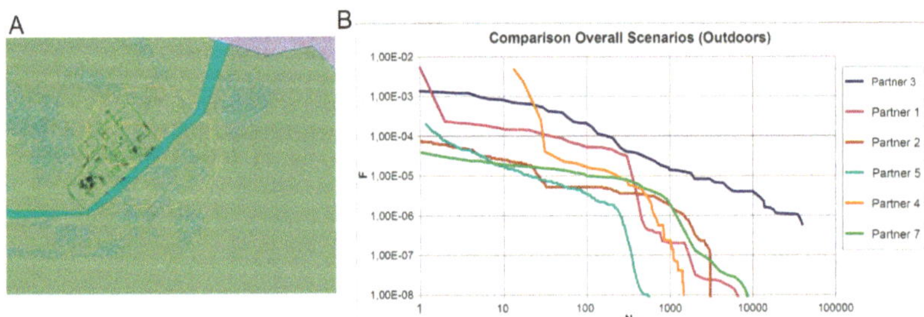

Fig. 12 (A) Largest and smallest individual risk contour. (B) Group risk *F–N* curves calculated by the teams. *After Lauridsen, K., Kozine, I., Markert, F., Amendola, A., Christou, M., & Fiori, M. (2002).* Assessment of uncertainties in risk analysis of chemical establishments. The ASSURANCE project. *Final summary report, Risø National Laboratory, Roskilde, Denmark, 52 p., http://www.risoe.dk/rispubl/sys/syspdf/ris-r-1344.pdf.*

Table 6 Factors contributing to uncertainty in QRA results.

Uncertainty factor	**Importance**
Differences in the qualitative analysis	**
Factors relating to frequency assessment:	
Frequency assessments of pipeline failures	***
Frequency assessments of loading arm failures	****
Frequency assessments of pressurized tank failures	****
Frequency assessments of cryogenic tank failures	***
Factors relating to consequence assessment:	
Definition of the scenario	*****
Modeling of release rate from long pipeline	***
Modeling of release rate from short pipeline	*
Release time (i.e., operator or shutdown system reaction time)	***
Choice of light, neutral, or heavy gas model for dispersion	****
Differences in dispersion calculation codes	***

More stars mean a larger contribution.
After Lauridsen, K., Kozine, I., Markert, F., Amendola, A., Christou, M., & Fiori, M. (2002). *Assessment of uncertainties in risk analysis of chemical establishments. The ASSURANCE project.* Final summary report, Risø National Laboratory, Roskilde, Denmark, 52 p., http://www.risoe.dk/rispubl/sys/syspdf/ris-r-1344.pdf.

A final EU benchmark BEQUER (Benchmarking Exercise in Quantitative Area Risk Assessment in Central and Eastern European Countries) has been conducted with the fully different objective to find out what spread could be seen in assessments of a fixed QRA by independent reviewers (Fabbri & Contini, 2009).

Because of the uncertainties in scenarios and failure rate data, with the launch of the 3rd Seveso Directive in 2003, the EU founded the European Working Group on Land Use Planning (EWGLUP) with, a.o., the task to collect scenarios as well as failure rate and modeling data (Christou, Gyenes, & Struckl, 2011); the activity was later absorbed by the Major Accident Hazards Bureau of EC JRC, Ispra.

In the United States, EPA (Environmental Protection Agency) did not have these problems as the Risk Management Program rule RMP 40 CFR Part 68 requires only safe distances given specified endpoint concentrations for a worst-case accident (full rupture of inventory containment) or an alternative scenario including protection layers. At the time, use of prescribed dispersion software was a must, a requirement that is meanwhile relaxed.

4. Highlighting of recent risk assessment methods and tools

The structure of risk assessment packages did not change. As a result of worldwide efforts, the source term and consequence models became much finer tuned and more accurate. Recently, the presumption that very large fuel clouds can detonate causing enhanced destructive force, became proven (Chamberlain et al., 2019), but enabling conditions and safety distances have still to be sorted out. The major changes, though, are in handling uncertain information and data. Uncertainty can be distinguished in *aleatory* uncertainty due to randomness in data, e.g., by inadequate repeatability of measurements, and *epistemic* uncertainty due to insufficient knowledge, e.g., in number and quality of data. It is in particular the latter that is addressed in the recent approaches. The latter varies from unawareness of any information, to various degrees of imprecise information. These degrees are (1) Interval analysis (uniform distribution with all values considered equally likely); (2) Possibility theory, consisting of a set of elements each with a likelihood value (Zadeh, 1978; relation with Fuzzy set approach); (3) Evidence theory (Dempster-Shafer), which specifies a limited number of focal elements with each a measure of credibility or basic belief assignment, together summing to 1; (4) Probabilities of elements in a fully developed structure embodied by a probability density function (see further Helton & Johnson, 2011).

Apart from the quest of improved hazard identification and characterization, various techniques developed under the heading of machine learning and artificial intelligence with applications in the realm of big data and analytics have recently been applied to risk assessment. This enabled higher accuracy of estimates and sharper confidence limits, but also picking up weak warning signals in case of operational risk assessment. Examples will be given of selected papers treating various topics. This chapter is not, however, a review, and the literature overview that attempts to cover all important aspects is not complete. Near the end of this section, a paper about the many flaws that afflicted QRA studies will be referred to as well as one on requirements with respect to QRA validity evaluation.

4.1 Hazard identification and scenario definition

In the ARAMIS project, Delvosalle, Fievez, Pipart, and Debray (2006) tried scenario definition a different way. They did not apply HAZOP, but they developed a two-step approach. The first step employs a set of generic fault and event trees for typical equipment and hazardous materials processes covering the whole of process industry leading without barriers to worst-case major accident events. This was called MIMAH (methodology for the identification of major accident hazards). The second step is taking account of safety barriers, failure frequencies, and various possible consequences leading to MIRAS scenarios (methodology for the identification of reference accident scenarios). Not having seen any feed-back on success or failure, one can argue that, even if it looks perfect, generic information will never be able to cover all possibilities that can occur in real practice.

Cameron et al. (2017) reviewed HAZOP procedure improvements, such as PFFM (Process Flow Failure Mode analysis, MacGregor, 2013), making use of process simulation models, such as Aspen HYSYS (Janošovský, Danko, Labovský, & Jelemenský, 2019), and (semi-) automated HAZOP developments. The latter applied expert systems, digraph or Petri net simulation. Taylor (2017) presented his own developments of automated modules and experiences over 40 years and concluded that parallel to HAZOP's social interaction, automation will have a role to provide "consistent or standardised terminology, better recording of consequences, a significant improvement in completeness, and a much larger knowledge base than is usually available to even the best HAZOP team." And once identified, digitized cause consequence paths can help in operational risk assessment and alarm management to trace or infer causes and also in new design. However, reasoning according to algorithms is different from the holistic way humans think, so the "creative thinking" of HAZOP shall be kept, but the automation will make tasks much more effective.

A system approach to hazard identification is essential. Leveson (2011) proposed a socio-technical system (STS) approach enabling an accident investigation technique STAMP (System-Theoretic Accident Model and Processes) and STPA (System-Theoretic Process Analysis) as a hazard identification tool. These are based on the principle that safety is emergent and a control problem. So, on all levels of the STS control loops can be defined, which with four questions regarding action appropriateness and its timing, it can indicate all possibilities of what can go wrong. Indeed, various researcher experiences learn that by this way hazard identification becomes much more complete, but due to lack of software the tool is very labor-intensive. Helpful in thinking to establish a causal structure, though, are System Dynamics Modeling and FRAM (Functional Resonance Analysis Method); the former is developed by Forrester (1961) at MIT, the latter by Hollnagel (2012). System dynamics consists of multiple causal loops and presents an overview of how changes interact; for a chlorine unloading example, see Garbolino, Chery, and Guarnieri (2016). FRAM consists of hexagonal elements with the functions: Input (I); Output (O); Resources (R) consumed; Constraints/controls (C); Preconditions (P); Time (T) at the vertices. The system is modeled by connecting functions, without details how exactly processes are within the element. The modeling can be at various levels of abstraction. Smith, Veitch, Khan, and Taylor (2017) show a propane feed control FRAM structure. Both system dynamics and FRAM software are available (FMV, 2016; Richmond, 2005, respectively).

Yet, another way of making use of a system approach is BLHAZID (Blended Hazid), which will enable automation (Cameron et al., 2017; Németh & Cameron, 2018). Basis is an electronic P&ID in the data standard ISO 15926 (2009) format. Based on this standard, a generic data model and a Reference Data Library for process plants have been developed. Digitization and Industrial Internet of Things is the future (Lee, Cameron, & Hassall, 2019); BLHAZID responds to that. The system is about a structure of plant, people and procedures, and capabilities delivering functions and intended targets (see Fig. 13. The blending refers to the combination of function-driven HAZOP and component-driven FMEA. Functions can fail and propagate effects. There are interactions among equipment, components, and the streams. This Socio-Technical System structure shapes the underlying causal model, the principle of which is depicted in Fig. 14 and an example in Fig. 15 showing detail for an ordinary pipe segment.

For various interconnected subsystems, this approach leads to build up of a knowledge base of causes, implications, and causal pathways. Then, after reading the P&ID, the whole is decomposed automatically into subsystems linking inputs to internal subsystems and mass and energy inventories, to outputs, all with three levels of aggregation, hence down to detail. Component types are automatically recognized and linked with the knowledge base. Given a functional component, or equipment mode failure, search algorithms are used to track down causes and implications for creating causal graphs. The latter can act as diagnostic tools for operators, and can be integrated with plant information systems and further developed. It opens up means for dynamic operational risk assessment.

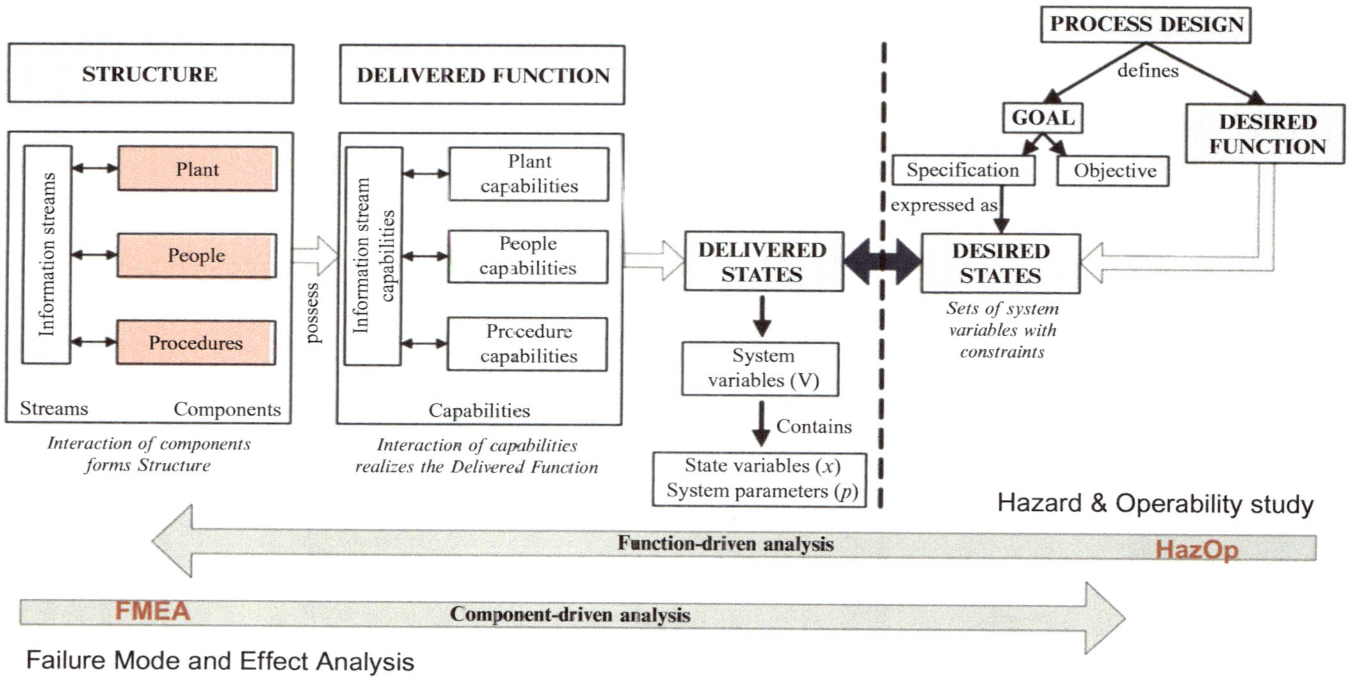

Fig. 13 A functional systems framework. *Adapted after Seligmann, B.J., Németh, E., Hangos, K.M., & Cameron, I.T. (2012). A blended hazard identification methodology to support process diagnosis.* Journal of Loss Prevention in the Process Industries, 25, *746–759.*

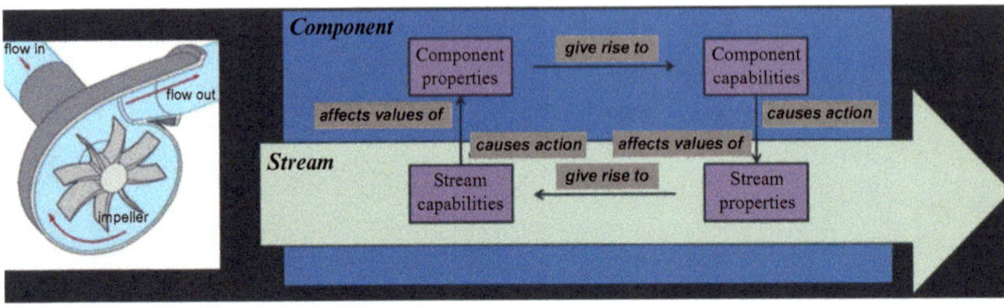

Fig. 14 Interaction of equipment and flow that shapes the causal model for failure and failure propagation. *After Németh, E. & Cameron, I. (2018). Multi-level failure, causality and hazard insights via knowledge based systems.* Mary Kay O'Connor Process Safety Center (MKOPSC), 21st Annual International Symposium October 23–25, *College Station, TX.*

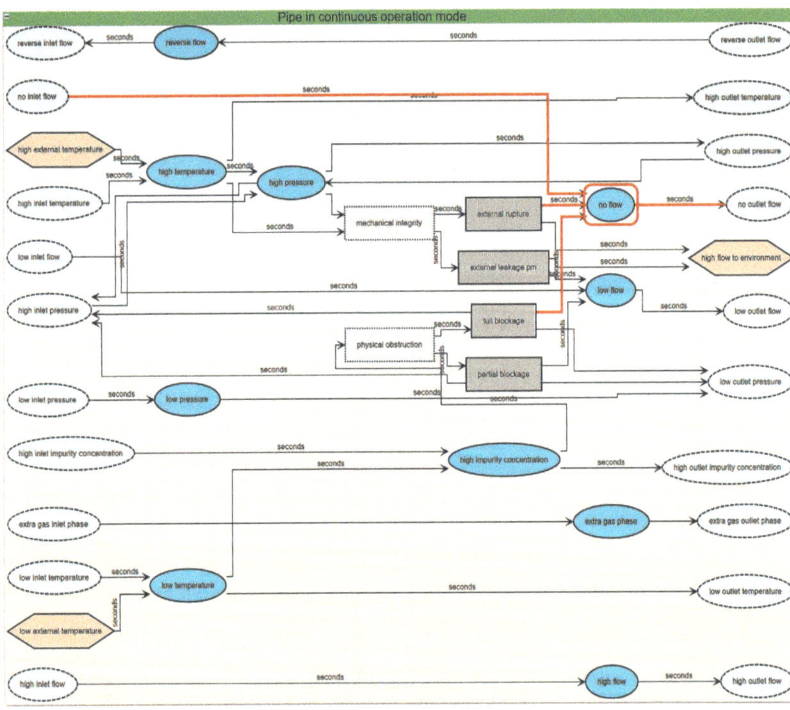

Fig. 15 Detailed causal model of a pipe in continuous flow operation mode including order of magnitude time constants of effect following the cause. For illustration, by going with the mouse over the state of "no flow" in the graph possible causes and implications are highlighted. *After Németh, E. and Cameron, I. (2018). Multi-level failure, causality and hazard insights via knowledge based systems. Mary Kay O'Connor Process Safety Center (MKOPSC), 21st Annual International Symposium October 23–25, College Station, TX.*

4.2 Mathematical statistical tools, machine learning, and artificial intelligence techniques: Causal networks, pattern recognition, weak abnormal/warning signal detection

Over the past century many statisticians have provided us with a richness of mathematical-statistical tools. As an illustration consider the collection with regard to continuous data streams under "Statistical" in Fig. 16, adapted from the review of Alauddin et al. (2018). These tools will bound uncertainty, discover trends in data, find useful signals in random clutter, or determine with a certain confidence a parameter value in observations obscured by randomness. Most of this was based on linear correlation.

Given an abundance of data, the tools developed over the last 4 decades in the framework of Machine Learning (ML) and Artificial Intelligence (AI) (Fig. 16 under "Hybrid" and "AI based") offer capabilities to create causal networks, that build from sometimes hidden root causes up to observable failures. And further, the tools provide capabilities to determine trends in collections of observations, hence as sophisticated non-linear curve fitting, and to recognize patterns. In general, the techniques enabled reducing uncertainty and to produce something useful from apparently confusing information. At first, in the development stage these techniques had mainly static data in focus, but later they were applied to continuous streams of observed data, which are subject to random error. In fact, this occurs the same way as mathematical-statistics based techniques in signal processing and process control, such as Kalman filtering, which have already been in use for quite some time. After exposure to historical process data, called training and testing, on-line applications in process control to detect faults and to diagnose causes becomes possible, thus performing dynamic operational risk assessment.

Most popular of techniques became BN and ANN (Artificial Neural Network, often shortened to NN) leading the branches under Hybrid and AI-based, respectively. These two and related ones will be briefly summarized in Sections 4.2.1 and 4.2.2.

4.2.1 Bayesian networks (BN), and petri net (PN)

BN, also called Bayesian Belief Networks, as it can handle imprecise subjective information, relates causes and their effects in a structure of nodes connected by directed arcs (acyclic directed graph) probabilistically (for any distribution function). Due to its built-in mathematics (software) one can infer the probability of an outcome of a complex structure and even the other way around given an observed outcome diagnose the (adjusted)

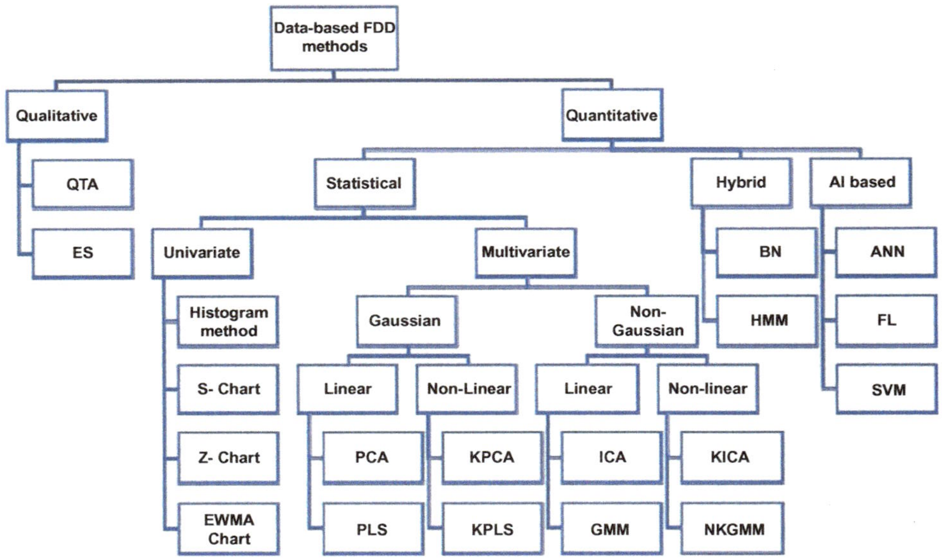

LEGEND

FDD = Fault Detection and Diagnosis
QTA = Qualitative Trend Analysis
ES = Expert Systems
S-Chart = Standard deviation chart
Z-Chart = three perspectives over time
EWMA = Exp. Weighted Moving Average
PCA = Principal Component Analysis
PLS = Partial Least Squares
K – Kernel cluster
ICA = Independent Component Analysis
GMM = Gaussian Mixture Model
NKGMM = Non-linear Kernel GMM
BN = Bayesian Network
HMM = Hidden Markov Model
ANN = Artificial Neural Network
FL = Fuzzy Logic
SVM = Support Vector Machine

Fig. 16 Classes of techniques to enable in process control fault detection and diagnosis. *Reprinted (adapted) with permission from Alauddin, M., Khan, F., Imtiaz, S., & Ahmed, S. (2018). A bibliometric review and analysis of data-driven fault detection and diagnosis methods for process systems.* Industrial & Engineering Chemistry Research, 57, *10719 – 10735. Copyright 2018 American Chemical Society.*

probability of a cause. BNs make use of the Bayesian theorem that states that joining older information (as prior distribution) with newly gained (likelihood) yields an update (as posterior distribution). BN is also Markovian in the sense that nodes (as effect) are only influenced by the causes they are directly connected with and not by ones higher up. All this made BNs in the past decade very popular in risk assessment. History, pioneers, and mathematical formulation and other details will be explained in Section 4. Meanwhile, variants developed are object-oriented BN (OOBN), which is just a software enabled feature to connect subsystem structures to a main one and so keeping overview, and dynamic BN (DBN) by introducing a time function, e.g., a reliability degradation. Applications of BN in risk assessment, including security, have become plentiful. Given data, algorithms, such as the K2 one, produce the most probable BN structure model (Cooper & Herskovits, 1992; Lerner & Malka, 2011). Schwarz (1978) developed the basis for the Bayesian Information Criterion (BIC) to judge the quality by the likelihood of the model derived. Khakzad, Amyotte, Cozzani, Reniers, and Pasman (2018) presents an application of BIC on BNs.

Not mentioned in Fig. 16 but also finding applications in dynamic risk assessment is *Petri net* (PN), which basically is a logistics network simulation with (token) storage places connected via conditional transition points. Petri nets have been used much in network traffic problems. The timed version, in which transitions are fired at pre-specified times, is much more flexible in temporal dynamics than DBN. It can be made stochastic and solved by MCMC, while tokens can be of different type (colored). Creating a net is usually more effort-intensive than in case of BN. Petri net is used much by reliability engineers for modeling dependability. So, is it used successfully in SIL verification, e.g., Zhao, Malasse, and Buchheit (2019). IEC 61508 (2010) recommends Petri net as a tool to analyze control loops, while IEC 62551 (2012) described standard shapes, other symbols, and typical network representations of the methodology. Kabir and Papadopoulos (2019) reviewed BN and PN applications.

4.2.2 Artificial neural network (ANN) and support vector machine (SVM)

In an AIChE perspective paper, Venkatasubramanian (2019) described ANN applications in chemical engineering including safety. Although the term had already been coined in 1956, it took a long time and many attempts before ANN became a practical tool that after training can recognize patterns. Venkatasubramanian starts with mentioning expert systems

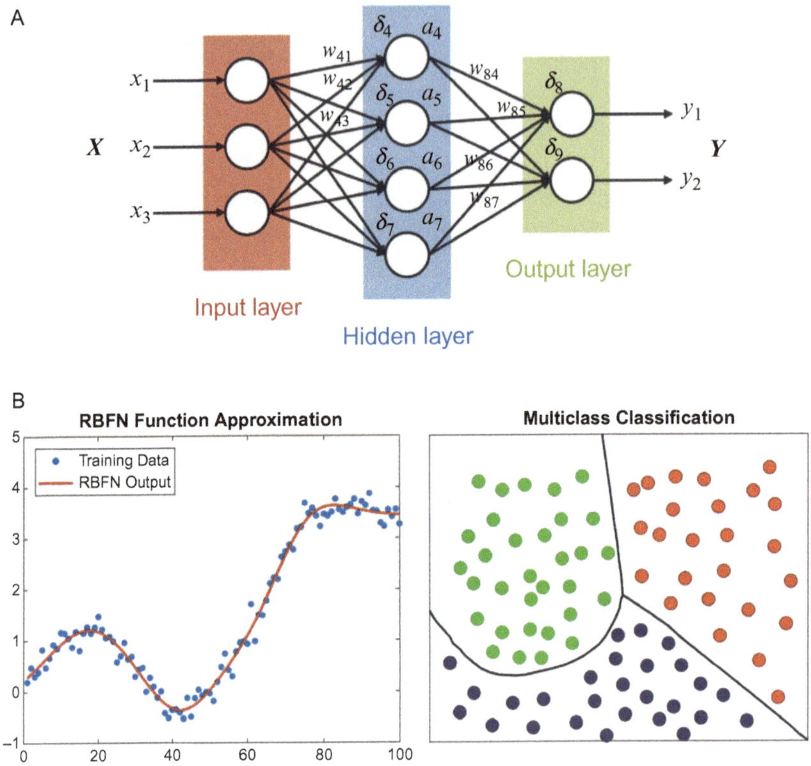

Fig. 17 (A) Feedforward neural net architecture (input *x*, weight *w*, backpropagated error *δ*, real output number *a*, output *y*); (B) Examples of Non-linear fit (RBFN = Radial Basis Function Network) and Multiclass Classification. Lee, Shina, and Realff (2018) wrote an overview article on Machine Learning techniques for process systems engineering. *After Venkatasubramanian, V. (2019). The promise of artificial intelligence in chemical engineering: Is it here, finally?* AICHE Journal, 65 *(2), 466–478; Figure 1, p.469, based on https://medium.com/@curiousily/tensorflow-for-hackers-part-iv-neural-network-from-scratch-1a4f504dfa8; https://neustan.wordpress.com/2015/09/05/neural-networks-vs-svm-where-when-and-above-all-why/; http://mccormickml.com/2015/08/26/rbfn-tutorial-part-ii-function-approximation/.*

(1980s until 1995) in which knowledge was cast into If-Then-Else rules, which have also been used in HAZOP automation attempts. In principle an attractive concept but it turned out to be too laborious and cumbersome. In the next phase the neural network (see Fig. 17) appeared as we know it now, because the backpropagation learning algorithm adapting weights had become practical. For detailed insight and understanding of how it works, Rojas (1996) is recommended. Meanwhile there are many variants. Supervised learning means that the system is trained by sets of corresponding input-output data. In case of unsupervised learning the system itself must find a relation.

There are yet only few applications of ANN in risk assessment. The group of Adedigba, Khan, and Yang (2017) published one on dynamic failure of a process system, taking as example the Tennessee Eastman (TE) process. It requires, of course, training data including all possible faults; in the example, 20 faults were simulated. Interesting is also Wu and Zhao (2018) describing a deep convolutional NN for feature extraction process, here applied to fault detection and diagnosis (FDD). A problem of ANN is so-called over-fitting. There are a few methods to prevent this and one of these is Bayesian regularization (the combination is shortened as BRANN), which is an effective algorithm to keep mean square errors low, hence minimizing estimation errors. The application is on saving runs by CFD code by first calculating the result of a number of representative scenarios. Shi, Khan, Zhu, Li, and Chen (2018) describes the procedure in detail for CFD code FLACS and dispersion of a cloud from an offshore platform. Input to the ANN is a set of scenario parameters: leak rate and direction, wind speed and direction. Output is the volume of a cloud as a function of time, not the geometry. A few hundred runs must be made for development and another few hundred as a test set. In a follow-on article Shi, Zhu, Khan, and Chen (2019) used the same technique and the same CFD code to fit the flammable volume and the overpressure-time history at various ignition points. Comparison with actual CFD outcomes was good. By introducing leak and ignition probabilities, risk can be derived. Shi, Li, et al. (2019) made further improvement by combining BRANN with Artificial Bee Colony (ABC) algorithm, for solving multidimensional and multimodal optimization problems.

Support vector machine (SVM) in its simplest form is a supervised learning algorithm able to classify data in two different groups by making use of a Kernel function, of which various types are available, e.g., Gaussian (Cortes & Vapnik, 1995). Taqvi, Tufa, Zabiri, Maulud, and Uddin (2018) applied SVM in FDD of a continuous process data stream from a distillation; Onel, Kieslich, Guzman, Floudas, and Pistikopoulos (2018) monitored a batch process for FDD.

4.3 Low-frequency event data: Hierarchical Bayesian analysis, WinBUGS, MCMC

A common problem in risk assessment is the lack of sufficient data. As regards equipment failure rate data, this is mainly due to lack of monitoring and registering failures, in particular when these are rare and that are just the large consequence–low-frequency major accident events, which are of high interest in risk assessment. To alleviate this problem, the two-stage hierarchical Bayesian analysis (HBA) or modeling (HBM) has been developed (Bier and Yi, 1995; Siu & Kelly, 1998). Alarms, safety barrier breaches, near misses, and other so-called precursors occur more frequently, and observation of the frequency of those precursors over a number of years and also possibly together of other plants in the sector will help. The relation between an initiating event and these observed occurrences can be modeled in an event tree.

The HBA procedure on the observed data is clearly described by Yu, Khan, and Veitch (2017), both deriving the branch probabilities of the event tree and the desired most probable rare event frequency. The first stage is applying Bayes theorem to derive an informative posterior distribution based on the observation data as likelihood and a non-informative prior governed by hyperparameters α and β. The theorem can be solved analytically when prior and likelihood are conjugate distributions, reason for Siu and Kelly (1998) to propose a gamma prior, while barrier breaches can be represented by a binomial distribution and an initiating or failure event by a Poisson distribution. As step 1 (see Fig. 18) the likelihood of the data given the hyperparameters must be derived by computing the expectation of the likelihood

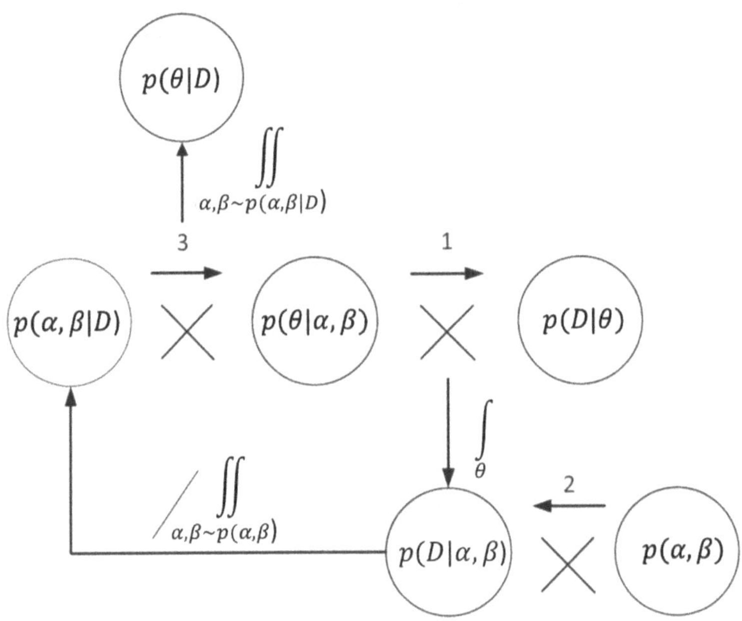

Fig. 18 HBA solving procedure according to Yu et al. (2017). 1–3 refer to the steps mentioned in the text.

function of the parameter of interest θ (e.g., event rate). In case of fault or event tree the latter is being jointly modeled, given the hyperparameters. In step 2 the Bayes theorem provides the posterior hyperparameter values. As step 3 the posterior distribution of θ given data, $p(\theta \mid D)$, is computed by marginalizing α and β out. Integrals are solved numerically by applying Markov Chain Monte Carlo incorporated in WinBUGS (Lunn, Jackson, Best, Thomas, & Spiegelhalter, 2012) or OpenBUGS, Rodionov, Kelly, and Uwe-Klugel (2009), where BUGS refers to Bayesian inference Using Gibbs Sampling. Beside the case in Yu et al. (2017), various other example cases are published recently (El-Gheriani, Khan, Chen, & Abbassi, 2017; Khakzad, Khan, & Paltrinieri, 2014; Yang, Khan, Lye, & Amyotte, 2015).

4.4 Rare process fault events: Maximum likelihood maximization of entropy (MLME) and hidden Markov model (HMM)

Ahooyi, Soroush, Arbogast, Seider, and Oktem (2014) proposed MLME and in a related paper Ahooyi, Arbogast, Oktem, Seider, and Soroush (2014) just Maximization of Entropy. The first applies MLME to determine a multivariate probability density function (pdf) of discrete random variables that maximizes simultaneously likelihood and entropy. The method needs historical data sets to train the computational system. A pdf is derived, of which the moments match with those of the random variable data set, and which represents of all pdfs possible the one with the highest uncertainty (maximum noise and thus entropy). For the final solution of identifying a rare event a BN for fault predicting and diagnosis is applied. The second method is based on determining multivariate probability mass functions (pmf). These papers were followed by one on the "rolling pin" method as an alternative to BN (Ahooyi, Arbogast, & Soroush, 2015a) and a second one to capture a probability density function (pdf) of a continuous random variable (Ahooyi, Arbogast, & Soroush, 2015b).

Rebello, Yu, and Ma (2019) developed an alternative approach applying HMM and BN. In a normal Markov process, states can be observed, but in an HMM they cannot, therefore it is called hidden. HMM is applied on sequential data, hence in case of a process the time series of its variable values. HMM has again many variants (Murphy, 2002) and found many applications, e.g., in speech recognition. In risk assessment it can be applied in fault detection and diagnosis in combination with a system model BN. HMM itself can be represented as the simplest DBN; given a transition matrix it converts discrete measurement values into probabilities of the possible hidden states (e.g., binary functional or faulty) of an equipment component applying the Viterbi algorithm. These probabilities are updating prior probabilities of the component state of the system BN. In case of fault the system BN guides to possible causes.

4.5 Expert elicitation: Fuzzy set, evidence theory, evidential network, probabilistic classical method

Expert interviewing and asking their judgment is another way to provide information given the lack of observed failures and other data albeit that considerable uncertainty remains. Below the three main approaches are summarized, although other, often more recent off-shoots exist:

- Zadeh's fuzzy set and logic
- The Dempster-Shafer Theory of Evidence
- Cook's classical method

4.5.1 Fuzzy set and logic

Zadeh (1965) developed fuzzy sets and logic; Klir and Yuan (1995) showed their applicability; Zimmermann (2001) explained how to work with fuzzy sets. The fuzzy approach became popular because the set, the fuzzy number, appeals to the way humans can express imprecision of an attribute (property or value) and the logic how different attributes, qualitative or quantitative, can be combined to a result, while set and logic can be processed by a computer. In risk assessment the approach is used to estimate values or to express even linguistic grades of consequence severity or event frequency. Zadeh's original thinking was that in classifying types it is often only possible to describe characteristics or criteria by a "continuum of grades of membership." In other words, the class or fuzzy set A is characterized by a membership function associating each point of A on the real line (X) between the extremes x_a to x_b, with a membership value (μ) in the interval [0, 1]. At x_0, $\mu = 1$, while at x_a and x_b, $\mu = 0$. The membership function can have any shape, but this is often assumed to be a triangle, trapezoid, or Gaussian, as shown in Figs. 3 and 19, left. Fuzzy sets of different experts can be weighted and averaged. Arithmetic operations with fuzzy numbers are based on the extension principle and follow certain rules.

Inference of fuzzy statements, e.g., linguistic variables, can be performed using fuzzy logic. A conditional statement can be inferred, such as IF $x = A$ THEN $z = C$; or an aggregation: IF $x = A$ AND $y = B$ THEN $z = D$, where logic AND can also be OR or NOT. The equalities before the THEN statement are called antecedents, and there after consequents. All can be numeric

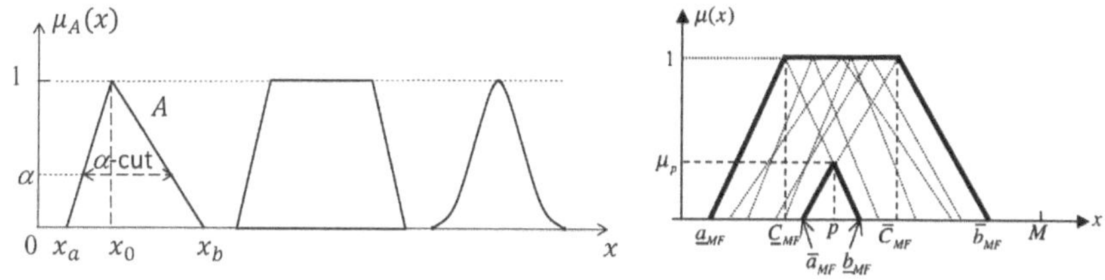

Fig. 19 *Left*: Fuzzy set shapes: e.g., triangle, trapezoid, Gaussian, on universe of discourse *X*. *Right*: Type-2 Footprint of Uncertainty, FoU, in which one can recognize composing triangle sets, which represent the intervals indicated by the experts, see Liu and Mendel (2008); the outer and inner FoU envelope are treated as fuzzy sets, index MF means membership function. The whole is a Type-2 fuzzy set.

or linguistic. An AND statement means intersection or minimum of the membership functions of the two sets; OR means union or maximum. The consequent can be a fuzzy set spanning the extremes of both antecedents on the universe (Mamdani-type), or a function and even a constant (Sugeno-type). The Mamdani-type result shall be de-fuzzified to determine a centroid: the center of area or the center of gravity. Fuzzy logic statements found widespread application in machine controls, but has also many other applications, e.g., in a risk matrix when event consequence severity and frequency are combined to a risk acceptance level.

Applications in risk assessment tools, such as fault tree and others have been reviewed by Kabir and Papadoupoulos (2018). Early-on, Markowski and Mannan (2008) applied the technique to develop a fuzzy risk matrix. There are also quite a few papers in which fuzzy numbers are inputs to BN, e.g., Yazdi and Kabir (2017) and Zarei, Khakzad, Cozzani, and Reniers (2019).

Fuzzy numbers defined above are Type-1 fuzzy set. However, in view of the uncertainty appearing when multiple experts express independently a judgment in a linguistic term, Zadeh (1975) proposed also Type-2 fuzzy set, of which the membership function itself is fuzzy. This makes it rather complex as uncertainty can now only be represented in three dimensions. By simplifying inputs to numerical intervals instead of fuzzy numbers, uncertainty can be reduced, though, to two dimensions, because interval information can be converted to triangle fuzzy numbers on the same universe by equating mean and standard deviation. In case of Interval Type-2 fuzzy set the combined fuzziness is represented by a Footprint of Uncertainty (FoU), see Fig. 19, right. Among other, in the late 1990s and the first decade of this century Mendel (2017) broadened and deepened this concept of "computing with words." It can be applied when a multiple of experts express their judgment first in a linguistic term but are asked then to indicate their idea of the linguistic term as an interval on a continuous numerical scale. Uncertainty due to differences in expert opinion is retained that way. So, for example Type-2 is useful in converting a linguistic risk matrix into a numerical one and to harmonize differences in risk matrix formats (Pasman, Hong, & Rogers, 2019). Application of theory and processing requires effort.

4.5.2 Dempster-Shafer theory of evidence (DST)

In the late 1960s, early 70s Dempster (1967) and Shafer (1976) trying to capture the uncertainty of human statements about an event, fact, or value proposed as an extension to the Bayesian approach, belief functions and evidential reasoning. DST encompasses belief, doubt, plausibility, disbelief, and ignorance, as depicted in Fig. 20 according to Rakowsky (2007).

Shafer explained it in simple words. Person A, if requested, makes a statement whether an event occurred, the interviewer can estimate person A's reliability, called *mass* (m_A) or *basic belief assignment* and compare it with person B (m_B), whom the same question was asked. In case both agree, the answer can be trusted more depending on reliabilities, or rather the product of unreliability values. If they do not agree, trust will decrease. If they agree, the joint reliability (m_{AB}) (AND, intersection, minimum) according to the Dempster combination rule is: $m_{AB}=1-(1-m_A)(1-m_B)$. In case of conflict this becomes: $m_{AB}=m_A(1-m_B)/(1-m_A m_B)$.

Dempster's combination rule can be considered as updating just as in the Bayes theorem:

$$m_{12}(A|\,A\neq\emptyset)=\frac{1}{1-k}\sum_{B\cap C=A} m_1(B)m_2(C);m_{12}(\emptyset)=0;k$$
$$=\sum_{B\cap C=\emptyset} m_1(B)m_2(C)$$

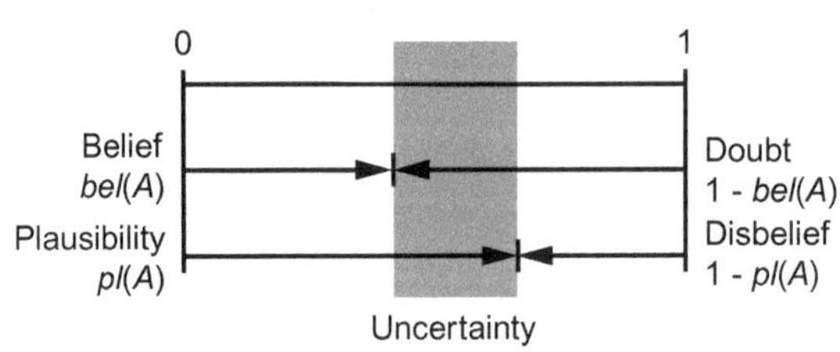

Fig. 20 Representation of belief and plausibility measures according to Rakowsky (2007).

The question can also be about the probability of an event. In case this is followed by a second question to the same person about the probability the event does not occur, the sum of both probabilities does not necessarily need to add to 1 (but the reliability mass of the person and the unreliability do sum to 1). If answers of different experts are much in conflict the rule does not produce a satisfactory answer and instead Yager's rule (Yager, 1987) must be applied.

The theory assumes that n hypotheses or information pieces of a data source, e.g., experts' estimates on states or events, may constitute a set of elements, A, formally called a frame of discernment or of disjoint states represented by a power set Ω with a number j of sub-sets of in total 2^n. To each sub-set a mass is assigned $m(A_j)$ measuring the degree of belief in claim A_j, while the sum of element masses $\sum_{A_j \subseteq \Omega} m(A_j) = 1$. The set contains beside uncertainty terms one is aware of, the sets of mixed information pieces, also a $\varnothing$ null set of uncertainty one is unaware of. If the latter is absent, that sub-set$=0$. All sub-sets $A_j > 0$ are called focal elements. The theory further defines a belief function $bel(A) = \sum_{A_j \subseteq A;\ Aj \neq \varnothing} m(A_j)$ and a plausibility function $pl(A) = \sum_{A_j \cap A \neq 0} m(A_j)$, the latter containing not only the belief terms but also the uncertainty ones, not being the $\varnothing$ null set. This will be clear from an example with three pieces of information: $A = \{\{\varnothing\}, \{A_1\}, \{A_2\}, \{A_3\}, \{A_1, A_2\}, \{A_1, A_3\}, \{A_2, A_3\}, \{A_1, A_2, A_3\}\}$, so that $bel(A_1, A_2) = m(A_1) + m(A_2) + m(A_1, A_2)$ and $pl(A_1, A_2) = m(A_1) + m(A_2) + m(A_1, A_2) + m(A_1, A_3) + m(A_2, A_3) + m(A_1, A_2, A_3)$.

Rakowsky (2007) applied the DST mathematics to reliability. The simplest case is one component with binary states: S functional and F failed. The powerset Ω in that case has four terms, of which the first one is here empty as the last one expresses all uncertainty: $A = \{\{\varnothing\}, \{S\}, \{F\}, \{S, F\}\}$. Upon request, an expert estimates the probability of the component staying functional over the next year to be 0.6, hence $bel(S) = 0.6$, but when asked to estimate the probability of failure he answers 0.25. If the expert is fully trusted (100%) the answers can be taken as the masses, which add up to 1, so $\{S,F\}$ will be 0.15, or $Pl(S) = 0.75$. This reasoning can be expanded to more states and more components forming a fault or an event tree. The whole can then be structured in an evidential network represented by a BN in which the uncertainty is propagated as a separate state (Khakzad, 2019; Simon & Weber, 2009; Simon, Weber, & Evsukoff, 2008). A fault tree application is presented by Curcurù, Galante, and La Fata (2012, 2013); an FMECA one by Certa, Hopps, Inghilleri, and La Fata (2017).

4.5.3 Probabilistic classical method of structured expert judgment

Cooke (1991) laid the foundation for this method. The rather intricate procedure described by Cooke and Goossens (2008), here summarized and oversimplified, is that experts are asked to provide estimates of values with probability to be the true value of $\leq$5%, >5% and $\leq$50%, >50% and $\leq$95%, and >95%, forming four inter-quantile intervals (or bins) as an impression of the perceived width of the answer distribution. If desired, elicitation can be refined by asking more quantile fractions. The data, so obtained are statistically processed into a calibration score of each expert, while the width of the distribution the expert provided produces an information score. If the latter is wide it may indicate the expert is uncertain, and when it is very narrow it may be an overconfidence sign. Next, based on the expert scores a so-called Decision maker is attributing weights, which can be done according to three different methods. Despite many trials, it is not fully clear which method is best. The method's preparation and processing require considerable effort. A recent application of the method with respect to dispersion model parameters for the climate panel has been described by Oppenheimer, Little, and Cooke (2016).

4.6 Dependence and correlation, copulas

One of the issues that did not get much attention in risk assessment yet, is the correlation or interdependence of variables, e.g., so-called basic faults in fault trees, or in case of process variables, e.g., reactor temperature and pressure. If common cause failures occur there is common dependency on a condition, e.g., availability of a utility, but it can also be that failure of a component leads to early wear out of others so that their failure rate increases dramatically. This can have a large effect on the reliability of a system. Fréchet correlation bounds for random variables A and B in AND or OR position are: $P(A\cup B) = [\max(0, a+b-1), \min(a, b)]$ and $P(A\cap B) = [\max(\mathrm{a}, \mathrm{b}), \min(1, \mathrm{a}+\mathrm{b})]$, in which $a = P(A)$ and $b = P(B)$. The actual values in case of correlation lie within the intervals (Ferson et al., 2004, section 2.3).

To describe mathematically the correlation of two variables a copula can be applied. Nelsen (2006) treats copulas in detail and explains the basics as follows. The cumulative distributions of random variables X and Y, $F(x) = P[X \leq x]$ and $G(y) = P[Y \leq y]$, respectively, and a joint distribution function $H(x, y) = P[X \leq x, Y \leq y]$. Each pair of real numbers (x, y) produces the numbers $F(x)$, $G(y)$,and $H(x, y)$ all in intervals [0,1]. In the unit square $[0, 1] \times [0, 1]$ there is therefore a point $(F(x), G(y))$, which corresponds with point $H(x, y)$ in [0, 1]. This correspondence between all points is a copula function. Central in copula theory is Sklar's theorem (Nelsen, 2006) that states: Let H be a joint distribution function with margins F and G. Then there exists a copula C such that for all x, y in domain $\overline{\mathrm{R}}$, $H(x, y) = C(F(x), G(y))$. If F and G are continuous, then C is unique, otherwise, C is uniquely determined on $\mathrm{Ran}F \times \mathrm{Ran}G$, the cartesian product of the ranges of the marginal cdf's. There are quite a few different copula functions depending on functions F and G; even an empirical copula can be derived based on observed pairs (Ferson et al., 2004, section 3.8). The Gaussian or multivariate normal copula is used much. It is derived by inverting Sklar's theorem: $C(u, v) = \Phi_r(\Phi^{-1}(u), \Phi^{-1}(v))$,in which $\Phi_r(x, y) = \frac{1}{2\pi\sqrt{1-r^2}} \int_{-\infty}^{x} \int_{-\infty}^{y} \exp\left(\frac{(-s^2 - 2rst + t^2)}{2(1-r^2)}\right) dsdt$, is the joint cumulative distribution function for the standard bivariate normal distribution with Pearson correlation coefficient r, $0 < r < 1$, while Φ^{-1} is the inverse cumulative distribution function of the standard univariate normal distribution function. In case of non-linear rank correlation, the Spearman rho or Kendall's tau correlation coefficient shall be used. Once formulated the copula function can be incorporated in BN (Elidan, 2010).

Applications in (dynamic) risk assessment have been presented by among other Hashemi, Khan, and Ahmed (2016) and Guo, Khan, and Imtiaz (2018, 2019).

4.7 Simulation and gamification

Process simulation for education and training has been found in the past of insufficient quality with respect to reality content. However, due to the vastly improved computer capabilities, e.g., by the introduction of Immersive Virtual Environments (Colombo & Golzio, 2016; Nazir & Manca, 2015), process simulation has changed and it has now become widely accepted. Simulation introduces the dynamics in real time, which for operator training and learning to control developing mishap will be a valuable tool. It certainly will have a role in hazard identification and design improvement. A recent review has been presented by Marcano, Haugen, Sannerud, and Komulainen (2019).

From a different direction evolved the development of serious games, which in combination with process simulation offers an environment in which trainees, also as a team, by the game element become more deeply involved and motivated. Patriarca, Falegnami, De Nicola, Villani, and Paltrinieri (2019) applied gamification for measuring a team's resilience-based early warning indicators and demonstrated the concept in an ammonia leak emergency response action. Din and Gibson (2019) treat in detail the application of games to improve occupational safety by better design and determined the effectiveness of the effort. They also collected interesting student responses. Jousselme, Pallotta, and Locke (2018) had a serious game played to obtain insight on rather psychologically oriented aspects of a player's belief, threat assessment, and decision making in risky environments and that all with imperfect information.

4.8 Game theory

Game theory is of a different nature. It originates in operations research in optimization of adversary positions and conditions, given both sides' limitations of investment, time and effort. It applies much to terrorist plant attack and cyber-attack. There are many publications on game theory, but few applied to process industry risk management; a recent paper by Zhang and Reniers (2018) provides besides an extension of the game theory also an overview of previous work. Musman and Turner (2018) describe and detailed their work on a game-oriented approach to hedge against cyber-attack.

4.9 Combinations of tools

A combination of tools may solve a problem; in quite some cases a combination with BN is used, because the BN algorithms solve readily the probabilistic computations. However, apart from the fact that there are other less frequently used techniques not mentioned here, there is no systematic guidance for what combinations to use, so we shall limit this section to a few examples.

Ferdous, Khan, Sadiq, Amyotte, and Veitch (2013) and earlier papers applied for constructing a bowtie Type-1 Fuzzy set in combination with DST to overcome the lack of failure and event probability data and model uncertainty. Experts provided fuzzy and DST basic assignment inputs on data and dependencies. For tackling the uncertainty in a security assessment, Misuri, Khakzad, Reniers, and Cozzani (2018) applied DST and Credal sets, which are evaluated using BN. A Credal set is another way to handle imprecise probabilities (Piatti, Antonucci, & Zaffalon, 2010) and consists of a closed convex set of probability measures that must sum to 1, but which can vary within that constraint. Under certain conditions, such sets can be evaluated by BN.

4.10 Making use of big data

Digitization of the industry, the merging of data and the processing by the Analytics tools described in this section holds the promise to provide people with information they never had before. Weaknesses of risk assessment methodology, such as incomplete hazard identification and lack of failure and other initiating event data, should be strongly reduced due to abundance of own historical plant data. Given a risk level, enterprise resource planning and cost figures will enable optimizing safety barriers and maintenance of system components. And in a dynamic fashion periodical updates of leading and lagging process safety performance data should relate to operational performance level and failure rates. Low-frequency events with an effect on risk level, such as issue of work permits, occurrence of simultaneous maintenance or other operations, and occasional groups of visitors, can be recorded. Vibration of pipelines, excessive equipment energy consumption, gas detection, degradation of barrier function, and other anomalies will also have its effect on risk level. All these data together with continuous monitoring of the processes and system and organizational behavior, should be able to output instantaneous risk and provide operational early warning. Lee et al. (2019) provides an excellent overview of what system approach, digitization shall mean for the improvement of operational performance and reduction of process safety accidents by the availability of data and models. Process plant and simulator will evolve to "digital twins."

4.11 Uncertainty analysis, deep uncertainty

We have to be aware that there are few certainties in life, and that one certainty is that nothing in risk assessment is certain. This can be demonstrated with an example about the "safety case." Apart from terminology confusion, the safety case in this context "claims" that a product, which at malfunction may kill the user, is safe given an upper failure rate of $1\cdot10^{-5}$/year, SI = 5. Known examples are a pacemaker or a heart valve. Graydon and Holloway (2017) analyzed various methods used by 12 teams in different parts of the world to underpin and proof the claim. It turned out that they were able to find in each case a counterexample in which the method output was implausible, and hence, proved to be not fully trustworthy (see also Pasman & Rogers, 2018).

On the other hand, the recently developed tools described above gave us much more grip on uncertainty. There are additional tools, which are not much used yet in risk assessment and that are tackling deep uncertainty.

Cox (2012) gave an overview summarized by Pasman and Rogers (2018). To name a few techniques applicable in ever deeper uncertainty: multiple priors, robust optimization, multiple predictive models, resampling, adaptive boosting, Bayesian model averaging, and (model-free) reinforcement learning.

4.12 Maturity model/independent evaluation

Rae, Alexander, and McDermid (2014) presented a unique paper in which they reported about all the types of flaws a risk assessment may be afflicted with. The assurance they obtained their inventory is complete was by consulting three sources: academic literature describing errors made in risk assessments, published peer reviews of risk assessments, and the author's experience. Of these, the peer reviews have been of 9 well respected mostly government related bodies that had screened QRA studies. The flaws are all spelled out and classified in four levels:

1. *Unrepeatable*: the record/report/description/documentation is incomplete, and it is not possible to reconstruct the assessment.
2. *Invalid*: the effects of flaws in the assessment are larger than the underlying uncertainties in the events investigated.
3. *Valid but inaccurate*: rigor in selecting data is insufficient.
4. *Accurate* but challengeable.

The paper has the objective to expose the flaws so that one can avoid those flaws, and that maturity of the assessments increases.

Lathrop and Ezell (2017) treated the issue of how to independently evaluate a QRA. They devised a scheme, as shown in Fig. 21, in which analysts' domain and that of reviewers and decision makers are strictly separated, although according to IRGS (2005), that also advocates separation, deliberation between parties is open. Ahead of a decision the final assessment is reviewed by an independent third party on content and by stakeholders with respect to the trust they can have in the assessment. Trust is the most important constituent of the risk communication process. Results of the reviews are passed on to the decision makers.

4.13 Future perspective/resilience assessment

Risk is a probabilistic concept. If according to an assessment a certain consequence event can occur with a certain frequency, it can happen today or not in a plant's lifetime. On the other hand, as we have seen, due to lack of knowledge and spread in values, the risk value is rather uncertain itself;

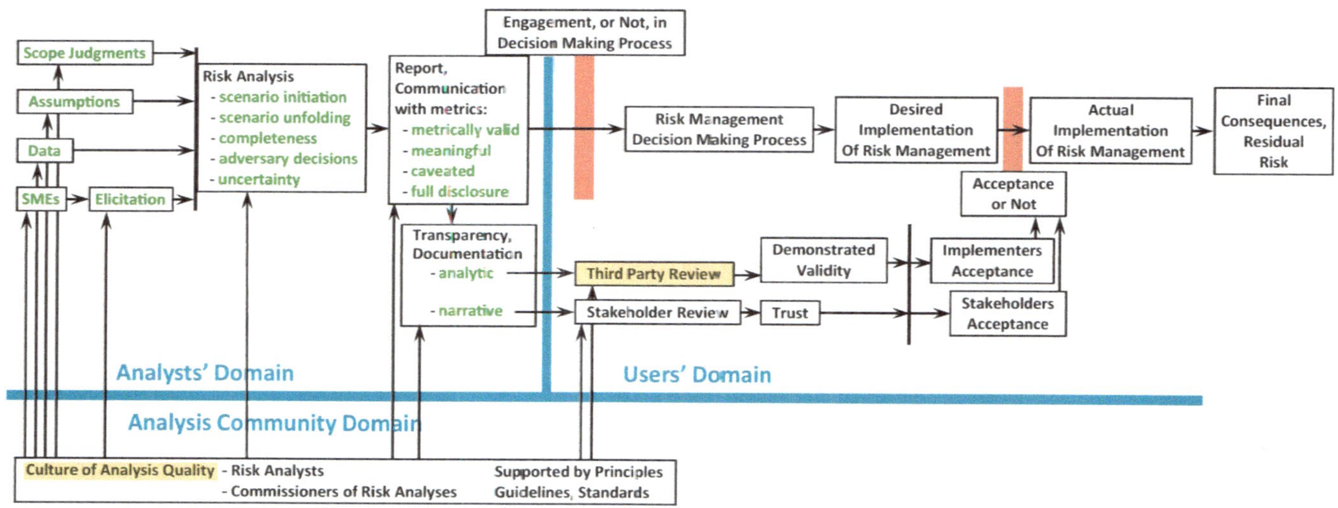

Fig. 21 Flow scheme of a risk assessment with clearly separate analyst's and user's domain as developed by Lathrop and Ezell (2017) (the highlighting of boxes "Culture of Analysis Quality" and Third Party Review" was added by Pasman & Rogers, 2018).

so even on a comparative basis one must be cautious with conclusions. The developments sketched in this chapter give much hope that uncertainty in risk figures can be significantly reduced and that it will be possible in continuous processes to monitor dynamic operational risk (risk dashboard). However, a guarantee on completeness of major hazard scenarios is not possible. Even, if we would be aware of all possible scenarios triggered by the equipment itself, due, of course, to a large variety of root causes, then still unexpected external threats can emerge. Hedging against the unexpected is possible by a sufficient level of resilience. It means that management has thought about preparing for disaster and keeping a possible dip in output as shallow and short as possible by taking measures to strengthen and nourish the organization safety culture.

Measures shall be based on sound risk assessments and can consist of early warning signal receiving arrangements, a preparedness attitude of all staff, plasticity of mind meaning flexible thinking of possible temporary solutions without creating other risks, effective emergency response force and further recovery, and contingency planning. The latter will have the objective to lower and shorten effects of business interruptions. This can be with respect to materials supply, attracting additional staff, satisfying customers with products from elsewhere, and sufficient reserve cash. It will be all to strengthen business continuity (Jain, Pasman, Waldram, Pistikopoulos, & Mannan, 2018).

4.14 Decision-making support models

Although not an organic part of risk assessment but, yet, each risk assessment is followed by risk decision making or what to do about the risk. Shall the risk be reduced by further investment or can we accept it as it is? Cost is an important factor: Reniers and Noël van Erp (2016) on safety economics and with respect to plant profitability, and even in more detail, Mellichamp (2019), may provide answers to questions arising. However, for acceptance also risk communication and risk perception are important. As mentioned, trust is crucial, and although risk is often interpreted as the product of consequence severity and event frequency, in general it holds that the larger the potential consequence, acceptable expected frequency will be smaller than the product indicates.

Decision-making methods are several, their collective name is Multi-Criteria Decision Making (MCDM), which can refer to continuous or discrete decision problems. The latter are more frequent, and their specific name is Multi-Attribute Decision Making (MADM), but in practice they are often called MCDM too. A decision problem consists of determining an order of preference (priority) among alternatives (products, activities) based on one or a set of criteria, against which the alternatives must be weighed.

In case for a decision various aspects or attributes of an item must be weighed by a number of participants in a decision team, various decision aiding models are available. Saaty's (1990, 2008) Analytic Hierarchy Process (AHP) is one of the most known. It is based on comparing against a criterion the pairwise importance of attributes, expressed as an ordinal importance scale of 1–9 and followed by matrix operations to derive a result. Weaknesses are possible group think and attributes that cannot be considered sufficiently independent. The former should be avoided by session organization, the latter by the Analytic Network Process (ANP) (Saaty, 2008) taking account of dependencies. An example is presented by Khakzad, Reniers, and Van Gelder (2017). Fuzzifying comparison decisions as proposed by Chang (1996) should, according to Saaty and Tran (2007), be avoided as it does not contribute. However, Chan, Sun, and Chung (2019) make clear when FAHP should be preferred over AHP. On the other hand, AHP can also be misused due to biased participants' opinions while pretending a rational decision, in fact a weakness adhering to any expert judgment. A striking example is given by Hazelrigg (2012). An overview of various other decision aids, such as TOPSIS and quite some others, is given in the review of multi-criteria decision-making methods by Velasquez and Hester (2013).

Zeraei (2015, 2016) developed the Best-Worst Method. In this method the decision maker identifies the best (most desirable or most important) criterion and the worst one first. Determine pairwise the preference of the best over all the other criteria, and the other way around, of all other ones over the worst. Derive the weights by applying the minmax operation and determine the consistency ratio. So, Best-Worst resembles AHP, but due to the preliminary selection it results in slightly better results measured in terms of the consistency ratio and in ordinal consistency. In addition, BWM came out best in total deviations and the conformity test, all as shown in an example of four smart phones selection and six criteria. Later, Zeraei (2016) extended the method to obtain multiple optimum solutions by an interval method, or if preferred a unique solution by linear programming.

If for a decision under uncertainty several attributes must be brought together in a causal relationship, the simplest is a Binary Decision Diagram based on Booleans (see, e.g., Andersen, 1999). In the 1970s

DEMATEL (Decision-making trial and evaluation laboratory) has been developed (for a review, see Si, You, Liu, & Zhang, 2018). By some matrix operations this method is able to split attributes into a cause group and an effect group revealing complex relations between both. It can also be used with fuzzy inputs. Much more refined will be the use of BN. Once one knows the cause-effect structure, one can bring together deterministic value nodes, nodes with continuous probability distributions, and ones with discrete distributions, where possibly values of criteria or attributes have been determined by judgment of the decision team members. Examples are given by Fenton and Neil (2019). Value estimates can have the form of a fuzzy set.

In later chapters of this volume these topics will be explained in more detail.

5. Conclusion

The conclusion can be brief. Over a period of 50 years risk assessment developed to a discipline of its own with many aspects. Because of the wide range of sciences feeding risk assessment methodology, sub-specialisms arose. However, for a long time it was applied only on static situations. In particular the last decade, though, due to information technology developments, computerization and digitization generating data, while enabling ML and AI techniques to process and analyze data streams, dynamic risk assessment has become a reality. In the ever more complex plant and product operations the latter will be of great support to help keep processes safer.

References

Adedigba, S. A., Khan, F., & Yang, M. (2017). Dynamic failure analysis of process systems using neural networks. *Process Safety and Environmental Protection, 111*, 529–543.

Ahooyi, T. M., Arbogast, J. E., Oktem, U. G., Seider, W. D., & Soroush, M. (2014). Estimation of complete discrete multivariate probability distributions from scarce data with application to risk assessment and fault detection. *Industrial & Engineering Chemistry Research, 53*, 7538–7547.

Ahooyi, T. M., Arbogast, J. E., & Soroush, M. (2015a). Applications of the rolling pin method. 1. An efficient alternative to Bayesian network modeling and inference. *Industrial & Engineering Chemistry Research, 54*, 4316–4325.

Ahooyi, T. M., Arbogast, J. E., & Soroush, M. (2015b). Rolling pin method: Efficient general method of joint probability modeling. *Industrial & Engineering Chemistry Research, 54*, 20191–20203.

Ahooyi, T. M., Soroush, M., Arbogast, J. E., Seider, W. D., & Oktem, U. G. (2014). Maximum-likelihood maximum-entropy constrained probability density function estimation for prediction of rare events. *AICHE Journal, 60*(3), 1013–1025.

Alauddin, M., Khan, F., Imtiaz, S., & Ahmed, S. (2018). A bibliometric review and analysis of data-driven fault detection and diagnosis methods for process systems. *Industrial & Engineering Chemistry Research, 57*, 10719–10735.

Amendola, A., Contini, S., & Ziomas, I. (1992). Uncertainties in chemical risk assessment: Results of a European benchmark exercise. *Journal of Hazardous Materials, 29*, 347–363.

Andersen, H. R. (1999). *An introduction to binary decision diagrams. In Lecture notes for efficient algorithms and programs.* https://www.cmi.ac.in/~madhavan/courses/verification-2011/andersen-bdd.pdf.

ANSI/ISA-84.00.01-2004 Parts 1–3 (IEC 61511 Mod), functional safety: Safety instrumented systems for the process industry sector, n.d. https://webstore.ansi.org/Standards/ISA/ISA8400012004PartIEC61511Mod-1643810.

Baybutt, P. (2015). A critique of the hazard and operability (HAZOP) study. *Journal of Loss Prevention in the Process Industries, 33*, 52–58.

Brighton, P. W. M., Byrne, A. J., Cleaver, R. P., Courtiade, P., Crabol, B., Fitzpatrick, R. D., et al. (1994). Comparison of heavy gas dispersion models for instantaneous releases. *Journal of Hazardous Materials, 36*, 193–208.

Britter, R. E., & McQuaid, J. (1988). *Workbook on the dispersion of dense gases. In HSE Contract Research Report No.17/1988.* https://ramp.nrc-gateway.gov/system/files/880b796981b948c77be66c0b848483a1/Britter_McQuaid%20%28CRR%2017_1988%29.pdf.

Cameron, I., Mannan, S., Németh, E., Park, S., Pasman, H., Rogers, W., et al. (2017). Process hazard analysis, hazard identification and scenario definition: Are the conventional tools sufficient, or should and can we do much better? *Process Safety and Environmental Protection, 110*, 53–70.

Cameron, I. T., & Raman, R. (2005). Process systems risk management. In *Process systems engineering*. Vol. 6Elsevier Academic Press. ISBN 0-12-156932-2.

Canvey. (1978/1981). *Canvey: An investigation of potential hazards from operations in the Canvey Island/Thurrock area. Health and Safety Executive.* First report 1978 H.M. Stationery. ISBN-10: 011883200X, ISBN-13: 978-0118832007; Second report 1981.

Casal, J. (2017). *Evaluation of the effects and consequences of major accidents in industrial plants* (2nd ed.). ElsevierISBN: 978-0-444-63883-0.

CCPS. (2000). *Guidelines for chemical process quantitative risk analysis* (2nd ed.). Center for Chemical Process Safety of AIChE, Wiley. ISBN 0-8169-0720-X.

Certa, A., Hopps, F., Inghilleri, R., & La Fata, C. M. (2017). A Dempster-Shafer theory-based approach to the failure mode, effects and criticality analysis (FMECA) under epistemic uncertainty: application to the propulsion system of a fishing vessel. *Reliability Engineering and System Safety, 159*, 69–79.

Chamberlain, G. A., Oran, E., & Pekalski, A. (2019). An analysis of severe vapour cloud explosions and detonations in the process industries. *Chemical Engineering Transactions, 77*, 853–858.

Chan, H. K., Sun, X., & Chung, S.-H. (2019). When should fuzzy analytic hierarchy process be used instead of analytic hierarchy process? *Decision Support Systems, 125*, 113114. 1–10.

Chang, D.-Y. (1996). Applications of the extent analysis method on fuzzy AHP. *European Journal of Operational Research, 95*, 649–655.

Christou, M., Gyenes, Z., & Struckl, M. (2011). Risk assessment in support to land-use planning in Europe: Towards more consistent decisions? *Journal of Loss Prevention in the Process Industries, 24*, 219–226.

CIA, Chemical Industries Association. (1977). *A guide to hazard and operability studies.* Chemical Industry Safety and Health Council of the Chemical Industries Association.

Colombo, S., & Golzio, L. (2016). The plant simulator as viable means to prevent and manage risk through competencies management: Experiment results. *Safety Science, 84*, 46–56.

Cooke, R. M. (1991). *Experts in uncertainty, opinion and subjective probability in science*. New York—Oxford: Oxford University Press. ISBN 0-19-506465-8.
Cooke, R. M., & Goossens, L. L. H. J. (2008). TU Delft expert judgment data base. *Reliability Engineering and System Safety*, *93*, 657–674.
Cooper, G. F., & Herskovits, E. (1992). A Bayesian method for the induction of probabilistic networks from data. *Machine Learning*, *9*, 309–347.
Cortes, C., & Vapnik, V. N. (1995). Support-vector networks. *Machine Learning*, *20*(3), 273–297. http://www.springerlink.com/content/k238jx04hm87j80g/.
COVO. (1982). *COVO Commission, risk analysis of six potentially hazardous industrial objects in the Rijnmond area, a pilot study. A report to the Rijnmond Public Authority*. Central Environmental Control Agency, Schiedam, The Netherlands, 1981 Dordrecht, The Netherlands: D. Reidel Publishing Co. ISBN 90 277 1393 6.
Cox, A. (2012). Confronting deep uncertainties in risk analysis. *Risk Analysis*, *32*(10), 1607–1629.
Crawley, F., Preston, M., & Tyler, B. (2000). *HAZOP: Guide to best practice*. Published by Institution of Chemical Engineers, © 2000 European Process Safety Centre. ISBN 0 85295 427 1.
Curcurù, G., Galante, G. M., & La Fata, C. M. (2012). Epistemic uncertainty in fault tree analysis approached by the evidence theory. *Journal of Loss Prevention in the Process Industries*, *25*, 667–676.
Curcurù, G., Galante, G. M., & La Fata, C. M. (2013). An imprecise Fault Tree Analysis for the estimation of the Rate of OCcurrence Of Failure (ROCOF). *Journal of Loss Prevention in the Process Industries*, *26*, 1285–1292.
de Dianous, V., & Fiévez, C. (2006). ARAMIS project: a more explicit demonstration of risk control through the use of bow-tie diagrams and the evaluation of safety barrier performance. *Journal of Hazardous Materials*, *130*(3), 220–233.
Delvosalle, C., Fievez, C., Pipart, A., & Debray, B. (2006). ARAMIS project: A comprehensive methodology for the identification of reference accident scenarios in process industries. *Journal of Hazardous Materials*, *130*, 200–219.
Dempster, A. P. (1967). Upper and lower probabilities induced by a multivalued mapping. *The Annals of Mathematical Statistics*, *38*(2), 325–339.
Din, Z. U., & Gibson, G. E. (2019). Serious games for learning prevention through design concepts: An experimental study. *Safety Science*, *115*, 176–187.
DoD. (1949). *MIL-P-1629-procedures for performing a failure mode effect and critical analysis*. U.S. Department of Defense. 9 November 1949.
Duijm, N. J., Carissimo, B., Mercer, A., Bartholome, C., & Giesbrecht, H. (1997). Development and test of an evaluation protocol for heavy gas dispersion models. *Journal of Hazardous Materials*, *56*, 273–285.
Ebeling, C. E. (2019). *An introduction to reliability and maintainability engineering* (2nd ed.). Waveland Press, Inc. ISBN-13: 978-1478637349, ISBN-10: 147863734X, Chapter 6.
El-Gheriani, M., Khan, F., Chen, D., & Abbassi, R. (2017). Major accident modelling using spare data. *Process Safety and Environmental Protection*, *106*, 52–59.
Elidan, G. (2010). Copula Bayesian networks. In *Advances in neural information processing systems (NIPS 2010), Vancouver* (pp. 559–567).
Fabbri, L., & Contini, S. (2009). Benchmarking on the evaluation of major accident-related risk assessment. *Journal of Hazardous Materials*, *162*, 1465–1476.
Fenton, N., & Neil, M. (2019). *Risk assessment and decision analysis with Bayesian networks* (2nd ed.). Boca Raton, FL 33487-2742, USA: CRC Press, Taylor & Francis Group. ISBN 978-1-138-03511-9.
Ferdous, R., Khan, F., Sadiq, R., Amyotte, P., & Veitch, B. (2013). Analyzing system safety and risks under uncertainty using a bow-tie diagram: An innovative approach. *Process Safety and Environmental Protection*, *91*, 1–18.

Ferson, S., Hajagos, J., Berleant, D., Zhang, J., Tucker, W. T., Ginzburg, L., et al. (2004). *Dependence in Dempster-Shafer theory and probability bounds analysis. Applied Biomathematics for Sandia National Laboratories under Contract No. <19094>*. https://ualr.edu/jdberleant/papers/Sandia04.pdf.

FMV. (2016). *FRAM model visualizer*. http://functionalresonance.com/FMV/index.html.

Forrester, J. W. (1961). *Industrial dynamics*. Cambridge, MA: MIT Press.

Garbolino, E., Chery, J.-P., & Guarnieri, F. (2016). A simplified approach to risk assessment based on system dynamics: An industrial case study. *Risk Analysis*, *36*(1), 16–28.

Graydon, P. J., & Holloway, C. M. (2017). An investigation of proposed techniques for quantifying confidence in assurance arguments. *Safety Science*, *92*, 53–65.

Guo, C., Khan, F., & Imtiaz, S. (2018). Risk assessment of process system considering dependencies. *Journal of Loss Prevention in the Process Industries*, *55*, 204–212.

Guo, C., Khan, F., & Imtiaz, S. (2019). Copula-based Bayesian network model for process system risk assessment. *Process Safety and Environmental Protection*, *123*, 317–326.

Haddon, W. (1973). Energy damage and the 10 countermeasure strategies. *The Journal of Trauma*, *13*, 321–331.

Hanna, S. R., Chang, J. C., & Strimaitis, D. G. (1993). Hazardous gas model evaluation with field observations. *Atmospheric Environment Part A General Topics*, *27*(15), 2265–2285.

Hashemi, S. J., Khan, F., & Ahmed, S. (2016). Multivariate probabilistic safety analysis of process facilities using the Copula Bayesian Network model. *Computers and Chemical Engineering*, *93*, 128–142.

Havens, J., Walker, H., & Spicer, T. (2012). Bhopal atmospheric dispersion revisited. *Journal of Hazardous Materials*, *233–234*, 33–40. https://doi.org/10.1016/j.jhazmat.2012.06.055.

Hazelrigg, G. A. (2012). *Fundamentals of decision making for engineering design and systems engineering*. Appendix C.2.3., Neill Corporation, ISBN 978-0-984997602.

Helton, J. C., & Johnson, J. D. (2011). Quantification of margins and uncertainties: Alternative representations of epistemic uncertainty. *Reliability Engineering and System Safety*, *96*, 1034–1052.

Hollnagel, E. (2012). *FRAM: The functional resonance analysis method: Modelling complex socio-technical systems*. Farnham, Surrey, UK: Ashgate Publishing Ltd. ISBN-13: 978 1409445517 and ISBN-10: 1409445518.

Houston, D. E. L. (1971). New approaches to the safety problem. *Institution of Chemical Engineers Symposium Series*, *34*, 210–216.

IEC 60812:2018 International Standard. (2018). *Analysis techniques for system reliability—Procedure for failure mode and effects analysis (FMEA)*. Geneva 20, Switzerland: International Electrotechnical Commission.

IEC 61025:2006 International Standard. (2006). *Fault tree analysis (FTA)*. Geneva 20, Switzerland: International Electrotechnical Commission.

IEC 61508: 2010 (E). (2010). *Functional safety of electrical/electronic/programmable electronic safety-related systems—Part 1: General requirements (IEC 61508-1:2010, IDT), and parts 2–7* (2nd ed.). Geneva, Switzerland: International Electrotechnical Commission.

IEC 61511: 2016 (E). (2016). *Functional safety—Safety instrumented systems for the process industry sector, Parts 1–3* (2nd ed.). Geneva, Switzerland: International Electrotechnical Commission.

IEC 62443-4-2:2019. (2019). *Security for industrial automation and control systems—Part 4-2: Technical security requirements for IACS components*. Geneva 20, Switzerland: International Electrotechnical Commission.

IEC 62551: 2012-10. (2012). *Analysis techniques for dependability—Petri net techniques*. Geneva, Switzerland: International Electro technical Commission. Ed.1.0.

IRGS. (2005). *White paper on risk governance: Towards an integrative approach*. CH-1219 Geneva, Switzerland: International Risk Governance Council. https://irgc.org/publications/core-concepts-of-risk-governance/.

ISO 15926. (2009). *Industrial automation systems and integration—Integration of life-cycle data for process plants including oil and gas production facilities*. CP 401—1214 Vernier, Geneva, Switzerland: International Organization for Standardization, ISO Central Secretariat. 11 Parts.

Jain, P., Pasman, H. J., Waldram, S., Pistikopoulos, E. N., & Mannan, M. S. (2018). Process resilience analysis framework (PRAF): A systems approach for improved risk and safety management. *Journal of Loss Prevention in the Process Industries*, *53*, 61–73.

Janošovský, J., Danko, M., Labovský, J., & Jelemenský, Ľ. (2019). Software approach to simulation-based hazard identification of complex industrial processes. *Computers & Chemical Engineering*, *122*, 66–79.

Jarvis, R., & Goddard, A. (2016). *An analysis of common causes of major losses in the onshore oil, gas and petrochemical industries: Implications for insurance risk engineering surveys*. Lloyd's Market Association. Version 1.0, September. (Search web on the title of the report).

Jousselme, A.-L., Pallotta, G., & Locke, J. (2018). Risk Game: Capturing impact of information quality on human belief assessment and decision making. *International Journal of Serious Games*, *5*(4), 23–44.

Lunn, D., Jackson, C., Best, N., Thomas, A., & Spiegelhalter, D. (2012). *The BUGS book—A practical introduction to bayesian analysis*. CRC Press/Chapman and Hall. ISBN-13: 978-1584888499, ISBN-10: 1584888490. https://www.mrc-bsu.cam.ac.uk/software/bugs/the-bugs-project-winbugs/.

Kabir, S., & Papadopoulos, Y. (2019). Applications of Bayesian networks and Petri nets in safety, reliability, and risk assessments: A review. *Safety Science*, *115*, 154–175.

Kabir, S., & Papadoupoulos, Y. (2018). A review of applications of fuzzy sets to safety and reliability engineering. *International Journal of Approximate Reasoning*, *100*, 29–55.

Kaplan, S., & Garrick, B. J. (1981). On the quantitative definition of risk. *Risk Analysis*, *1*(1), 11–27.

Khakzad, N. (2019). System safety assessment under epistemic uncertainty: Using imprecise probabilities in Bayesian network. *Safety Science*, *116*, 149–160.

Khakzad, N., Amyotte, P., Cozzani, V., Reniers, G., & Pasman, H. J. (2018). How to address model uncertainty in the escalation of domino effects? *Journal of Loss Prevention in the Process Industries*, *54*, 49–56.

Khakzad, N., Khan, F., & Paltrinieri, N. (2014). On the application of near accident data to risk analysis of major accidents. *Reliability Engineering and System Safety*, *126*, 116–125.

Khakzad, N., Reniers, G., & Van Gelder, P. (2017). A multi-criteria decision-making approach to security assessment of hazardous facilities. *Journal of Loss Prevention in the Process Industries*, *48*, 234–243.

Khan, F. I., Sadiq, R., & Husain, T. (2002). Risk-based process safety assessment and control measures design for offshore process facilities. *Journal of Hazardous Materials*, *A94*, 1–36.

Kletz, T. A. (1971). Hazard analysis – A quantitative approach to safety. *Institution of Chemical Engineers Symposium Series*, *34*, 78–81.

Klir, G. J., & Yuan, B. (1995). *Fuzzy sets and fuzzy logic: Theory and applications*. Upper Saddle River: Prentice Hall.

Lathrop, J., & Ezell, B. (2017). A systems approach to risk analysis validation for risk management. *Safety Science*, *99*(Pt. B), 187–195.

Lauridsen, K., Kozine, I., Markert, F., Amendola, A., Christou, M., & Fiori, M. (2002). *Assessment of uncertainties in risk analysis of chemical establishments. The ASSURANCE project. Final summary report*. Roskilde, Denmark: Risø National Laboratory.52. http://www.risoe.dk/rispubl/sys/syspdf/ris-r-1344.pdf.

Lawley, H. G. (1974). Operability studies and hazard analysis. *Chemical Engineering Progress*, *70*(4), 45–57.

Lee, J., Cameron, I., & Hassall, M. (2019). Improving process safety: What roles for digitalization and industry 4.0? *Process Safety and Environmental Protection*, *132*, 325–339. Article in press. https://doi.org/10.1016/j.psep.2019.10.021.

Lee, J. H., Shina, J., & Realff, M. J. (2018). Machine learning: Overview of the recent progresses and implications for the process systems engineering field. *Computers and Chemical Engineering*, *114*, 111–121.

Lees, F. P. (1996). Loss prevention in the process industries (2nd ed.). In (Vol. 3) Butterworth-Heinemann, © Reed. ISBN 0 7506 1547 8. Appendices 7 and 8.

Lerner, B., & Malka, R. (2011). Investigation of the K2 algorithm in learning Bayesian network classifiers. *Applied Artificial Intelligence*, *25*, 74–96.

Leveson, N. G. (2011). *Engineering a safer world, systems thinking applied to safety*. The MIT Press608. ISBN-10:0-262-01662-1 ISBN-13:978-0-262-01662-9.

Liu, F., & Mendel, J. M. (2008). Encoding words into interval type-2 fuzzy sets using an interval approach. *IEEE Transactions on Fuzzy Systems*, *16*(6), 1503–1521.

MacGregor, R. J. (2013). Assess hazards with process flow failure modes analysis. *Chemical Engineering Progress*, *109*(3), 48–56.

Mannan, M. S. (2015). *Lees' loss prevention in the process industries: Hazard identification, assessment, and control* (4th ed.). Elsevier. ISBN-13: 978-0123971890, ISBN-10: 0123971896, Chapter 8.

Marcano, L., Haugen, F. A., Sannerud, R., & Komulainen, T. (2019). Review of simulator training practices for industrial operators: How can individual simulator training be enabled? *Safety Science*, *115*, 414–424.

Markowski, A. S., & Mannan, M. S. (2008). Fuzzy risk matrix. *Journal of Hazardous Materials*, *159*, 152–157.

Mellichamp, D. A. (2019). Profitability, risk, and investment in conceptual plant design: Optimizing key financial parameters rigorously using NPV%. *Computers and Chemical Engineering*, *128*, 450–467.

Mendel, J. M. (2017). *Uncertain rule-based fuzzy logic systems: Introduction and new directions* (2nd ed.). Springer. ISBN 978-3-319-51369-0.

Misuri, A., Khakzad, N., Reniers, G., & Cozzani, V. (2018). Tackling uncertainty in security assessment of critical infrastructures: Dempster-Shafer Theory vs. Credal Sets Theory. *Safety Science*, *107*, 62–76.

Murphy, K. P. (2002). *Dynamic bayesian networks: Representation, inference and learning*. Ph.D. Dissertation. Berkeley: University of California. https://www.cs.ubc.ca/∼murphyk/Thesis/thesis.pdf

Musman, S., & Turner, A. J. (2018). A game oriented approach to minimizing cybersecurity risk. *International Journal of Safety and Security Engineering*, *8*(2), 212–222.

Nazir, S., & Manca, D. (2015). How a plant simulator can improve industrial safety. *Process Safety Progress*, *34*(3), 237–243.

Nelsen, R. B. (2006). *An introduction to Copulas* (2nd ed.). Springer. ISBN-10: 0-387-28659-4.

Németh, E., & Cameron, I. (2018). Multi-level failure, causality and hazard insights via knowledge based systems. In *Mary Kay O'Connor Process Safety Center (MKOPSC), 21st Annual International Symposium October 23–25, College Station, TX.*

Onel, M., Kieslich, C. A., Guzman, Y. A., Floudas, C. A., & Pistikopoulos, E. N. (2018). Reprint of: Big data approach to batch process monitoring: Simultaneous fault detection and diagnosis using nonlinear support vector machine-based feature selection. *Computers and Chemical Engineering*, *116*, 503–520.

OpenBUGS, Rodionov, A., Kelly, D., & Uwe-Klugel, J. (2009). *Guidelines for analysis of data related to ageing of nuclear power plant components and systems. In JRC scientific and technical reports*. Cambridge University, Medical Research Council, Biostatistics Unit, EUR 23954 EN, ISSN 1018-5593; see for project BUGS. http://www.mrc-bsu.cam.ac.uk/bugs/ while OpenBUGS can be downloaded from. https://www.mrc-bsu.cam.ac.uk/software/bugs/openbugs/.

Oppenheimer, M., Little, C. M., & Cooke, R. M. (2016). Expert judgement and uncertainty quantification for climate change. *Nature Climate Change*, *6*, 445–451. https://doi.org/10.1038/NCLIMATE2959.

Pasman, H. J. (2015). *Risk analysis and control for industrial processes—gas, oil and chemicals. A system perspective for assessing and avoiding low-probability, high consequence events.* Butterworth Heinemann, Copyright © 2015 Elsevier Inc. ISBN:978-0-12-800057-1.

Pasman, H. J., Hong, Y., & Rogers, W. J. (2019). Can judgements by different experts of a hazard event risk be reconciled? An on first sight exotic, yet practical solution!. In *Loss prevention 2019 symposium, Delft, the Netherlands, June 16–19, Paper #2* (p. 6).

Pasman, H. J., & Rogers, W. J. (2018). How trustworthy are risk assessment results, and what can be done about the uncertainties they are plagued with? *Journal of Loss Prevention in the Process Industries*, *55*, 162–177.

Patriarca, R., Falegnami, A., De Nicola, A., Villani, M. L., & Paltrinieri, N. (2019). Serious games for industrial safety: An approach for developing resilience early warning indicators. *Safety Science*, *118*, 316–331.

Petersen, K. E. (1994). European model evaluation activity. *Journal of Loss Prevention in the Process Industries*, 7(2), 130–132.

Piatti, A., Antonucci, A., & Zaffalon, M. (2010). *Building knowledge-based systems by Credal networks: a tutorial. In A. R. Baswell (Ed.), Advances in mathematics research.* Vol. 11New York: Nova Science Publishers. http://people.idsia.ch/~zaffalon/papers/2010nova-cns.pdf.

Rae, A., Alexander, R., & McDermid, J. (2014). Fixing the cracks in the crystal ball: A maturity model for quantitative risk assessment. *Reliability Engineering & System Safety*, *125*, 67–81.

Rakowsky, U. K. (2007). Fundamentals of the Dempster-Shafer theory and its applications to reliability modeling. *International Journal of Reliability, Quality and Safety Engineering*, *14*(6), 579–601.

Rasmussen, N. (1975). *Reactor safety study. An assessment of accident risks in U. S. commercial nuclear power plants. Executive summary. WASH-1400 (NUREG-75/014).* Rockville, MD, USA: Federal Government of the United States, U.S. Nuclear Regulatory Commission.

Rebello, S., Yu, H., & Ma, L. (2019). An integrated approach for real-time hazard mitigation in complex industrial processes. *Reliability Engineering and System Safety*, *188*, 297–309.

Reniers, G. L. L., & Noël van Erp, H. R. (2016). *Operational safety economics: A practical approach focused on the chemical and process industries.* Wiley, ISBN 978-1-118871126.

Richmond, B. (2005). *An introduction to systems thinking.* ISEE Systems, Lebanon, NH STELLA®. Systems thinking, experimental learning laboratory with animation developed by Barry Richmond 1985, see https://www.iseesystems.com/.

Rojas, R. (1996). *Neural networks, a systematic introduction.* Berlin: Springer-Verlag. https://www.springer.com/us/book/9783540605058.

Saaty, T. L. (1990). How to make a decision: The analytic hierarchy process. *European Journal of Operational Research*, *48*, 9–26.

Saaty, T. L. (2008). The analytic hierarchy and analytic network measurement processes: Applications to decisions under risk. *European Journal of Pure and Applied Mathematics*, *1*(1), 122–196.

Saaty, T. L., & Tran, L. T. (2007). On the invalidity of fuzzifying numerical judgments in the Analytic Hierarchy Process. *Mathematical and Computer Modelling*, *46*, 962–975.

Salvi, O., & Debray, B. (2006). A global view on ARAMIS, a risk assessment methodology for industries in the framework of the SEVESO II directive. *Journal of Hazardous Materials*, *130*, 187–199. followed by 10 more papers on pages 200-299.

Schwarz, G. (1978). Estimating the dimension of a model. *The Annals of Statistics*, *6*(2), 461–464.

Shafer, G. (1976). *A mathematical theory of evidence.* Princeton University Press.

Shi, J., Khan, F., Zhu, Y., Li, J., & Chen, G. (2018). Robust data-driven model to study dispersion of vapor cloud in offshore facility. *Ocean Engineering*, *161*, 98–110.

Shi, J., Li, X., Khan, F., Chang, Y., Zhu, Y., & Chen, G. (2019). Artificial bee colony Based Bayesian Regularization Artificial Neural Network approach to model transient flammable cloud dispersion in congested area. *Process Safety and Environmental Protection, 128*, 121–127.

Shi, J., Zhu, Y., Khan, F., & Chen, G. (2019). Application of Bayesian Regularization Artificial Neural Network in explosion risk analysis of fixed offshore platform. *Journal of Loss Prevention in the Process Industries, 57*, 131–141.

Si, S.-L., You, X.-Y., Liu, H.-C., & Zhang, P. (2018). *DEMATEL technique: A systematic review of the state-of-the-art literature on methodologies and applications.* Hindawi Mathematical Problems in Engineering. ID 3696457, 33 pages https://doi.org/10.1155/2018/3696457.

Simon, C., & Weber, P. (2009). Evidential networks for reliability analysis and performance evaluation of systems with imprecise knowledge. *IEEE Transactions on Reliability, 58*(1), 69–87.

Simon, C., Weber, P., & Evsukoff, A. (2008). Bayesian networks inference algorithm to implement Dempster Shafer theory in reliability analysis. *Reliability Engineering and System Safety, 93*, 950–996.

Siu, N. O., & Kelly, D. L. (1998). Bayesian parameter estimation in probabilistic risk assessment. *Reliability Engineering & System Safety, 62*(1), 89–116.

Smith, D., Veitch, B., Khan, F., & Taylor, R. (2017). Understanding industrial safety: Comparing Fault tree, Bayesian network, and FRAM approaches. *Journal of Loss Prevention in the Process Industries, 45*, 88–101.

Suokas, J., & Rouhiainen, V. (1989). Quality control in safety and risk analyses. *Journal of Loss Prevention in the Process Industries, 2*, 67–77.

Taqvi, S. A., Tufa, L. D., Zabiri, H., Maulud, A. S., & Uddin, F. (2018). Multiple fault diagnosis in distillation column using multikernel support vector machine. *Industrial & Engineering Chemistry Research, 57*, 14689–14706.

Taylor, J. R. (2017). Automated HAZOP revisited. *Process Safety and Environmental Protection, 111*, 635–651.

Taylor, J. R. (2020). Organisational failure analysis for industrial safety. *ASCE-ASME Journal of Risk and Uncertainty in Engineering Systems Part B: Mechanical Engineering, 6*(1), 011006. https://doi.org/10.1115/1.4044945.

UK HSE. (2017). *Review of consequence model evaluation protocols for major hazards under the EU SAPHEDRA platform.* UK Health and Safety Executive. RR1099 www.hse.gov.uk/research/rrpdf/rr1099.pdf.

Velasquez, M., & Hester, P. T. (2013). An analysis of multi-criteria decision making methods. *International Journal of Operations Research, 10*(2), 56–66.

Venkatasubramanian, V. (2019). The promise of artificial intelligence in chemical engineering: Is it here, finally? *AICHE Journal, 65*(2), 466–478. Figure 1, p.469, based on https://medium.com/@curiousily/tensorflow-for-hackers-part-iv-neural-network-from-scratch-1a4f504dfa8. https://neustan.wordpress.com/2015/09/05/neural-networks-vs-svm-where-when-and-above-all-why/. http://mccormickml.com/2015/08/26/rbfn-tutorial-part-ii-function-approximation/.

Vesely, W., Stamatelatos, M., et al. (2002). lead author *Fault tree handbook with aerospace applications.* NASA Office of Safety and Mission Assurance. https://kscddms.ksc.nasa.gov/Reliability/Documents/Fault_Tree_Handbook_with_Aerospace_Applications_August_2002.pdf.

Visser, J. (1998). Developments in HSE management in oil and gas exploration and production. In A. Hale & M. Baram (Eds.), *Safety management: The challenge of change* (pp. 43–65). Oxford, UK: Pergamon. (Chapter 3).

Watson, H. A. (1961). Launch control safety study, section VII. In (Vol. 1)Murray Hill, NJ: Bell Labs.

Wu, H., & Zhao, J. (2018). Deep convolutional neural network model based chemical process fault diagnosis. *Computers and Chemical Engineering, 115*, 185–197.
Yager, R. (1987). On the Dempster-Shafer framework and new combination rules. *Information Sciences, 41*, 93–137.
Yang, M., Khan, F., Lye, L., & Amyotte, P. (2015). Risk assessment of rare events. *Process Safety and Environmental Protection, 98*, 102–108.
Yazdi, M., & Kabir, S. (2017). A fuzzy Bayesian network approach for risk analysis in process industries. *Process Safety and Environmental Protection, 111*, 507–519.
Yu, H., Khan, F., & Veitch, B. (2017). A flexible hierarchical Bayesian modeling technique for risk analysis of major accidents. *Risk Analysis, 37*(9), 1668–1682.
Zadeh, L. A. (1965). Fuzzy sets. *Information and Control, 8*, 338–353.
Zadeh, L. A. (1975). The concept of a linguistic variable and its application to approximate reasoning. *Information Sciences, 8*, 199–249.
Zadeh, L. A. (1978). Fuzzy sets as a basis for the theory of possibility. *Fuzzy Sets and Systems, 1*, 3–28. reprinted 1999, Fuzzy sets and systems, 100 supplement, 9–34.
Zarei, E., Khakzad, N., Cozzani, V., & Reniers, G. (2019). Safety analysis of process systems using Fuzzy Bayesian Network (FBN). *Journal of Loss Prevention in the Process Industries, 57*, 7–16.
Zeraei, J. (2015). Best-worst multi-criteria decision-making method. *Omega, 53*, 49–57.
Zeraei, J. (2016). Best-worst multi-criteria decision-making method: Some properties and linear model. *Omega, 64*, 126–130.
Zhang, L., & Reniers, G. (2018). Applying a Bayesian Stackelberg game for securing a chemical plant. *Journal of Loss Prevention in the Process Industries, 51*, 72–83.
Zhao, X., Malasse, O., & Buchheit, G. (2019). Verification of safety integrity level of high demand system based on Stochastic Petri Nets and Monte Carlo Simulation. *Reliability Engineering and System Safety, 184*, 258–265.
Zimmermann, H.-J. (2001). *Fuzzy set theory—And its applications* (4th ed.). Springer Science & Business Media. ISBN 978-94-010-3870-6.

Further reading

CCPS. (2014). *Guidelines for initiating events and independent protection layers in layer of protection analysis*. Center for Chemical Process Safety of AIChE, Wiley. ISBN-13: 978-0470343852, ISBN-10: 0470343850.
IEC 61882:2016 International Standard. (2016). *Hazard and operability studies (HAZOP studies)—Application guide*. Geneva 20, Switzerland: International Electrotechnical Commission.

CHAPTER FOUR

(mis)Using Bayesian networks for dynamic risk assessment

Nima Khakzad*
School of Occupational and Public Health, Ryerson University, Toronto, ON, Canada
*Corresponding author: e-mail address: nima.khakzad@ryerson.ca

Contents

1. Introduction

Process systems and operations are highly complicated, interdependent, and dynamic where the probability of malfunction of equipment or occurrence probability of undesired events can change with time. Temporal change of probabilities can take place due to degradation, aging or maintenance of mechanical and electrical equipment or because of temporal dependencies among the components of systems. Dynamic gates

Methods in Chemical Process Safety, Volume 4
ISSN 2468-6514
https://doi.org/10.1016/bs.mcps.2020.03.001

such as the functional dependency (FDEP) gate and the priority AND (PAND) gate are some examples of temporal dependencies among the components of a system which require a predefined order of failures of the components to result in a system failure.

As a result of temporal changes of probabilities, the failure probability of the system and accordingly the calculated risk of system failure may change over time. This has been the cornerstone of dynamic risk assessment in the recent years (Khakzad, Yu, Paltrinieri, & Khan, 2016). The risk calculated using conventional and mostly static techniques, such as fault tree analysis, provides the analysts with a snapshot of the risk within a limited time framework without taking account of future changes of the risk components. Clearly enough, adopting risk management strategies or making risk-based decisions merely based on such a risk snapshot would quite often result in suboptimal and cost-ineffective (if not totally useless) risk management strategies.

As a daily life example of the difference between conventional and dynamic risk assessments, consider a case when you decide to take a few-day road trip in the winter season starting on Wednesday Feb. 5th until Saturday Feb. 8th and are wondering if you should equip your car with winter tires or not. A conventional risk assessment of weather condition on Tuesday Feb. 4th would give a zero probability of snowfall on Wednesday Feb. 5th and no information on the coming days. As such, you will have no choice but to extend the same zero probability of snowfall to the coming days (Fig. 1).

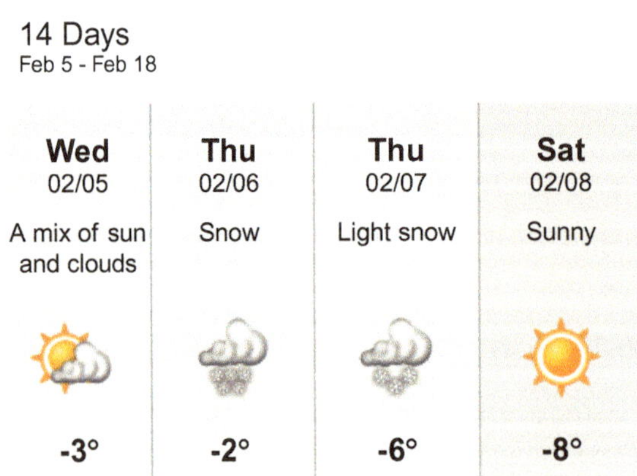

Fig. 1 Today being Tuesday, Feb. 4, a conventional risk assessment of weather condition would predict a zero snowfall within a limited time horizon, that is, until Wednesday Feb. 5. A dynamic risk assessment would however provide a broader picture of the weather condition, showing the probability of snowfall on the coming days.

On the other hand, a dynamic risk assessment would assign probabilities of snowfall to the coming Thursday and Friday as well. Obviously, the decision as to install winter tires on your car would only be credible based on the result of the dynamic risk assessment.

The issue of dynamic risk assessment is more crucial in the hazardous industries such as chemical and process industries as the risk of an unwanted release of hazardous chemicals or energy (in the form of fire and explosion) could be catastrophic.

In recent years, Bayesian network (BN) has widely been used in safety assessment and management of process systems and facilities. BN is a graphical probabilistic technique based on Bayes' theorem for reasoning under uncertainty. BN can be used either in a forward analysis to predict the probability of unknown variables (e.g., the failure probability of a heat exchanger) based on known variables (e.g., the degradation or failure of the heat exchanger's components) or in a backward analysis to update the probability of known variables given some evidence using the Bayes' theorem (e.g., to update the failure probability of the components given the failure of the heat exchanger).

Due to its flexible graphical representation and robust reasoning engine, BN has proved as a promising technique for safety assessment of a wide variety of complex systems and events (Khakzad, Khan, & Amyotte, 2011). The fact that BN can work with a variety of probabilities such as crisp probabilities, fuzzy probabilities (Zarei, Khakzad, Cozzani, & Reniers, 2019), and imprecise probabilities (Khakzad, 2019a) has further helped promote its application in safety assessment and risk analysis. Two features of BN, i.e., its capability in modeling dependencies (not temporal dependencies but mostly common-cause failures) and probability updating, have been extensively used in system safety and reliability assessment. However, the probability updating feature of BN, although very effective in fault diagnostic analysis, has usually been misused in the context of dynamic risk assessment.

The previous work on the application of BN to dynamic risk assessment all but has overlooked the fact that the probability updating feature of BN merely provides the analyst's with updated prior beliefs in the same limited time horizon as provided by the conventional risk assessment. In other words, by employing the probability updating the conditional probabilities used to populate the conditional probability tables (CPTs) of BN would not change whatsoever, leaving the analyst literary with the same model structure and prior probabilities to reason about the future of the system of interest.

The other key feature of BN, known as *probability adapting* or sequential learning, is much more relevant to dynamic risk assessment but has surprisingly received much less attention than probability updating. Unlike probability updating, probability adapting enables the analyst to explicitly take account of new information to revise the BN parameters (conditional probabilities) as time passes and more information and knowledge becomes available.

In the past decade, an extension of BN, known as dynamic Bayesian network (DBN), has started to be employed in modeling and safety assessment of dynamic systems and processes. In its simplest form, a DBN can be made by replicating a BN at several time slices over a limited time horizon. Every two sequential time slices identify a time interval, where the BN at the beginning of a time interval can be used to predict the probabilities over that time interval. Accordingly, the outputs of the BN in the beginning of the time interval are used as the inputs of the BN in the end of the time interval with the possibility of having different conditional probability tables for the BNs. This way, a DBN relies on the collective ability of a number of sequential BNs each capable of predicting over a limited time interval with a reasonable degree of accuracy. That being said, the shorter the length of the time intervals the higher the number of BNs and thus the higher the accuracy of the predictions (reasonably assuming that the temporal probabilities would not change too dramatically over a short time interval). The more sensitive the temporal probabilities to time the shorter time intervals would be required to obtain a good accuracy.

There are tones of literature on applications of BN and DBN to system modeling and safety assessment. Therefore, in the present chapter, instead of replicating commonplace applications of BN and DBN, we have tried to emphasize those aspects of BN and DBN which have received less attention or have long been misused in the context of dynamic risk assessment.

In Section 2, the fundamentals of Bayes' theorem, BN and DBN are briefly revisited, and an application of DBN, as an intuitive dynamic technique for dynamic safety (risk) assessment, is demonstrated via a simple example. A common misapplication of DBN is also discussed. In Section 3, it is demonstrated that BN—and in particular its probability updating feature—is not an appropriate technique for doing dynamic risk assessment, and instead a probability adapting should be performed if one insists on using BN for this purpose. A tutorial on the application of probability adapting to dynamic risk assessment is also presented in this section. Section 4 reviews some risk importance measures such as Birnbaum importance measure and mutual information and gives tips on the ones more appropriate to dynamic risk assessment. Section 5 presents the key outcomes of the study and concludes the chapter.

2. Bayesian inference

2.1 Bayes' theorem

A major task in any dynamic risk assessment is to predict a prior probability for a variable of interest X, update the prior probability given new information (also known as evidence E), and more importantly, predicting the changes of the probability over an arbitrary time horizon with regard to environmental and operational changes. The prior probability distribution can be estimated in several ways including statistical analysis of historical data or data collected from inspection and condition monitoring, deductive reasoning by means of quantitative risk analysis techniques (e.g., fault tree), or expert judgment. The updating of the prior probability distribution is usually carried out using Bayes' theorem:

$$f_1(X|E) = \frac{f_0(X)\,L(E|X)}{\int f_0(X)\,L(E|X)\,dX} \tag{1}$$

where $f_1(X|E)$ is the updated (posterior) probability distribution of X given the evidence E; $f_0(X)$ is the prior probability distribution of X; $L(E|X)$ is the likelihood function of X or alternatively the probability distribution of the evidence given the parameter X. In the case of having discrete probability values instead of probability distributions, Eq. (1) can be simplified as:

$$P(X|E) = \frac{P(X)\,L(E|X)}{\sum_X P(X)\,L(E|X)} \tag{2}$$

Since in Eqs. (1) and (2) the denominator is used to normalize the posterior probability distribution, in many applications it suffices to employ Eq. (1) in the form of $f_1(X|E) \propto f_0(X)\,L(E|X)$, implying the posterior probability distribution is proportionate to the product of the prior and the likelihood probability distributions.

For example, consider a case where the prior probability of failure time (t) of a component follows a beta distribution $f_0(t) \propto t^{\alpha}(1-t)^{\beta}$ while the likelihood function is estimated using a binomial distribution

$L(E|t) \propto t^m (1-t)^{n-m}$. α and β are the parameters of beta distribution whereas m and n are the parameters of binomial distribution, representing that the component has failed m times out of being demanded n times. As a result, the posterior probability distribution of t will be $f_1(t|E) \propto t^{\alpha+m} (1-t)^{\beta+n-m}$, which is a beta distribution with the updated parameters $\alpha' = \alpha + m$ and $\beta' = \beta + n - m$.

Bayes' rule can be integrated with conventional risk assessment techniques such as fault tree, event tree, and bow-tie diagram so that cumulative knowledge (e.g., the number of failures of a component recorded over a time period) could be used to sequentially update the probabilities (Kalantarnia, Khan, & Hawboldt, 2009; Khakzad, Khakzad, & Khan, 2014; Khakzad, Khan, & Amyotte, 2012; Khakzad, Khan, & Paltrinieri, 2014; Meel & Seider, 2006). Although effective in probability updating, Bayes' rule cannot alone be employed for predicting the temporal changes of the probability of a random variable of interest.

2.2 Conventional Bayesian network

Bayesian network BN=(G, θ) (Pearl, 1988) is a directed acyclic graph for knowledge representation and probabilistic inference. G is the structure of the graph in which the random variables are presented as nodes while the dependencies among the random variables are denoted as directed arcs connecting the nodes (Fig. 2).

The graph G satisfies Markovian condition in that each variable in G is independent of its non-descendants given its immediate parents. As a result, the associated joint probability distribution of the random variables can be factorized as the multiplication of conditional probabilities of the nodes (variables) given their parents as:

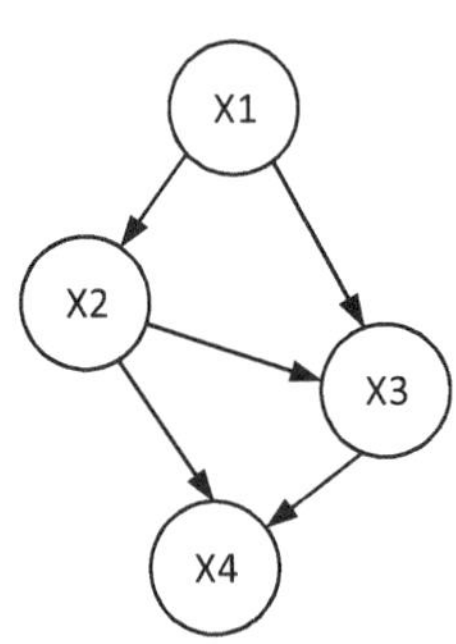

Fig. 2 A typical Bayesian network.

$$P(U) = \prod_{i=1}^{n} P(X_i|Pa(X_i)) \tag{3}$$

where $P(U)$ is the joint probability distribution of variables, and $Pa(X_i)$ is the parent set of variables X_i. The conditional probabilities $\theta = P(Xi \mid Pa(Xi))$ are known as the network parameters which can either be elicited from subject matter experts or be learned from data. Considering the BN in Fig. 2, the joint probability of the variables can be presented as: $P(X1, X2, X3, X4) = P(X1)\ P(X2 \mid X1)\ P(X3 \mid X1, X2)\ P(X4 \mid X2, X3)$.

BN can be used for predictive analysis (prediction of the symptom based on the cause), e.g., $P(X4 = x_4 \mid X1 = x_1)$, and for diagnostic analysis (diagnosis of the cause given the symptom), e.g., $P(X4 = x_4|X1 = \overline{x_1})$. Having the marginal probabilities of the root nodes (nodes with no parents, e.g., X1 in Fig. 2) and the conditional probabilities of child nodes (e.g., X2–X4 in Fig. 2), a number of exact inference algorithms such as bucket elimination or variable elimination (Dechter 1996) and junction tree (Jensen, 1996; Lauritzen & Spiegelhalter, 1988) as well as approximate inference techniques such as belief propagation (also known as sum-product message passing) (Pearl, 1982), Monte Carlo Markov Chain (MCMC) (Cheng & Druzdzel, 1999), dynamic discretization and MCMC (Fenton & Neil, 2019) and rank correlation (Hanea & Kurowicka, 2008) can be used to calculate the marginal probabilities of the child nodes.

The main application of BN is to carry out diagnostic or backward analysis where prior probabilities could be updated given a new set of observations or evidence. BN directly takes advantage of Bayes' rule in Eq. (2) to update the probabilities as:

$$P(U|E) = \frac{P(U)P(E|U)}{\sum_U P(U)P(E|U)} = \frac{P(U,E)}{\sum_U P(U,E)} \tag{4}$$

In safety risk analysis of chemical and process facilities, the probability updating feature of BN can effectively be employed to identify the root causes or failure modes contributing the most to the failure of a system of interest (see risk importance measures in Section 5).

The advantages of BN over other conventional techniques such as fault tree, event tree, and bow-tie diagram have been thoroughly discussed in previous work (Khakzad et al., 2011; Khakzad, Khan, & Amyotte, 2013a, 2013b). Nevertheless, it is still a common approach in safety assessment first to model a system (or event) using fault tree or bow-tie diagram and then to convert it into an equivalent BN. Despite its advantages over the conventional techniques, conventional BN is not a suitable technique for dynamic risk assessment when the sequence of failures matters or when the prior probabilities may change over time.

For illustrative purposes, consider the pumps system in Fig. 3 consisting of three pumps M1, M2, and M3. For the system to operate, either M1 or M2 and M3 should be operational.

M2 is the spare of M1 and becomes operational only if M1 fails. M2 is also assumed not to fail while in dormant (idle) state and at time of activation (probability of failure on demand = 0). However, due to a lesser capacity of M2 compared to M1, M3 would be under more load when M2 is operational and M1 is failed or dormant, leading to different failure probabilities of M3 depending on whether M1 is operational or M2. Besides, when M2 is activated, M1 could be repaired to be back to an operational state in case M2 fails and vice versa. Further, for the sake of simplicity, assume that the failure probability of a repaired pump, whether M1 or M2, would be the same as before the failure.

To calculate the system failure (operation) probability over time using BN approach, assume the following conditional probabilities based on the foregoing system description:

(1) P(M1 = failed | M2 = not working) = 0.1
(2) P(M2 = failed | M1 = not working) = 0.1
(3) P(M3 = failed | M1 = working, M2 = not working) = 0.1
(4) P(M3 = failed | M1 = not working, M2 = working) = 0.2
(5) P(M3 = failed | M1 = not working, M2 = not working) = 0.3
(6) P(M3 = failed | M1 = working, M2 = working) = 0.1
(7) P(M1 = idle | M2 = working, M1 = failed) = P(M2 = idle | M1 = working, M2 = failed) = 0.6.

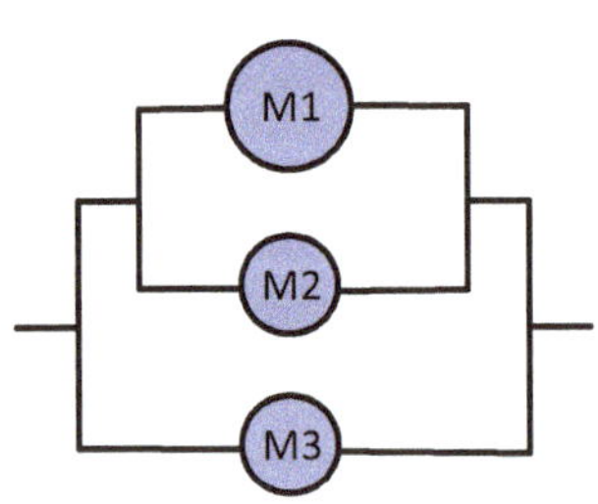

Fig. 3 A parallel system of pumps. M1 and M3 are operational while M2 is dormant. If M1 fails, M2 is activated. For the system to be operational, M1 and M3 or M2 and M3 should be operational.

In the foregoing conditional probabilities, the state of "not working" refers to one of the states of "idle" or "failed." It should be noted that these probabilities are simplistic and merely for illustrative purposes. As we will see later, these conditional probabilities are not constant and should be calculated given the failure rates and repair rates of the pumps—whether constant or time-dependent—and time intervals used in the DBN modeling.

For instance, P(M1 = ready to work | M2 = working, M1 was under repair) depends on the failure rate of M2 and repair rate of M1 and how much time would be available from the activation to the failure of M2 which is equal to the time available for the repair of M1. After demonstrating the modeling of the pumps system using these simplistic probabilities, we will briefly demonstrate how failure/repair rates can be used to calculate time-dependent conditional probabilities more realistically and accurately.

The conditional probabilities, no matter how simplistic, reflect mutual cause-effect relationships among the system components, that cannot be modeled via a conventional BN. For instance, the conditional probabilities in line 7 imply the dependence of the state of M1 on the state of M2 and vice versa, which should be realized in the BN with an arc from M1 to M2 and an arc from M2 to M1. Obviously, such a mutual cause-effect relationship cannot be modeled in a conventional BN since BN is an acyclic graph and M1 → M2 → M1 would form a cycle (Fig. 4).

To avoid the cycle in modeling the foregoing system using conventional BN, one would have no choice but, for instance, to simplify the model by ignoring the dependence of M1 on M2 as depicted in Fig. 5. As can be seen, the possibility of M1 undergoing repair while M2 is operational and vice

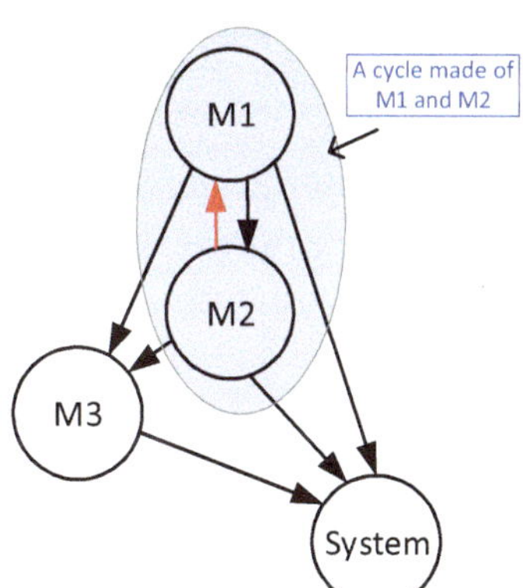

Fig. 4 A directed graph that cannot be considered as a BN due to the cycle made of M1 and M2.

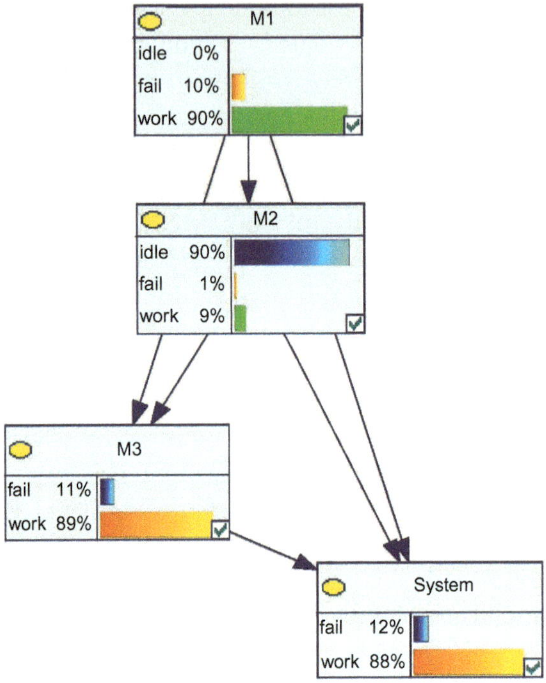

Fig. 5 Simplified BN for modeling the pumps system, in which the possibility of M1 and M2 undergoing repair has been overlooked.

Table 1 Conditional probability table of node M2 in the BN of Fig. 5.

M1→	Idle	Failed	Working
M2 = Idle	0	0	1
M2 = Failed	0.1	0.1	0
M2 = Working	0.9	0.9	0

versa, M2 undergoing repair after it fails and the repaired (idle) M1 is back to operation cannot be modeled using the oversimplified BN in Fig. 5. For the sake of clarity, the conditional probabilities of M2 and M3 are presented in Tables 1 and 2.

In the next section, we will show how DBN can be utilized to mode the reciprocal dependency between M1 and M2, and how it may affect the calculation of probabilities.

2.3 Dynamic Bayesian network

Dynamic Bayesian network (DBN) offers a flexible structure and unique modeling techniques to intuitively model temporal dependencies via a

Table 2 Conditional probability table of node M3 in the BN of Fig. 5.

M1→	**Idle**			**Failed**			**Working**		
M2→	**Idle**	**Failed**	**Working**	**Idle**	**Failed**	**Working**	**Idle**	**Failed**	**Working**
M3 = Failed	0.3	0.3	0.2	0.3	0.3	0.2	0.1	0.1	0.1
M3 = Working	0.7	0.7	0.8	0.7	0.7	0.8	0.9	0.9	0.9

probabilistic framework. The reasoning and learning engine of DBN allows time dependencies and complex interactions among the components of a system to be considered. This attribute of DBN helps it outperform conventional BN in likelihood modeling and safety analysis of stochastic systems. Generally speaking, two different approaches have been developed for DBN: (i) interval-based approach and (ii) instant-based approach. In this chapter we consider the second approach, i.e., instant-based DBN, as it provides more flexibility and less computational complexity than the first approach. For some applications of interval-based DBNs see Boudali and Dugan (2005), Khakzad, Khan, and Amyotte (2013c), Khakzad, Khan, Amyotte, and Cozzani (2014).

In the instant-based DBN the time line is divided into a finite number of time instants (time slices). Identical BN structures (such as the one depicted in Fig. 2) are then generated at each time instant and connected to each other by means of temporal arcs (red arcs in Fig. 6). In a DBN, the arcs connecting the BNs of different time slices are known as inter-arcs while the arcs connecting the nodes of a BN in the same time slice (the black arcs in Fig. 6) are known as intra-arcs.

Using the instant-based DBN, a node at i^{th} time slice can be conditionally dependent not only on its parents at the same time slice but also on its parents and even itself at the previous time slice. When the temporal transition rates (or transition probabilities) are constant, only two time slices can be considered in the modeling of DBN (Khakzad, 2015; Montani, Portinale, Bobbio, & Codetta-Raiteri, 2008; Neapolitan, 2003). As such, the joint probability distribution of a set of random variables at time $t+\Delta$ can be expanded as:

$$
\begin{aligned}
P(U^{t+\Delta}) &= P\left(X_1^{t+\Delta}, X_2^{t+\Delta}, \ldots, X_n^{t+\Delta}\right) \\
&= \prod_{i=1}^{n} P\left(X_i^{t+\Delta} \middle| X_i^t, Pa\left(X_i^t\right), Pa\left(X_i^{t+\Delta}\right)\right)
\end{aligned} \tag{5}
$$

where X_i^t and $X_i^{t+\Delta}$ are the copies of X_i in two consecutive time slices with a time interval of Δ; $Pa(X_i^t)$ and $Pa(X_i^{t+\Delta})$ are the parent sets of X_i at the

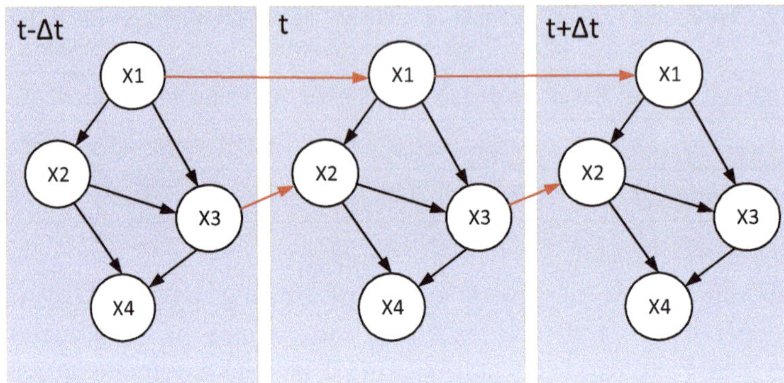

Fig. 6 Schematic of an instant-based dynamic Bayesian network by replicating similar Bayesian networks at consecutive time slices. The arcs in color red represent temporal arcs.

foregoing time slices. For example, in Fig. 6, the conditional probability of the copy of X2 at $t+\Delta$ can be calculated as $P(X2^{t+\Delta}|X1^{t+\Delta}, X3^{t})$. Instant-based DBN can effectively be employed to failure analysis of systems which may degrade over time or to capture the influence of maintenance on the failure probability of systems.

To demonstrate the outperformance of DBN over conventional BN, consider the previous pumps system in Fig. 3. Assuming that the conditional probabilities would not change over time, only for illustrative purposes, Fig. 7 depicts a DBN developed over three time slices for calculating the temporal changes of the failure probabilities of the pumps and the entire system.

As can be seen, in the first time slice, there is an arc from M1 to M2 to show that the state of M2 depends on the state of M1; it is owing to the fact that M2 would be in an idle or operational state depending on whether M1 is in operational or failed state, respectively. In the second time slice, however, there is an arc from M2 in the first time slice to M1 in the second time slice as the state of M1 (i.e., a failed M1 can be restored to an idle M1) would depend on whether M2 has been operational, providing enough time to repair M1. To make a comparison between the DBN and the BN of the same system, the conditional probability tables of nodes M1 and M2 in the second time slice, i.e., Copy of M1 and Copy of M2, are presented in Tables 3 and 4. It is worth mentioning that the CPTs of nodes M3 and System in both the BN and the DBN would remain unchanged.

Some BN modeling software such as GeNIe (2019) enables the modeler to present and model the DBN in a compact form since replicating the DBN

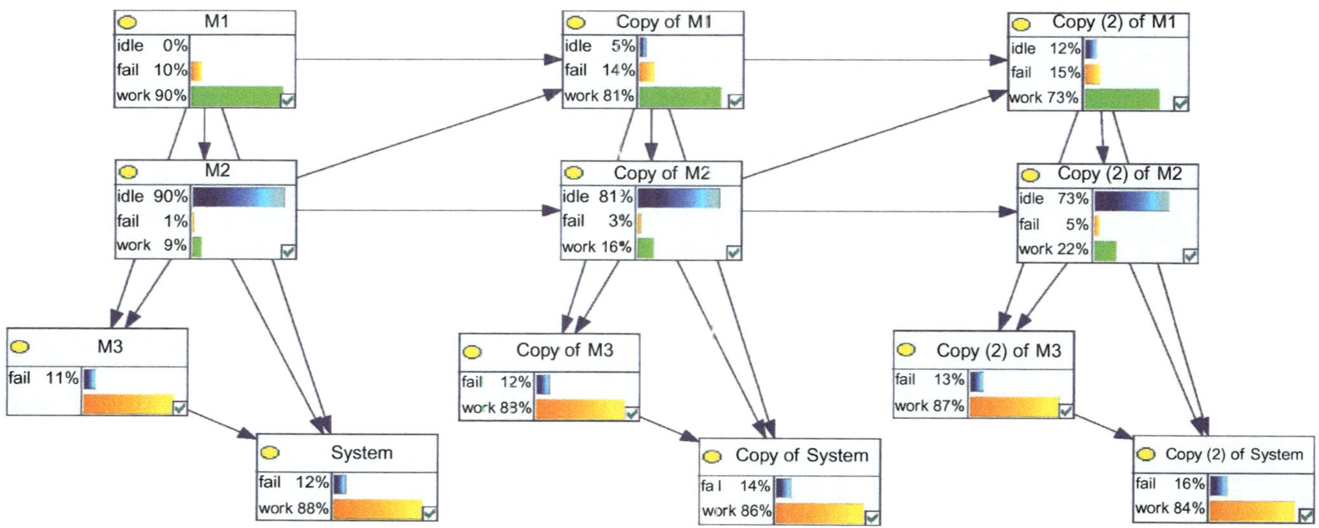

Fig. 7 DBN model of the pumps system. Only three time slices have been depicted for brevity.

Table 3 Conditional probability table of node M1 in the second time slice, i.e., Copy of M1, in the DBN of Fig. 7.

M1 →	**Idle**			**Failed**			**Working**		
M2 →	**Idle**	**Failed**	**Working**	**Idle**	**Failed**	**Working**	**Idle**	**Failed**	**Working**
Copy of M1 = Idle	0	0	1	0	0	0.6	0	0	0
Copy of M1 = Failed	0.1	0.1	0	1	1	0.4	0.1	0.1	0.1
Copy of M1 = Working	0.9	0.9	0	0	0	0	0.9	0.9	0.9

Table 4 Conditional probability table of node M2 in the second time slice, i.e., Copy of M2, in the DBN of Fig. 7.

Copy of M1 →	**Idle**			**Failed**			**Working**		
M2 →	**Idle**	**Failed**	**Working**	**Idle**	**Failed**	**Working**	**Idle**	**Failed**	**Working**
Copy of M2 = Idle	0	0	0	0	0	0	1	0.6	1
Copy of M2 = Failed	0.1	0.4	0.1	0.1	1	0.1	0	0.4	0
Copy of M2 = Working	0.9	0.6	0.9	0.9	0	0.9	0	0	0

for more time slices in the form of Fig. 7 would be neither practical nor tractable. A compact representation of the foregoing DBN in software GeNIe has been displayed in Fig. 8, where labeled arcs denote the inter-arcs (dynamic arcs), and regular arcs denote the intra-arcs. The numbers inside the labels show the number of previous time slices to be considered for assessing the conditional probabilities. Normally, when the conditional probabilities are constant (e.g., when failure rates are constant, or alternatively when the probability distribution of failure times are exponential), only two sequential time slices would be considered for calculating the conditional probabilities, illustrated by an "1" labeled arc (e.g., the arc from M2 to M1).

In the previous example, we assumed that the conditional probabilities would remain constant regardless of the length of time intervals, which are time between two sequential BNs. Considering Fig. 6, such time interval has been denoted as $(t+\Delta t)-t=t-(t-\Delta t)=\Delta t$, although the presumption of

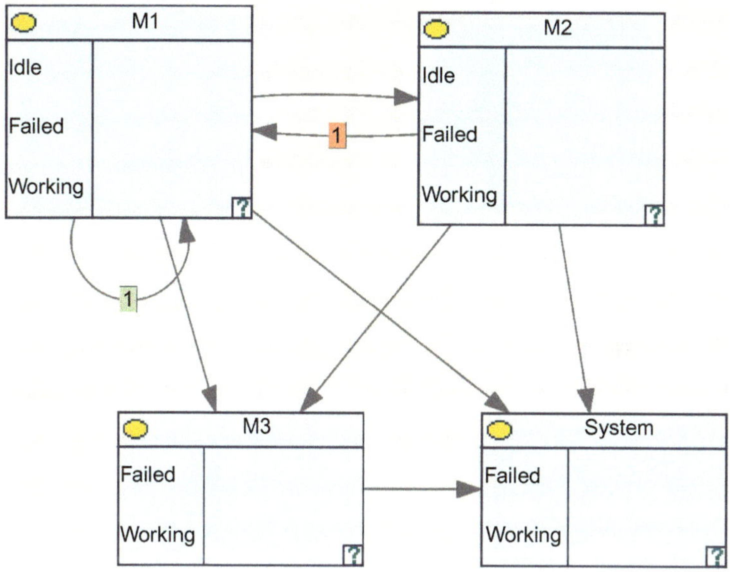

Fig. 8 Compact representation of the pumps DBN in GeNIe.

identical time intervals among sequential time slices is not necessary in developing DBNs (obviously it largely facilitates the computation). However, the length of time intervals is a key parameter in calculating the conditional probabilities as a longer time interval may imply a larger failure probability for an operating pump as well as a larger probability of getting back to operating condition for a failed pump being under repair.

As a result, the assumption of constant probabilities—be it failure probabilities or repair probabilities—is either realistic nor accurate. Besides, without specifying the length of time intervals, it is not possible to accurately interpret the temporal changes in the failure probability of the pumps system in Fig. 7. For instance, it is not clear how long it would take for the system's failure probability to increase from 12% in the first time slice to 16% in the second time slice: Would it take 2 hours, 2 days, 2 weeks, 2 months, or 2 years?

For illustration purposes, assume that both the time to failure and time to repair of pump M1 have exponential distributions. λ is the constant failure rate as the sole parameter in the exponential distribution of the time to failure, and μ is the constant repair rate as the sole parameter in the exponential distribution of the time to repair of pump M1. As a result, the failure and repair probability of M1 during a Δt time interval can be calculated, respectively, as Eqs. (6) and (7) (Ebeling, 1996):

$$P(M1^{t+\Delta t} = failed\,|M1^{t} = working) = 1 - e^{-\lambda.\Delta t} \quad (6)$$

$$P(M1^{t+\Delta t} = repaied\,|\,M1^{t} = failed) = 1 - e^{-\mu.\Delta t} \quad (7)$$

Having these conditional probabilities, the CPT of M1 in the DBN can be populated as Table 5. It is worth recalling that a repaired pump will stay in an idle state until activated upon the failure of the other pump (only for M1 and M2).

Aside from their application to dynamic safety (risk) assessment of complex systems, DBNs have been extensively utilized in modeling, decision making, and risk assessment of dynamic events such as domino effects (Khakzad, 2015), firefighting (Khakzad, 2018) and fire protection of process plants (Khakzad, Landucci, & Reniers, 2017), spread of wildfire in wildland-industrial interfaces (Khakzad, 2019b), protection of process plants against wildfires (Khakzad, Dadashzadeh, & Reniers, 2018), and optimal inspection intervals of buried pipelines (Abubakirov, Yang, & Khakzad, 2020), dynamic assessment of process systems resilience (Zinetullina, Yang, Khakzad, & Golamn, 2019), protection of process plants against imminent terrorist attacks (Khakzad & Reniers, 2019), failure assessment of subsea pipelines (Arzaghi et al., 2018), and vulnerability assessment of process plants to domino effects (Khakzad, Reniers, Abbassi, & Khan, 2016).

2.4 Misapplication of DBN to dynamic risk assessment

BN and DBN are both based upon Markovian property in the sense that the conditional probability distribution of future states of a node (random variable) depends only on the present state, not on the sequence of events that preceded it. As such, in DBN usually the conditional probabilities only within a single time interval (for two sequential BNs) are developed and thereafter replicated for the other consecutive time intervals. In other words, given the same time interval (Δt) between the sequential BNs, it is usually mistakenly assumed in the development of DBNs that always $P(X^{t+\Delta t}|X^{t}) = P(X^{t}|X^{t-\Delta t})$.

However, this presumption is only valid if the failure (occurrence) of X has an exponential distribution or alternatively a constant failure (occurrence) rate. If the failure rate is time dependent, which is the case in, among

Table 5 Conditional probability table of node M1 in the second time slice, i.e., "Copy of M1."

M1 →	**Idle**			**Failed**			**Working**		
M2 →	**Idle**	**Failed**	**Working**	**Idle**	**Failed**	**Working**	**Idle**	**Failed**	**Working**
Copy of M1 = Idle	0	0	1	0	0	$1-e^{-\mu.\ \Delta t}$	0	0	0
Copy of M1 = Failed	$1-e^{-\lambda.\ \Delta t}$	$1-e^{-\lambda.\ \Delta t}$	0	1	1	$e^{-\mu.\ \Delta t}$	$1-e^{-\lambda.\ \Delta t}$	$1-e^{-\lambda.\ \Delta t}$	$1-e^{-\lambda.\ \Delta t}$
Copy of M1 = Working	$e^{-\lambda.\ \Delta t}$	$e^{-\lambda.\ \Delta t}$	0	0	0	0	$e^{-\lambda.\ \Delta t}$	$e^{-\lambda.\ \Delta t}$	$e^{-\lambda.\ \Delta t}$

Given equal time intervals, the conditional probability table of M1 in the third time slice, i.e., "Copy(2) of M1" would be the same.

others, Weibull, normal, or lognormal distribution, $P(X^{t+\Delta t}|X^{t}) \neq P(X^{t}|X^{t-\Delta t})$ even given the same time intervals. For the sake of clarity, consider two cases:

(1) X is a component the time to failure (T_x) of which follows an exponential distribution, and

(2) Y is a component the time to failure of which (T_y) follows a Weibull distribution.

Regarding X, the probability that X fails before a specific time t is given as the cumulative distribution function of exponential distribution as (Ebeling, 1996):

$$P(T_X < t) = F_X(t) = 1 - e^{-\lambda t} \tag{8}$$

where λ is the failure rate of X, or the parameter of the exponential distribution. As such, given that X has been working at t, the failure probability of X within the next time interval Δt can be calculated as:

$$\begin{aligned} P(t < T_X < t + \Delta t \mid T_X > t) &= \frac{P(t < T_X < t + \Delta t)}{P(T_X > t)} \\ &= \frac{F_X(t + \Delta t) - F_X(t)}{1 - F_X(t)} = \frac{e^{-\lambda t} - e^{-\lambda(t+\Delta t)}}{e^{-\lambda t}} \\ &= 1 - e^{-\lambda.\Delta t} \end{aligned} \tag{9}$$

Likewise,

$$\begin{aligned} P(t - \Delta t < T_X < t \mid T_X > t - \Delta t) &= \frac{P(t - \Delta t < T_X < t)}{P(T_X > t - \Delta t)} \\ &= \frac{F_X(t) - F_X(t - \Delta t)}{1 - F_X(t - \Delta t)} = \frac{e^{-\lambda(t-\Delta t)} - e^{-\lambda t}}{e^{-\lambda(t-\Delta t)}} \\ &= 1 - e^{-\lambda.\Delta t} \end{aligned} \tag{10}$$

As can be seen via a comparison between Eqs. (9) and (10), as long as the length of time intervals is the same between the sequential BNs of a DBN, the conditional probabilities needed to populate the CPTs of an exponential variable (node) in the DBN would remain the same in all the BNs. This significantly facilitates the development of the DBN as it is not required to calculate the conditional probabilities separately in each time interval.

Regarding Y, the probability that Y fails before a specific time t is given as the cumulative distribution function of Weibull distribution as (Ebeling, 1996):

$$P(T_Y < t) = F_Y(t) = 1 - e^{-\left(\frac{t}{\gamma}\right)^k} \tag{11}$$

where γ and k are the parameters of Weibull distribution. As such, given that Y has been working at t, the failure probability of Y within the next time interval Δt can be calculated as:

$$\begin{aligned} P(t < T_Y < t + \Delta t\,)\ T_Y > t) &= \frac{P(t < T_Y < t + \Delta t)}{P(T_Y > t)} \\ &= \frac{F_Y(t + \Delta t) - F_Y(t)}{1 - F_Y(t)} \\ &= \frac{e^{-\left(\frac{t}{\gamma}\right)^k} - e^{-\left(\frac{t+\Delta t}{\gamma}\right)^k}}{e^{-\left(\frac{t}{\gamma}\right)^k}} = 1 - e^{-\left(\frac{t+\Delta t}{\gamma}\right)^k + \left(\frac{t}{\gamma}\right)^k} \end{aligned} \tag{12}$$

Similarly,

$$\begin{aligned} P(t - \Delta t < T_Y < t\,)\ T_Y > t - \Delta t) &= \frac{P(t - \Delta t < T_Y < t)}{P(T_Y > t - \Delta t)} \\ &= \frac{F_Y(t) - F_Y(\mathrm{t} - \Delta t)}{1 - F_Y(t - \Delta t)} \\ &= \frac{e^{-\left(\frac{t-\Delta t}{\gamma}\right)^k} - e^{-\left(\frac{t}{\gamma}\right)^k}}{e^{-\left(\frac{t-\Delta t}{\gamma}\right)^k}} = 1 \quad e^{-\left(\frac{t}{\gamma}\right)^k + \left(\frac{t-\Delta t}{\gamma}\right)^k} \end{aligned} \tag{13}$$

Having a glimpse at the conditional probabilities in Eqs. (12) and (13), despite very similar structures, one can see that even given the same length of time intervals between (t $-$ Δt, t) and (t, t $+$ Δt), these conditional probabilities would generally not be the same. In other words, the conditional probabilities needed to populate the CPTs of a Weibull variable (node) should separately be calculated for each time interval no matter if the length of time intervals is the same.

To make the discussion more concrete, let's make a numerical comparison between the two cases. Consider a $\lambda = 0.083$ for X and a $\gamma = 6$ and $k = 2$ for Y, and $\Delta t = 3$ for both cases. The numerical values of the exponential and Weibull distribution have been chosen so that equal failure paraxialities could be obtained for both X and Y in the first time interval (just to facilitate the comparison). The DBNs for calculating the failure probabilities of X and Y over three consecutive time intervals have been illustrated in Fig. 9.

The CPT used for component X in Fig. 9a would be the same for all the three time slices (BNs) as presented in Table 6 while those of component Y would be different for each time slice as presented in Tables 7–9.

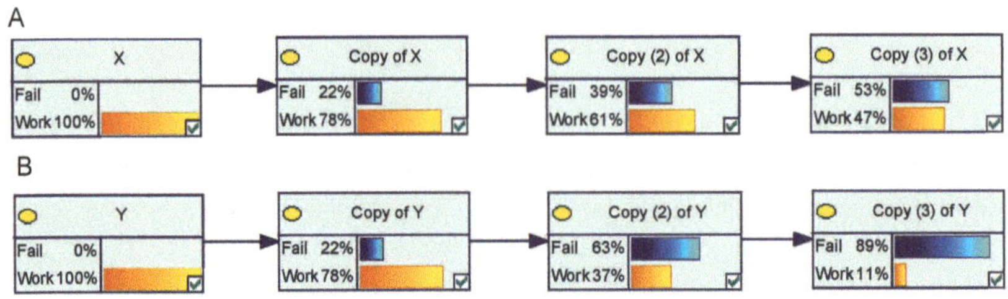

Fig. 9 DBNs for calculating the failure probabilities of (a) component X with an exponential distribution for its time to failure and (b) component Y with a Weibull distribution for its time to failure.

Table 6 CPT of component X in the second time slice, i.e., Copy of X, in Fig. 9a. The CPTs of X in the third and fourth time slices are also the same as the second time slice.

Copy of X ↓ X →	**Fail**	**Work**
Fail	1	0.22
Work	0	0.78

Table 7 CPT of component Y in the second time slice, i.e., Copy of Y, in Fig. 9b.

Copy of Y ↓ Y →	**Fail**	**Work**
Fail	1	0.22
Work	0	0.78

Table 8 CPT of component Y in the third time slice, i.e., Copy (2) of Y, in Fig. 9b.

Copy (2) of Y ↓ Copy of Y →	**Fail**	**Work**
Fail	1	0.53
Work	0	0.41

Table 9 CPT of component Y in the fourth time slice, i.e., Copy (3) of Y, in Fig. 9b.

Copy (3) of Y ↓ Copy (2) of Y →	**Fail**	**Work**
Fail	1	0.71
Work	0	0.29

3. Probability updating vs. probability adapting

3.1 Misapplication of probability updating to dynamic risk assessment

Probability adapting, also known as sequential learning, is different than probability updating which is a direct application of Bayes' rule. Through probability updating, evidence can be used to update our prior knowledge of the probabilities or relative frequencies of root events without changing the main parameters of the BN, i.e., the conditional probabilities used to populate the CPTs.

In other words, probability updating, which nowadays is mistakenly used as a means of dynamic risk analysis, has really nothing to do with dynamic analysis as it is incapable of using the information gained and recorded over time for revising (learning) the parameters of the BN.

To put it another way, given evidence as the occurrence of an event, probability updating cannot account for the number of occurrences of the event over time. Probability updating just tells us what would have been the change in the share of each root event if a consequent event were known to occur, no matter how many times. This type of gained knowledge, though valuable, cannot seem to be helpful in dynamic risk analysis where the cumulative observations over time and space should be considered in updating our priorly assigned probabilities and relative frequencies. To make it clearer, consider a simple example where a series system C comprises two binary components A and B.

The reliability of components A and B have been tested by the supplier, and their relative failure frequency in 100 tests have been identified as f(A fails) = 10/100 and f(B fails) = 30/100. If we have enough trust in the accuracy and comprehensiveness of the tests, we may take these relative frequencies as the maximum likelihood estimate of the failure probabilities of A and B as P(A fails) = 0.1 and P(B fails) = 0.3. The BN for predicting the failure frequency of C is presented in Fig. 10a in which nodes A and B are connected to node C via a deterministic OR gate (as an indication of series system).

As can be seen, the failure probability of C is calculated as 0.37 which can be interpreted as the relative frequency of 37 failures over 100 tests. Now, given the failure of C as evidence, the failure probabilities (frequencies) of A and B can be updated as P′(A fails) = P(A fails | C fails) = 0.27 and P′(B fails) = P(B fails | C fails) = 0.81, as shown in Fig. 10b.

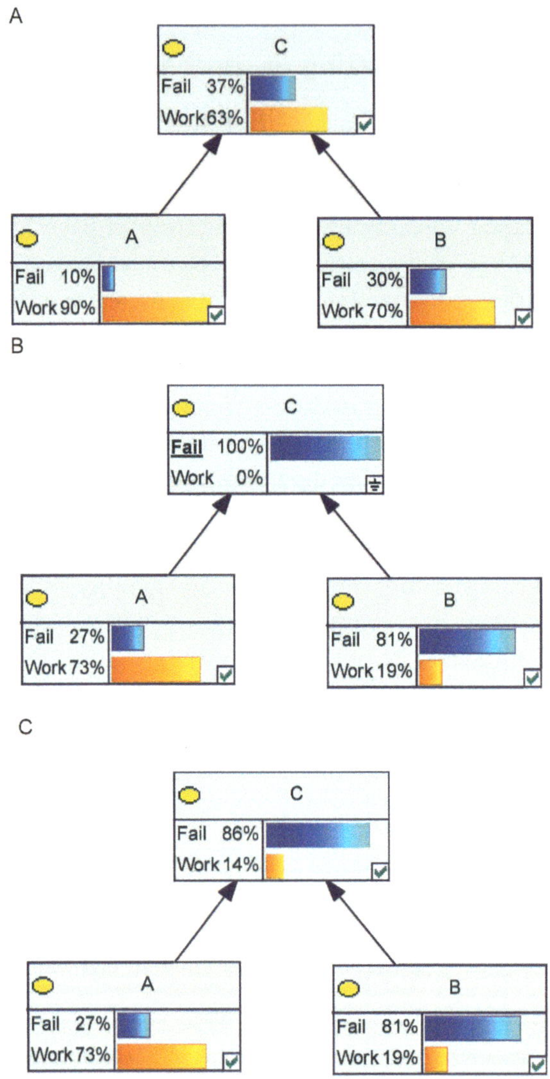

Fig. 10 (a) BN model of a series system C with two components A and B. (b) Updating the failure probabilities of A and B given the failure of C. (c) Using the updated probabilities calculated in (b) to update the failure probability of C.

These updated probabilities only tell we believe that the failure of C could have been 27% due to the failure of A, 81% due to the failure of B, and 22% (=27% × 81%) due to the failure of both A and B. However, if we mistakenly take $P'(\text{A fails}) = 0.27$ and $P'(\text{B fails}) = 0.81$

as the new (updated) probabilities of A and B for predicting the failure probability of C in the future (which is disturbingly the case in most applications of BN in safety and risk assessment), the updated failure probability of C in the future given that it has recently failed would be calculated as $P'(C\text{ fails}) = 0.86$, as shown in Fig. 10c.

However, this is contradicting with the notion of relative frequency which was first used to develop our prior believes about the failure probabilities of A, B, and C. In other words, given the previous 100 tests for reasoning about the failure frequencies of A and B (and C as well), observing C fail once would update the failure frequency of C as $f'(C\text{ fails}) = (37+1)/(100+1) = 0.376$, which is less than half the updated failure probability of C.

Repeating the same procedure for a case where C is observed to fail twice would wrongly result in updated probabilities of $P''(A\text{ fails}) = P(A\text{ fails}\,|\,C\text{ fails twice}) = 0.31$ and $P''(B\text{ fails}) = P(B\text{ fails}\,|\,C\text{ fails twice}) = 0.94$ and thus $P''(C\text{ fails}) = 0.96$ which is about more than twice the updated failure frequency of C as $f''(C\text{ fails}) = \frac{37+1+1}{100+1+1} = 0.382$. This simple example seems sufficient in illustrating the misapplication of probability updating to dynamic risk analysis and sequential updating of probabilities. Thus, the application of probability updating must be limited to fault diagnosis and measuring the importance of root events using the importance metrics discussed in Section 4.

3.2 Probability adapting

Probability adapting, which is also known as sequential learning, is a means for updating one's prior belief when more data (evidence) becomes available over time. We can use probability adapting only when the probabilities are relative frequencies. To make the discussion more concrete, consider a binomial experience such as tossing a coin with two outcomes, heads and tails. Depending on the shape and homogeneity of the coin, we may express our prior believes about the probability of the outcomes in $m = a + b$ trials as:

$$P(\text{heads}) = \frac{a}{m} \;\&\; P(\text{tails}) = \frac{b}{m} \tag{14}$$

After observing s heads and t tails in $n = s + t$ trials (evidence), our posterior probabilities would be:

$$P(\text{heads} \mid \text{st}) = \frac{a+s}{m+n} \;\&\; P(\text{tails} \mid \text{st}) = \frac{b+t}{m+n} \tag{15}$$

The posterior probabilities in Eq. (15) are called the *maximum a posterior probability* (MAP). It is worth noting that even if we feel like that the prior probability of heads and tails for a given coin would be equal, i.e., 50%, it matters if we choose a = b = 3 or a = b = 50 to express our prior believes, but why when both cases lead to P(heads) = P(tails) = 0.5?

First consider the case where we choose a = b = 3. This will lead to $P(\text{heads}) = \frac{3}{3+3} = 0.5$. After seeing 42 heads in 100 tosses (trials), our MAP for heads would be $P(\text{heads}|42, 58) = \frac{3+42}{6+100} = 0.42$. Now, consider the second case where a = b = 50. After seeing 42 heads in 100 tosses (trials), our MAP for heads would be $P(\text{heads} | 42, 58) = \frac{50+42}{100+100} = 0.46$. As can be seen, the sample size used to express our prior believes (6 and 100 in the first and second cases, respectively) determines how much we trust in our prior believes and would want the evidence to influence (update) our prior beliefs.

The guidelines for binary random variables (that easily can be extended to multinomial random variables) are as follows:

- a = b = 1: We would use these values when we have no knowledge at all regarding the prior probabilities (prior ignorance) and do not want to impose none of our beliefs on the learning algorithm. In other words, we would use these values when we want the objective data (evidence) to totally reflect on the posterior probabilities.
- a, b > 1: We would use these values when we want to impose our beliefs concerning the relative frequency on the learning algorithm. The larger the values of a and b, the more we believe in our prior probabilities (see the foregoing example).
- a, b < 1: We would use these values when we feel it is likely that the probability of one of the outcomes is high, but we are not sure which one! Like the previous guideline, we would use these values when we want to impose our beliefs concerning the relative frequency on the learning algorithm although such prior beliefs are quickly overwhelmed by data. We may take $a = b \approx 0$ if we are almost certain that the probability of one of the outcomes is very close to 1.

As an example, consider a two-node BN in Fig. 11 for parameter learning with each node having only two states as $X = (x_1, x_2)$ and $Y = (y_1, y_2)$.

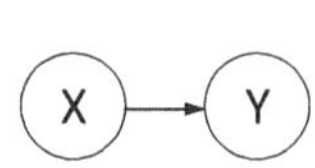

Fig. 11 BN with binary nodes.

In absence of any prior knowledge, we may attempt to represent our prior ignorance as to the value of all probabilities as:

$$P(x_1) = \frac{\|x_1\|}{\|x_1\| + \|x_2\|} = \frac{1}{1+1} = \frac{1}{2} \tag{16}$$

$$P(y_1|x_1) = \frac{\|y_1, x_1\|}{\|x_1\|} = \frac{1}{2} \tag{17}$$

$$P(y_1|x_2) = \frac{\|y_1, x_2\|}{\|x_2\|} = \frac{1}{2} \tag{18}$$

where $\|x_1\|$ represents the number of times x_1 has occurred; $\|y_1, x_1\|$ is the number of times both x_1 and y_1 have occurred together. In light of new data presented in Table 10, these prior probabilities can be updated as:

$$P(x_1|Data) = \frac{1+6}{1+1+10} = \frac{7}{12}$$

$$P(y_1|x_1, Data) = \frac{1+5}{2+6} = \frac{6}{8}$$

$$P(y_1|x_2, Data) = \frac{1+2}{2+4} = \frac{3}{6}$$

Obviously, $P(y_2|x_1, Data) = 1 - P(y_1|x_1, Data) = 1 - \frac{6}{8} = \frac{2}{8}$ and $P(y_2|x_2, Data) = 1 - P(y_1|x_2, Data) = 1 - \frac{3}{6} = \frac{3}{6}$.

Table 10 New data for the two-node BN.

Case	X	Y
1	x_1	y_1
2	x_1	y_1
3	x_1	y_1
4	x_1	y_1
5	x_1	y_1
6	x_1	y_2
7	x_2	y_1
8	x_2	y_1
9	x_2	y_2
10	x_2	y_2

3.3 Equivalent sample size in probability adapting

In the previous example, although it seems natural to set $\|x_1\|=1$, $\|y_1, x_1\|=1$ and $\|y_1, x_2\|=1$ to express our prior ignorance of all the conditional probabilities, such assignments are not consistent. For one thing, consider Eq. (16), in which the assessor's experience (prior knowledge) is equivalent to having seen x_1 occur one time and x_2 one time in two trials, i.e., $\|x_1\|=\|x_2\|=1$. However, in Eq. (17) the assessor's experience is equivalent to having seen y_1 occur one time out of the two times x_1 occurred, i.e., $\|y_1, x_1\|=1$ while $\|x_1\|=2$. This is not consistent though, because first we say x_1 occurred once and then we assume that it occurred twice! As illustrated in Neapolitan (2003), such inconsistency can result in odd posterior probabilities.

We could eliminate this inconsistency, for instance, by keeping $\|y_1, x_1\|=1$ but setting $\|x_1\|=\|x_2\|=2$, or by keeping $\|x_1\|=\|x_2\|=1$ but setting $\|y_1, x_1\| = \frac{1}{2}$. Either case would result in the same prior probabilities as in Eqs. (16)–(18) yet without the aforementioned inconsistency among the samples size. This modification to the samples size can be explained via the *prior equivalent sample size* (Neapolitan, 2003). For a BN, the prior equivalent sample size of each node can be identified as:

$$\textit{Prior equivalent sample size} = \frac{N}{p \cdot q} \tag{19}$$

where N is an arbitrary positive integer, p is the number of states of the node, and q is the number of instantiations (combinations of the states) of the parents of the node. Considering $N=2$, for the node X in Fig. 11 the prior equivalent sample size is $\frac{N}{p \cdot q} = \frac{2}{2 \times 1} = 1$. As can be seen, $p=2$ since X is a binary node $X=(x_1, x_2)$, and $q=1$ because X is a root node and does not have any parent(s) instantiations. As such, $\|x_1\|=\|x_2\|=1$. For the node Y, the prior equivalent sample size is $\frac{N}{p \cdot q} = \frac{2}{2 \times 2} = \frac{1}{2}$. In this case, $p=2$ since Y is a binary node $Y=(y_1, y_2)$, and $q=2$ because Y has only one parent, X, and the number of the instantiations of X is 2 (i.e., x_1 and x_2). As such, $\|y_1, x_1\| = \|y_1, x_2\| = \|y_2, x_1\| = \|y_2, x_2\| = \frac{1}{2}$, or in short $\|Y, X\| = \frac{1}{2}$. This way, the total equivalent sample size (the total number of experiments in which the values of X and Y are recorded) of the BN in Fig. 11 is 2.

Now consider the BN in Fig. 12 where X and Y are binary nodes, and Z is a ternary node. In this case, the prior equivalent sample sizes of X and Y are the same as before but that of node Z is $\frac{N}{p \cdot q} = \frac{2}{3 \times 4} = \frac{1}{6}$. It is

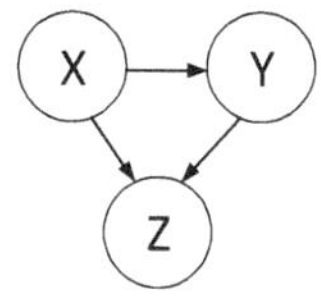

Fig. 12 BN with binary nodes X and Y and a ternary node Z.

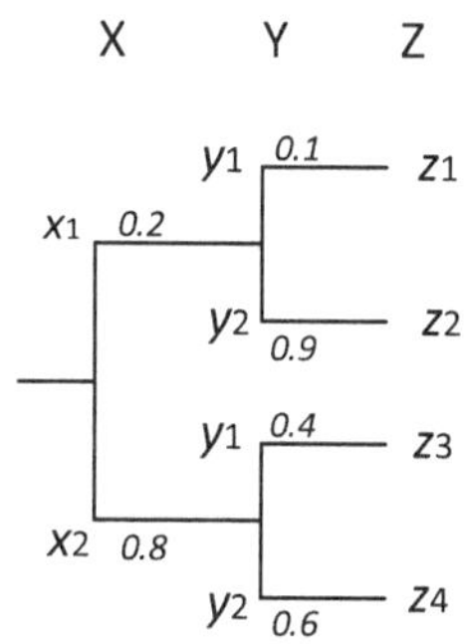

Fig. 13 Event tree diagram for modeling the outcomes z1–z4 of two sequential incidents X and Y.

because Z is a ternary node $p=3$ and the number of the instantiation of its parents q = (x_1, x_2) × (y_1, y_2) = 4. Therefore, $||Z, X, Y|| = \frac{1}{6}$.

3.4 Application of probability adapting to dynamic risk assessment

Probability adapting or sequential learning can effectively be applied to dynamic risk assessment, where the occurrence probability of incidents may change following the evidence becoming available over time. To illustrate one of such applications, consider the event tree diagram in Fig. 13. It is worth mentioning that the event tree can be modeled as the BN shown in Fig. 12 where X and Y are binary nodes and Z a quaternary node.

The assessor has expressed his belief about the occurrence probabilities of the events X and Y as relative frequencies shown in the figure for each branch: $P(x_1)=0.2$, $P(x_2)=0.8$, $P(y_1|x_1)=0.1$, $P(y_1|x_2)=0.4$, $P(y_2|x_1)=0.9$, and $P(y_2|x_2)=0.6$. In absence of any prior knowledge of such, the assessor could use the prior equivalent sample size approach to express their prior ignorance about the probabilities of X and Y. Having the probabilities of the states of X and Y, the probabilities of the states of

Z can simply be calculated using the event tree as $P(z_1)=0.2\times0.1=0.02$, $P(z_2)=0.2\times0.9=0.18$, $P(z_3)=0.8\times0.4=0.32$, and $P(z_4)=0.8\times0.6=0.48$.

Depending on how much confident the assessor is about the prior probabilities, he may use a, b > 1 or a, b < 1 to express his prior beliefs in the form of relative frequencies (see Section 3.2). In case the assessor is confident about his prior beliefs, he may use a relatively large sample size a + b = 100 (total number of experiments) to express the prior relative frequencies as $P(x_1)=\frac{20}{100}=0.2$, $P(x_2)=\frac{80}{100}=0.8$. As such, using the law of total probability, $P(y_1)=P(x_1)P(y_1|x_1)+P(x_2)P(y_1|x_2)=0.2\times0.1+0.8\times 0.4=0.34=\frac{34}{100}$, and similarly $P(y_2)=0.66=\frac{66}{100}$. Given the observations in Table 11, the maximum a posterior probability (MAP) of the variables can be calculated. For instance, regarding the observations made in the first year, the MAP of the states of Z can be calculated as:

$P(z_1|D1)=\frac{2+4}{100+6}=0.057$, $P(z_2|D1)=\frac{18+1}{100+6}=0.179$, $P(z_3|D1)=\frac{32+0}{100+6}=0.302$, and $P(z_4|D1)=\frac{48+1}{100+6}=0.46$. Further, since the occurrence of z_1 or z_3 implies the occurrence of y_1, the MAP of y_1 given the data in the first year can be calculated as $P(y_1|D1)=\frac{34+(4+0)}{100+6}=0.358$. Alternatively, the MAP of y_2 could be calculated using the number of times z_2 and z_4 have occurred as $P(y_2|D1)=\frac{66+(1+1)}{100+6}=0.642$. The same way, the MAP of x_1 (or x_2) can be calculated using the number of times z_1 and z_2 (or z_3 and z_4) have occurred. The dynamically varying MAPs of x_1, y_1, and z_3 have been presented in Fig. 14 for the sake of exemplification.

Table 11 Data observed over 3 years for the states of Z in the event tree.

D1: data in year 1	D2: data in year 2	D3: data in year 3
z_1	z_1	z_2
z_1	z_2	z_2
z_1	z_2	z_1
z_2	z_1	z_1
z_4	z_1	z_3
z_1	–	z_1
–	–	z_3
–	–	z_2

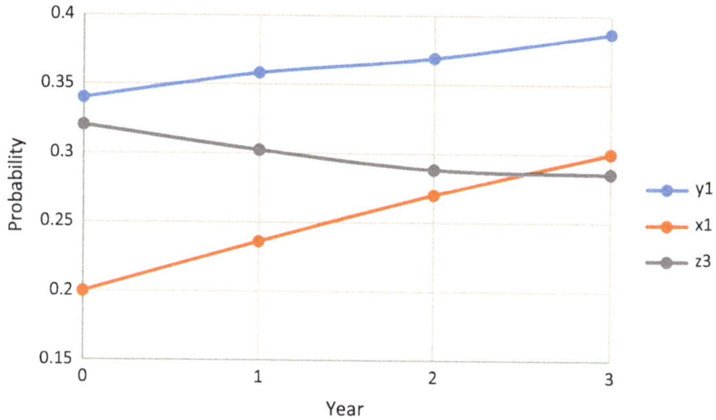

Fig. 14 Dynamic variation of the maximum a posterior probability of x1, y1, and z3.

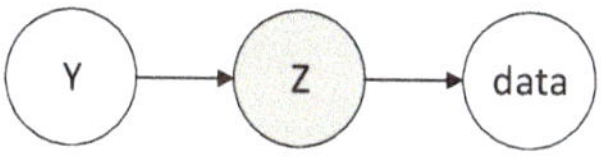

Fig. 15 Incapability of probability updating in adapting the probability of Y given the new observations in node "data." Given a series connection, instantiation of Z blocks the transfer of information back to Y.

It is worth mentioning that the application of Baye's rule to the observed data would be of no help in updating (adapting) the prior beliefs of X and Y. To make the discussion more concrete, consider the application of Bayes' rule to updating the probability of Y as $P(Y|Z) = \frac{P(Y).P(Z|Y)}{P(Z)}$.

Given observations of Z as the data, the above conditional probability would be modified as $P(Y|\,Z, data)$. Considering the serial connection among the Y, Z, and data nodes in a corresponding BN as shown in Fig. 15, $P(Y|\,Z,\ data) = P(Y|\,Z)$ since the instantiation of node Z would make Y conditionally independent of data. In other words, a conventional application of Bayes' rule would not be enough to adapt the probability of Y (and X).

4. Importance measures

One of the most important applications of Bayes' rule is in identification of critical components of a system. To demonstrate the metrics used for this purpose and to make a comparison among their results, consider a fault tree in Fig. 16. The fault tree has been purposely selected simple so

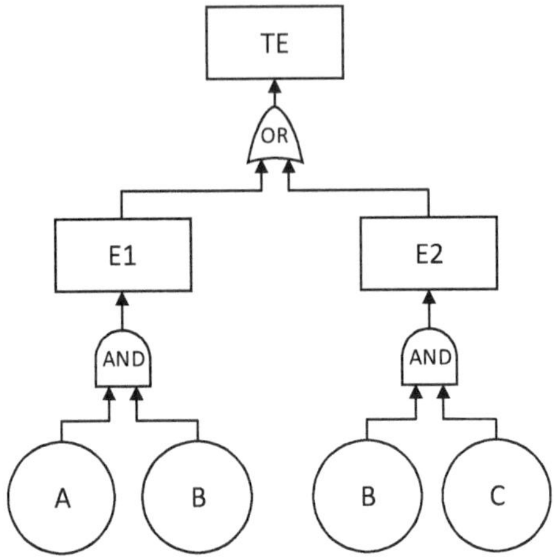

Fig. 16 A typical fault tree for risk importance measure analysis. B is the most critical component while A and C are equally the second most critical components.

that the critical components could be identified intuitively even without resorting to the risk importance metrics which will be discussed in the following sections.

With a glance on the fault tree, it is known that the tree has two minimal cut sets E1 and E2 since not only E1 and E2 are connected to the top event (TE) by means of an OR gate but they are connected to the root events by means of AND gates. Qualitatively, root events which show up in more minimal cut sets may be more critical than the other root events. As such, B, which is contributing to the both minimal cut sets may be the most critical component while A and C are equally as the second most critical components. In other words, the failure or occurrence of B (depending on whether it is a component or an event) is more important than A and C to the failure or occurrence of the TE. This could also be shown using Boolean algebra in Eq. (20) where B and (A ∪ C) are shown to contribute equally to the TE, implying that the share of B is more than that of A, C, or they both altogether.

$$TE = E1 \cup E2 = (A \cap B) \cup (B \cap C) = B \cap (A \cup C) \tag{20}$$

However, in the next sections we will demonstrate that such qualitative risk importance assessment would be reliable only if the failure probabilities of the root events are equal.

4.1 Birnbaum importance measure

Birnbaum (1969) proposed a metric for measuring the importance of system components with regard to the reliability (probability of non-failure) of the system. The Birnbaum Importance Measure of component i, BIM (i), is defined as the partial derivative of the system reliability R_{sys} with respect to the component reliability R_i:

$$BIM(i) = \frac{\partial R_{sys}}{\partial R_i} \tag{21}$$

BIM is a well-known measure in sensitivity analysis. If BIM(i) is large, a small change in the reliability of component i results in a comparatively large change in the reliability of the system. In case of fault tree analysis, the BIM(i) can equally be defined as:

$$BIM(i) = \frac{\partial P(TE)}{\partial P(i)} \tag{22}$$

where P(TE) is the probability of the top event, and P(i) is the failure probability of component i. Given the fault tree in Fig. 16, the probability of top event can be calculated as:

$$\begin{aligned} P(TE) &= P(B \cap (A \cup C)) = P(B).P(A \cup C) \\ &= P(B).\{P(A) + P(C) - P(A)P(C)\} \end{aligned}$$

As such, the BIM of the components can be calculated as:

$$BIM(A) = \frac{\partial P(TE)}{\partial P(A)} = P(B) - P(B)P(C)$$

$$BIM(B) = \frac{\partial P(TE)}{\partial P(B)} = P(A) + P(C) - P(A)P(C)$$

$$BIM(C) = \frac{\partial P(TE)}{\partial P(C)} = P(B) - P(B)P(A)$$

Given different failure probabilities for A, B, and C, the BIM of the components have been calculated and presented in Table 12. As can be noted, only in the case of components with equal failure probabilities do the results of BIM comply with the outcomes of qualitative importance measurement techniques (minimal cut sets or Boolean algebra). By changing the failure probabilities of the components, their criticality (importance) with regard to the system failure changes accordingly.

Table 12 Comparison of components' Birnbaum importance measures given different prior failure probabilities.

Case	P(A)	P(B)	P(C)	BIM(A)	BIM(B)	BIM(C)	Descending order of critical components (components in the brackets are equally critical)
1	0.20	0.20	0.20	0.16	0.36	0.16	B, (A, C)
2	0.20	0.45	0.20	0.36	0.36	0.36	All the same
3	0.20	0.50	0.20	0.40	0.36	0.40	(A, C), B
4	0.20	0.50	0.40	0.30	0.52	0.40	B, C, A

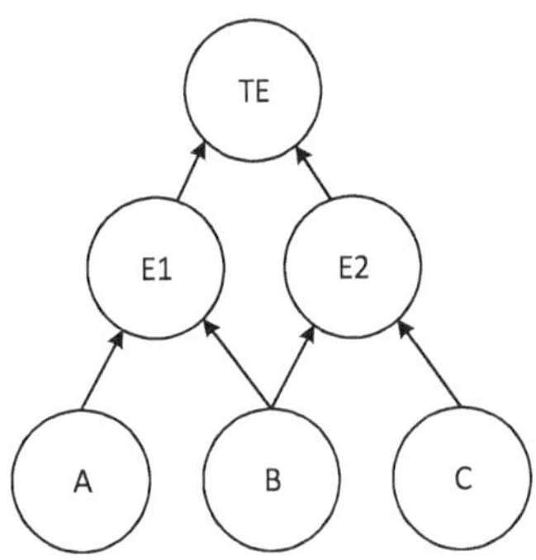

Fig. 17 Equivalent Bayesian network of the fault tree shown in Fig. 16.

BIM of component i can alternatively be defined as:

$$BIM(i) = R_{sys|i=1} - R_{sys|i=0} \tag{23}$$

where $R_{sys|i=1}$ is the system reliability when component i is in operational state ($R_i=1$), and $R_{sys|i=0}$ is the system reliability when component i is in failure state ($R_i=0$). Likewise, in case of fault trees:

$$BIM(i) = P(TE|i=0) - P(TE|i=1) \tag{24}$$

where $P(TE|i=0)$ is the system (top event) failure probability when component i is in the failure state, and $P(TE|i=1)$ is the system failure probability when component i is in the operational state. Given BIM as in Eq. (24), the equivalent BN of the fault tree, shown in Fig. 17, can be used for a quick identification of critical components.

To calculate the BIM of component i, the updated probability of the top event once for an instantiation of the component to its failure state and the other time for an instantiation of the component to its working state should be calculated. The difference between these two updated probabilities equals the BIM of the component. As an example, the BIM(A) for a case where P(A) = P(B) = P(C) = 0.2 has been calculated using the BNs shown in Fig. 18. In the BN of Fig. 18a the node A has been instantiated to its failure state so that P(A fails) = 1 while Fig. 18b shows the BN where A has been instantiated to its working state,

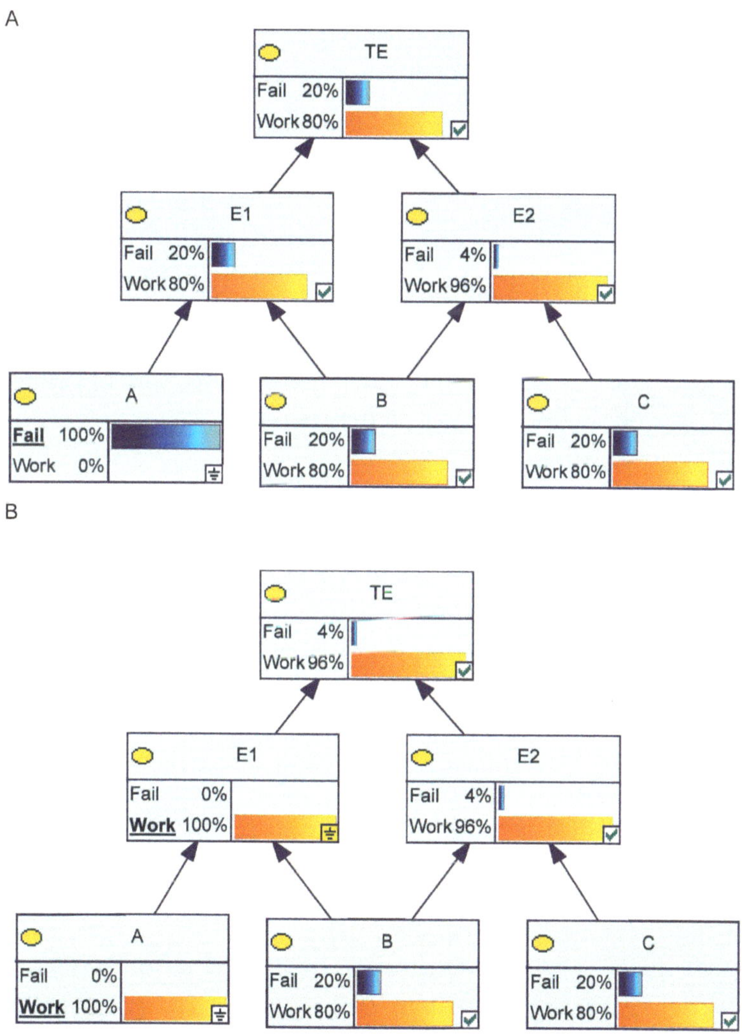

Fig. 18 BN used to calculate BIM(A). (a) Node A has been instantiated to its failure state: P(A fails) = 100%. (b) Node A has been instantiated to its working state: P(A works) = 100%.

P(A works) = 1. As can be seen from the updated probabilities of the TE, BIM(A) = P(TE | A fails) − P(TE | A works) = 0.2–0.04 = 0.16, the same as shown in Table 12.

The change in the BIM of components following the change in their probabilities (Table 12) points out the importance of updating the components' BIM in dynamic systems. In other words, in a dynamic system the failure probability of the components may change with time so does the BIM of the components.

4.2 Fussell-Veseley

Fussell-Veseley (FV) is a risk importance metric that measures the overall contribution of minimal cut sets containing a basic event of interest to the total risk (total minimal cut sets):

$$FV(i) = \frac{\sum P(MCS_i)}{\sum P(MCS)} \tag{25}$$

where FV(i) is the Fussell-Veseley risk importance measure of component i, MCS_i is one minimal cut set of the system containing component i, and MCS is any minimal cut set of the system. Considering the fault tree in Fig. 16, A∩B and B∩C are the only MCSs of the system. As such, the FV can be calculated as:

$$FV(A) = \frac{P(A \cap B)}{P(A \cap B) + P(B \cap C)} < 1$$

$$FV(B) = \frac{P(A \cap B) + P(B \cap C)}{P(A \cap B) + P(B \cap C)} = 1$$

$$FV(C) = \frac{P(B \cap C)}{P(A \cap B) + P(B \cap C)} < 1$$

As can be seen from the equations above, no matter what the failure probabilities, B is always the most critical component of the system. The FVs of the components given different failure probabilities of which are calculated and presented in Table 13.

As can be seen, FV metric is not so sensitive as is the BIM metric to changes in the failure probabilities of the components and yields similar results to the qualitative method of minimal cut sets. However, FV is more precise than the qualitative approach as it differentiates between the criticality of A and C when the failure probability of C becomes greater than that of

Table 13 Comparison of components' Fussell-Veseley importance measures given different prior failure probabilities.

Case	P(A)	P(B)	P(C)	FV(A)	FV(B)	FV(C)	Descending order of critical components (components in the brackets are equally critical)
1	0.20	0.20	0.20	0.50	1.00	0.50	B, (A, C)
2	0.20	0.45	0.20	0.50	1.00	0.50	B, (A, C)
3	0.20	0.50	0.20	0.50	1.00	0.50	B, (A, C)
4	0.20	0.50	0.40	0.33	1.00	0.66	B, C, A

A. In comparison with the BIM, the FV seems to be of less value in dynamic risk assessment as temporal changes in the failure probabilities do not reflect on the criticality of the respective components.

4.3 Ratio of variation

Ratio of variation (RV) of component i has been recently proposed by some researchers as a measure of component importance in Bayesian assessment of system safety (Zarei, Azadeh, Khakzad, Aliabadi, & Mohammadfam, 2017). The RV of component i is defined as the normalized difference of the posterior and prior probabilities of the component as:

$$RV(i) = \frac{P(i|TE) - P(i)}{P(i)} \tag{26}$$

where P(i) is the prior priority of component i, and P(i | TE) is the posterior probability of the component when the TE is instantiated to its failure (or occurrence) state. Fig. 19 illustrates the procedure for calculating the RV(A) where the prior failure probabilities of A, B and C are all equal to P(A) = P(B) = P(C) = 0.2.

Given the prior and posterior probabilities of A in Fig. 19a and b, the RV(A) can be calculated as $RV(A) = \frac{0.56-0.2}{0.2} = 1.80$. The RVs of the components given different failure probabilities of which are calculated and presented in Table 14. As can be seen, the results of RV analysis are very similar to those of BIM analysis except for case 2 where RV shows B as a slightly less critical component than A and C. (BIM for case 2 shows the same criticality for all the components). To be in safe side, we suggest using BIM instead of RV as the reasoning behind the development of the former sounds more robust, intuitive, and extensively tested.

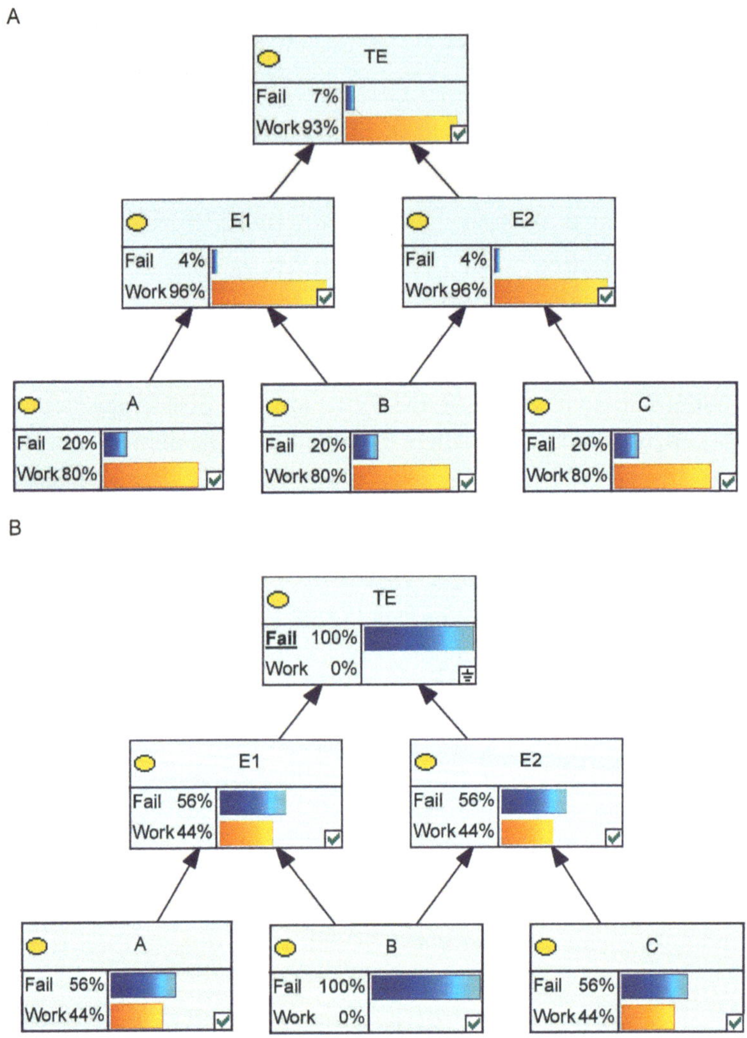

Fig. 19 Bn used to calculate the RV of component A. (a) BN before instantiating the TE to its failure state. Prior probability of A is $P(A)=0.2$. (b) BN after instantiating the TE to its failure state: $P(TE=fail)=1$. Posterior probability of A given the failure of TE is $P(A|TE)=0.56$.

4.4 Mutual information

Considering a random variable Z with probability mass function of P(Z) for the states of Z, the amount of uncertainty associated with Z can be measured using the concept of entropy H(Z) as:

Table 14 Comparison of components' ratio of variation given different prior failure probabilities.

Case	P(A)	P(B)	P(C)	RV(A)	RV(B)	RV(C)	Descending order of critical components (components in the brackets are equally critical)
1	0.20	0.20	0.20	1.80	4.00	1.80	B, (A, C)
2	0.20	0.45	0.20	1.80	1.22	1.80	(A, C), B
3	0.20	0.50	0.20	1.80	1.00	1.80	(A, C), B
4	0.20	0.50	0.40	0.90	1.00	0.93	B, C, A

$$H(Z) = -\sum_{z \in Z} P(z) \log P(z) \tag{27}$$

The conditional entropy of Z given another random variable as Y can be defined as:

$$H(Z|Y) = -\sum_{z \in Z, y \in Y} P(z, y)\ log \frac{P(z, y)}{P(y)} \tag{28}$$

Thus, the mutual information of Z and Y, I(Z,Y), can be defined as the reduction in the uncertainty of Z given the observation of Y:

$$I(Z, Y) = H(Z) - H(Z|Y) = \sum_{z \in Z, y \in Y} P(z, y)\ log \frac{P(z, y)}{P(z)P(y)} \tag{29}$$

Using the chain rule, the mutual information in Eq. (29) can be simplified as:

$$I(Z, Y) = \sum_{z \in Z, y \in Y} P(y)P(z|y)\ log \frac{P(z|y)}{P(z)} \tag{30}$$

In the above equations, natural logarithm can be used instead to avoid obtaining too small values for the mutual information. In the case of the fault tree analysis, the most critical component can be considered as a component with the highest mutual information shared with the TE. For instance, I(A, TE) can be calculated as:

$$I(A, TE) = \sum P(A)P(TE|A)\ ln\frac{P(TE|A)}{P(TE)} = P(A = fail)P(TE = fail\ |A = fail)\ ln\frac{P(TE = fail\ |A = fail)}{P(TE = fail)} + P(A = fail)P(TE = work\ |A = fail)\ ln\frac{P(TE = work\ |A = fail)}{P(TE = work)} + P(A = work)P(TE = fail\ |A = work)\ ln\frac{P(TE = fail\ |A = work)}{P(TE = fail)} + P(A = work)P(TE = work\ |A = work)\ ln\frac{P(TE = work\ |A = work)}{P(TE = work)}.$$

Given the BN, the above conditional probabilities can readily be calculated and plugged in Eq. (30) to calculate the I(A, TE). Not to mention that among the above conditional probabilities, only P(TE=fail|A=fail) and P(TE=fail|A=work) need to be calculated using the BN as the other two conditional probabilities can be calculated as the complements of these two conditional probabilities. Calculating P(TE=fail|A=fail) =0.20 and P(TE=fail|A=work)=0.04 using the BN for case 1 (i.e., P(A)=P(B)=P(C)=0.2), I(A, TE) can be calculated as $I(A, TE) = 0.2 \times 0.2 \times ln\frac{0.2}{0.07} + 0.2\times\ 0.8 \times ln\frac{0.8}{0.93} + 0.8 \times 0.04 \times ln\frac{0.04}{0.07} + 0.8 \times 0.96 \times ln\frac{0.96}{0.93} = 0.024$.

The mutual information of the components and the TE given different failure probabilities of the components are calculated and presented in Table 15. As can be noticed from a comparison between Tables 15 and 13, mutual information analysis results in the same order of critical

Table 15 Comparison of components' mutual information (with TE) given different prior failure probabilities of the components.

Case	P(A)	P(B)	P(C)	I(A, TE)	I(B, TE)	I(C, TE)	Descending order of critical components (components in the brackets are equally critical)
1	0.20	0.20	0.20	0.024	0.250	0.024	B, (A, C)
2	0.20	0.45	0.20	0.099	0.382	0.099	B, (A, C)
3	0.20	0.50	0.20	0.090	0.392	0.090	B, (A, C)
4	0.20	0.50	0.40	0.015	0.423	0.214	B, C, A

components as Fussell-Veseley analysis, and not so sensitive to the probability changes as BIM and RV analyses.

In addition to its application to criticality analysis, mutual information analysis can be used to infer the likelihood of low-probability events from the occurrence of relevant more frequent events (Khakzad, Khan, & Amyotte, 2015). To make the discussion more concrete, consider the event tree in Fig. 20.

In an event tree, considering an upper branch for the occurrence (or failure) of a respective event (or component), the end events z_1–z_3 are usually ordered from top to bottom based on the occurrence likelihood descendingly; that is, $P(z_1) > P(z_2) > P(z_3)$.

As can be seen from Fig. 20, the end events z_1, z_2, and z_3 have common contributing factors in the form of the same initiating event (IE) and the intermediate events X and Y. (X and Y could be safety measures whose failure or operation can result in different outcomes of the IE). As such, the occurrence of z_1 or z_2 would imply the occurrence of the contributing events IE, X and Y, which in turn could be used to update the occurrence probability of z_3.

Between z_1 and z_2, the one with the highest mutual information with z_3 would be expected to be more informative in the sense that its occurrence could decrease the uncertainty about the prospective occurrence of z_3. The mutual information of z_1 and z_3, for instance, can be calculated as:

$$I(z_1, z_3) = \sum P(z_1)\, P(z_3|z_1)\, log\, \frac{P(z_3|z_1)}{P(z_3)} \tag{31}$$

However, it should be noted that the mutual information in the current form would result in a value of zero since the end events of an event tree are

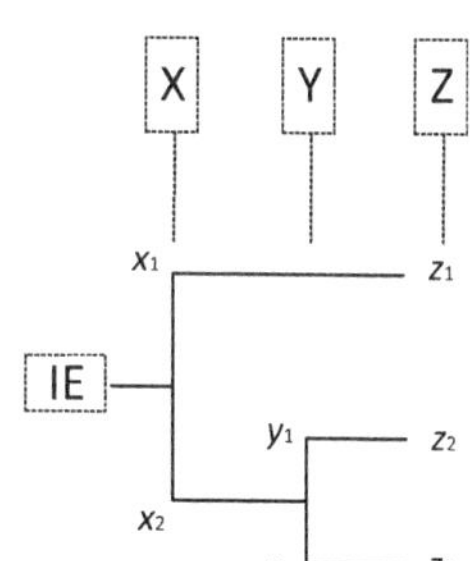

Fig. 20 Modeling the end events (z1–z3) of an initiating event (IE) and a sequence of other events (X and Y) as an event tree diagram. z1 and z3 are, respectively, the most and least likely end events.

mutually exclusive and their conditional probability distribution in the foregoing equation equals zero, that is, P $(z_3|z_1) = P(z_3|z_2) = P(z_2|z_1) = 0$.

To resolve this problem, assume that the occurrence probability of the initiating event IE can be defined as $P(IE) = \frac{a_1}{a_1 + a_2}$, showing that among $n = a_1 + a_2$ trials IE has occurred a_1 times. Similarly, assume that the probabilities of X and Y can also be defined as $P(x_1) = \frac{\|x_1\|}{\|x_1\| + \|x_2\|}$ and $P(y_1) = \frac{\|y_1\|}{\|y_1\| + \|y_2\|}$, respectively, where $\|x_1\|$, for instance, is the number of times the outcome x_1 has occurred in total number of occurrence of X, i.e., $\|X\| = \|x_1\| + \|x_2\|$. This way, the prior probability of z_3 at time t would be calculated as:

$$\begin{aligned} P(Z^t = z_3) &= P(IE^t = a_1).P(X^t = x_2).P(Y^t = y_2) \\ &= \frac{a_1}{a_1 + a_2} \cdot \frac{\|x_2\|}{\|x_1\| + \|x_2\|} \cdot \frac{\|y_2\|}{\|y_1\| + \|y_2\|} \end{aligned} \tag{32}$$

Now, assume that $Z = z_2$ is observed once at time t. The occurrence of z_2 would imply the occurrence of the $IE = a_1$ as well as $X = x_2$ and $Y = y_1$. As such, the prior probabilities of IE, X, and Y can be updated as:

$$P\left(IE^{t+1} = a_1 | Z^t = z_2\right) = \frac{a_1 + 1}{a_1 + a_2 + 1} \tag{33}$$

$$P\left(X^{t+1} = x_2 | Z^t = z_2\right) = \frac{\|x_1\| + 1}{\|x_1\| + \|x_2\| + 1} \tag{34}$$

$$P\left(Y^{t+1} = y_2 | Z^t = z_2\right) = \frac{\|y_2\|}{\|y_1\| + \|y_2\| + 1} \tag{35}$$

Having these updated probabilities, the updated probability of $P(Z^{t+1} = z_3 | Z^t = z_2)$ can be calculated as (see Eq. (32) for a comparison):

$$\begin{aligned} P\left(Z^{t+1} = z_3 | Z^t = z_2\right) &= P\left(IE^{t+1} = a_1 | Z^t = z_2\right) \\ &\quad .P\left(X^{t+1} = x_2 | Z^t = z_2\right).P\left(Y^{t+1} = y_2 | Z^t = z_2\right) \\ &= \frac{a_1 + 1}{a_1 + a_2 + 1} \cdot \frac{\|x_1\| + 1}{\|x_1\| + \|x_2\| + 1} \cdot \frac{\|y_2\|}{\|y_1\| + \|y_2\| + 1} \end{aligned} \tag{36}$$

As such, the mutual information of $Z^{t+1} = z_3$ and $Z^t = z_2$ can be calculated as:

$$I\left(z_3^{t+1}, z_2^t\right) = \sum P\left(z_2^t\right).P\left(z_3^{t+1}|z_2^t\right).\,log\,\frac{P\left(z_3^{t+1}|z_2^t\right)}{P\left(z_3^{t+1}\right)} \quad (37)$$

An innovative application of DBN and mutual information to risk-based decision making can be found in Khakzad (2018).

5. Summary

In this chapter, we highlighted and discussed some common misapplications of BN to dynamic risk assessment. We demonstrated that conventional BN is not an appropriate technique for dynamic risk assessment, and more importantly, its probability updating feature is quite often misused for this purpose. In other words, when it comes to dynamic risk assessment, conventional BN (alternatively called static BN or just BN) does not really have any advantage over other conventional techniques such as fault tree analysis except that it could more efficiently account for dependencies. (This is not really the main objective of dynamic risk assessment unless the dependencies are temporal dependencies, which is not the case in the conventional BN!)

The probability updating feature of BN was also shown to be usually misused in the context of dynamic risk assessment. It is mainly because modelers often mistake the updated probabilities for dynamically revised probabilities reflecting the future of the system of interest in the light of temporal changes. However, we showed via simple examples that this is not correct, and the application of probability updating must be limited to fault diagnosis and risk importance analysis.

On the other hand, the probability adapting feature of BN, also known as sequential learning or updating, was shown to be much more relevant to dynamic risk assessment where cumulative observations or knowledge obtained over time can be utilized to dynamically revise the probabilities.

DBN was also demonstrated to be an inherently more appropriate technique than BN for dynamic risk assessment. However, due care should be given to calculating and developing the conditional probability tables of the nodes in sequential time intervals since for a non-exponential variable (e.g., a component with a lognormal time to failure) such conditional tables would be different from one time slice to the other time slice.

Among risk importance measures used in dynamic risk assessment, we demonstrated that the Birnbaum importance measure shows the most sensitivity to the change of probabilities and is thus more dependable than other importance measures such as Fussel-Veseley, mutual information, and the ratio of variation when it comes to dynamic risk analysis. The results of the ratio of variation were also shown to be the closest to those of Birnbaum importance measure.

References

Abubakirov, R., Yang, M., & Khakzad, N. (2020). A risk-based approach to determination of optimal inspection intervals for buried oil pipelines. *Process Safety and Environmental Protection, 134*, 95–107.

Arzaghi, E., Abbassi, R., Garaniya, V., Binns, J., Chin, C., Khakzad, N., et al. (2018). Developing a dynamic model for pitting and corrosion-fatigue damage of subsea pipelines. *Ocean Engineering, 150*, 391–396.

Birnbaum, Z. (1969). On the importance of different components in a multicomponent system. In P. R. Krishnaiah (Ed.), *Multivariate analysis II* (pp. 581–592). New York: Academic Press.

Boudali, H., & Dugan, J. B. (2005). A discrete-time Bayesian network reliability modeling and analysis framework. *Reliability Engineering and System Safety, 87*, 337–349.

Cheng, J., & Druzdzel, M. J. (1999). Computational investigation of low-discrepancy sequences in simulation algorithms for Bayesian networks. In *Proceedings of the 16th conference on uncertainty in artificial intelligence* (pp. 72–81), Morgan Kaufmann.

Dechter, R. (1996). Bucket elimination: A unifying framework for probabilistic inference. In *Proceedings of the conference on uncertainty in artificial intelligence* (pp. 211–219), San Francisco: Morgan Kaufmann.

Ebeling, C. (1996). *An introduction to reliability and maintainability engineering*. New Delhi: TATA McGraw-Hill Edition.

Fenton, N., & Neil, M. (2019). *Risk assessment and decision analysis with Bayesian networks* (2nd ed.). Boca Raton, FL, USA: CRC Press, Taylor & Francis Group.

GeNIe, 2019 2.4.4601 Academic Installer. Decision Systems Laboratory, University of Pittsburg. Available online from https://download.bayesfusion.com/files.html?category=Academia, accessed 15.10.2019.

Hanea, A. M., & Kurowicka, D. (2008). Mixed non-parametric continuous and discrete Bayesian belief nets. In T. Bedford, J. Quigley, L. Walls, B. Alkali, A. Daneshkhah, & G. Hardman (Eds.), *Advances in mathematical modeling for reliability*. Amsterdam: IOS Press. ISBN: 978-1-58603-865-6.

Jensen, F. (1996). *An introduction to Bayesian networks*. UCL Press.

Kalantarnia, M., Khan, F., & Hawboldt, K. (2009). Dynamic risk assessment using failure assessment and Bayesian theory. *Journal of Loss Prevention in the Process Industries, 22*, 600–606.

Khakzad, N. (2015). Application of dynamic Bayesian network to risk analysis of domino effects in chemical infrastructures. *Reliability Engineering and System Safety, 138*, 263–272.

Khakzad, N. (2018). Which fire to extinguish first? A risk-informed approach to emergency response in oil terminals. *Risk Analysis, 38*(7), 1444–1454.

Khakzad, N. (2019a). System safety assessment under epistemic uncertainty: Using imprecise probabilities in Bayesian network. *Safety Science, 116*, 149–160.

Khakzad, N. (2019b). Modeling wildfire spread in wildland-industrial interfaces using dynamic Bayesian network. *Reliability Engineering & System Safety, 189*, 165–176.

Khakzad, N., Dadashzadeh, M., & Reniers, G. (2018). Quantitative assessment of wildfire risk in oil facilities. *Journal of Environmental Management, 223*, 433–443.

Khakzad, N., Khakzad, S., & Khan, F. (2014). Probabilistic risk assessment of major accidents: Application to offshore blowouts in the Gulf of Mexico. *Natural Hazards, 74*, 1759–1771.

Khakzad, N., Khan, F., & Amyotte, P. (2011). Safety analysis in process facilities: Comparison of fault tree and Bayesian network approaches. *Reliability Engineering and System Safety*, *96*, 925–932.

Khakzad, N., Khan, F., & Amyotte, P. (2012). Dynamic risk analysis using bow-tie approach. *Reliability Engineering and System Safety*, *104*, 36–44.

Khakzad, N., Khan, F., & Amyotte, P. (2013a). Dynamic safety analysis of process systems by mapping bow-tie into Bayesian network. *Process Safety and Environmental Protection*, *91*, 46–53.

Khakzad, N., Khan, F., & Amyotte, P. (2013b). Quantitative risk analysis of offshore drilling operations: A Bayesian approach. *Safety Science*, *57*, 108–117.

Khakzad, N., Khan, F., & Amyotte, P. (2013c). Risk-based design of process systems using discrete-time Bayesian networks. *Reliability Engineering and System Safety*, *109*, 5–17.

Khakzad, N., Khan, F., & Amyotte, P. (2015). Major accidents (gray swans) likelihood modeling using accident precursors and approximate reasoning. *Risk Analysis*, *35*(7), 1336–1347.

Khakzad, N., Khan, F., Amyotte, P., & Cozzani, V. (2014). Risk management of domino effects considering dynamic consequence analysis. *Risk Analysis*, *34*(6), 1128–1138.

Khakzad, N., Khan, F., & Paltrinieri, N. (2014). On the application of near accident data to risk analysis of major accidents. *Reliability Engineering and System Safety*, *126*, 116–125.

Khakzad, N., Landucci, G., & Reniers, G. (2017). Application of dynamic Bayesian network to performance assessment of fire protection systems during domino effects. *Reliability Engineering & System Safety*, *167*, 232–247.

Khakzad, N., & Reniers, G. (2019). Low-capacity utilization of process plants: A cost-robust approach to tackle man-made domino effects. *Reliability Engineering & System Safety*, *191*, 106114.

Khakzad, N., Reniers, G., Abbassi, R., & Khan, F. (2016). Vulnerability analysis of process plants subject to domino effects. *Reliability Engineering & System Safety*, *154*, 127–136.

Khakzad, N., Yu, H., Paltrinieri, N., & Khan, F. (2016). *Techniques of reactive frequency update: Bayesian inference. In F. Khan & N. Paltrinieri (Eds.), Dynamic risk analysis in the chemical and process industry* (pp. 51–61). UK: Elsevier. https://doi.org/10.1016/B978-0-12-803765-2.00005-6.

Lauritzen, S. L., & Spiegelhalter, D. J. (1988). Local computations with probabilities on graphical structures and their application to expert systems. *Journal of the Royal Statistical Society: Series B: Methodological*, *50*, 157–224.

Meel, A., & Seider, W. D. (2006). Plant-specific dynamic failure assessment using Bayesian theory. *Chemical Engineering Science*, *61*(21), 7036–7056.

Montani, S., Portinale, L., Bobbio, A., & Codetta-Raiteri, D. (2008). RADYBAN: A tool for reliability analysis of dynamic fault trees through conversion into dynamic Bayesian networks. *Reliability Engineering and System Safety*, *93*, 922–932.

Neapolitan, R. (2003). *Learning Bayesian networks* (1st ed.). Upper Saddle River, NJ: Prentice Hall, Inc.

Pearl, J. (1982). Reverend Bayes on inference engines: A distributed hierarchical approach. In *Proceedings of the second national conference on artificial intelligence* (pp. 133–136). Pittsburgh, PA. Menlo Park, California: AAAI-82. AAAI Press.

Pearl, J. (1988). *Probabilistic reasoning in intelligent systems: Networks of plausible inference*. ISBN: 978-0-08-051489-5. https://doi.org/10.1016/C2009-0-27609-4.

Zarei, E., Azadeh, A., Khakzad, N., Aliabadi, M., & Mohammadfam, I. (2017). Dynamic safety assessment of natural gas stations using Bayesian network. *Journal of Hazardous Materials*, *321*, 830–840.

Zarei, E., Khakzad, N., Cozzani, V., & Reniers, G. (2019). Safety analysis of process systems using Fuzzy Bayesian Network (FBN). *Journal of Loss Prevention in the Process Industries*, *57*, 7–16.

Zinetullina, A., Yang, M., Khakzad, N., & Golamn, B. (2019). Dynamic resilience assessment for process units operating in arctic environments. *Safety in Extreme Environments*. https://doi.org/10.1007/s42797-019-00008-3.

CHAPTER FIVE

Data driven methods

Yiyang Dai[a], Feifan Cheng[b], Hao Wu[b], Deyang Wu[b], Jinsong Zhao[b,*]
[a]School of Chemical Engineering, Sichuan University, Chengdu, Sichuan, PR China
[b]Department of Chemical Engineering, Tsinghua University, Beijing, PR China
[*]Corresponding author: e-mail address: jinsongzhao@mail.tsinghua.edu.cn

Contents

1. Importance of the data in risk assessment

Risk assessment techniques play a fundamental role in the full life cycle of a chemical process, including research, process design, facility siting, construction, operation, maintenance, emergency management, transportation as well as decommissioning.

Traditional risk assessment methods use a structured approach to identify possible hazards/threats, analyze their causes and consequences, and determine the risk levels in a qualitative or quantitative manner. Compared with qualitative risk assessment, quantitative risk assessment (QRA) can provide more effective decision support. As defined by ISO/IEC standard (ISO, 2009), QRA includes: establishment of the context, hazard identification, performance of the risk analysis, risk evaluation. The first step of QRA for chemical process is hazard identification, aiming to find the possible

Methods in Chemical Process Safety, Volume 4
ISSN 2468-6514
https://doi.org/10.1016/bs.mcps.2020.02.002

abnormal situations which may cause undesired consequences and describe the potential hazardous scenarios. In this step, several approaches may be employed, such as preliminary hazard analysis (PHA), failure modes and effects analysis (FMEA), fault tree analysis (FTA), Hazard and Operability study (HAZOP), etc. Frequency estimation and consequence estimation are the two most critical steps in QRA. For frequency estimation, a knowledge-based model such as fault trees, event trees (Necci, Antonioni, Bonvicini, & Cozzani, 2016), Petri nets (Kabir & Papadopoulos, 2019), and dynamic process simulation (Carlos, Fatine, Nelly, & Nadine, 2018) should be established at first to reflect the evolution of potential hazardous scenarios from cause to consequence. And the failure rates of every factor in the scenarios are used to calculate the final frequencies. Although some of the generic failure rates are available in commercial industrial databases, such as SERH (exida, 2015), OREDA (SINTEF & NTNU, 2015), PERD (CCPS/AIChE, 2007), and PDS (Håbrekke, Hauge, & Onshus, 2013), the evaluation result could be more reliable by using the data from the chemical plant's own mechanical integrity management database. The consequence estimation usually relies on chemical accident modeling tools such as the U.S. EPA's ALOHA and the DNV's PHAST to determine the severity level. In order to predict more precisely the consequence in a complex environment, computational fluid dynamics (CFD) models (Seike, Kawabata, & Hasegawa, 2017; Zhang, Liu, & Qiao, 2019) are often built. However, CFD models still need to be verified with a large amount of actual and/or experimental data.

As the actual environment is constantly changing, new risks could appear in practice, the static QRA cannot update and integrate the overall risk. So, it could only be applied during process design. To overcome this limit, dynamic risk assessment (DRA) was proposed (Villa, Paltrinieri, Khan, & Cozzani, 2016; Zio, 2018). Meel and Seider (2006, 2008) developed the first DRA approach, called Dynamic Failure Assessment (DFA). Near miss or accident data and real-time data from the processes was used to estimate the dynamic probabilities of accident sequences. Since, the risks of a chemical process change dynamically with its operating conditions, improved DRA approaches have been proposed to incorporate alarm data and/or condition monitoring data to predict real-time risks (Zeng & Zio, 2018; Zhu, Zhao, & Yang, 2014).

From qualitative risk assessment to QRA, and then to DRA, risk assessment is continuously meeting the high demand from the industry. In the meantime, the availability of large amount of data that could be used for risk assessment is also becoming more and more critical.

2. Source of the data

Traditional data for risk analysis of chemical processes is mainly gathered from labs and safety management system. DRA's data, however, usually comes from the automation system as shown in Fig. 1.

In the big data era, with the rapid development of computer and sensor technology, various types of data could be easily gathered from the Internet of Things (IoTs). Data from IoTs shown in Fig. 2 provides a new perspective for risk assessment.

(1) Process data.

Nowadays, chemical processes are usually equipped with DCS systems for better quality control and safety control. Process data is retrievable from DCS systems or process historians.

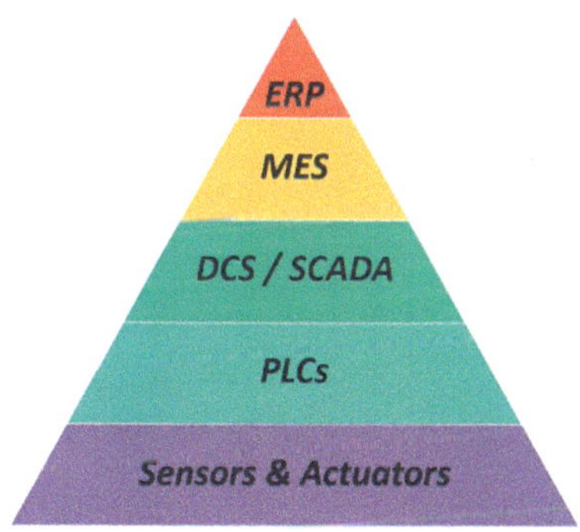

Fig. 1 Automation system pyramid in chemical plants.

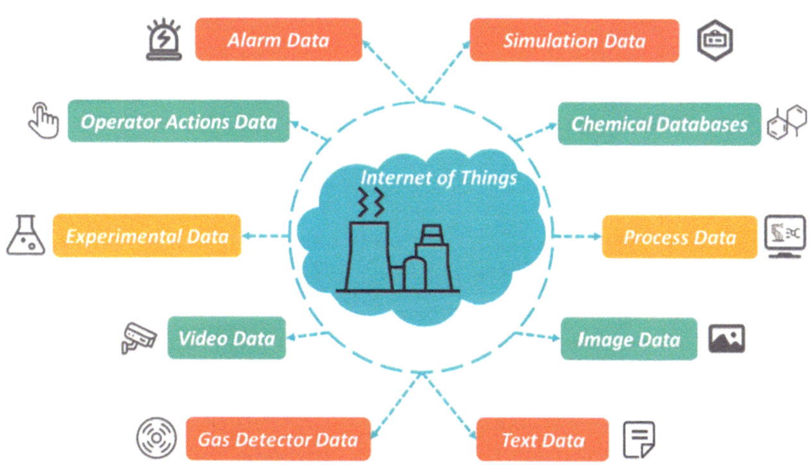

Fig. 2 Data from IoTs in a chemical plant.

(2) Alarm data.
With the development of DCS systems, setting up alarms to important process variables is becoming quite easy. Alarm data is an important data source to reflect the process risk.
(3) Experimental data.
With the laboratory information management system (LIMS) installed in more and more modern chemical plants and research institutes, experimental data is also becoming easily available for risk analysis of chemical products.
(4) Operator actions data.
When the process conditions are switched, or an unexpected situation occur, manual operation would instead automatic control to ensure the smooth operation of the process. The operator's actions on DCS would be saved in event summary documents, and the on-site actions would be manually recorded in the shift logs or other operation logs.
(5) Video data and image data.
In order to enhance the security and emergency management capability, more and more video cameras are installed in chemical plants. Sometime, there are hundreds of video cameras in a large chemical plant. The volume of the videos stored in a chemical plant could be hundreds of petabytes.
(6) Gas detector data.
Various flammable and toxic gas detectors are now available in the market. They are becoming a standard configuration of a chemical plant. Based on gas detector data, we can estimate the real-time severity of chemical leakage and verify chemical release models.
(7) Text data.
In chemical plants, a lot of information such as operation instructions, risk analysis reports, historical accident or near miss reports, and safety inspection records are stored in text documents. These text data are unstructured but contain much useful information for risk assessment.
(8) Chemical databases.
The various chemical hazardous properties directly affect the risks of chemical processes. With hundreds of millions of chemicals in the world, relevant chemical property data could be collected from existing chemical databases.
(9) Simulation data.
During the process design or retrofit, process simulation data is generated for mass balance, heat balance and cost analysis. With virtual reality technology deployed in the chemical industry, data from emergency simulators is becoming available.

3. Data driven methods for risk assessment and management

In the big data era, a large amount of data gathered from IoTs contains deep hidden knowledge that people don't master before. Compared with model driven methods, data driven methods are more suitable for mining the hidden knowledge from big data for risk assessment and management. The data driven methods can be roughly divided into two categories: statistical and artificial intelligent.

3.1 Statistical methods

The statistical methods can be further classified into univariate and multivariate methods. In the univariate methods, statistical analysis is performed for each variable individually to summarize the basic properties of the single piece of information. Measures of central tendency (mean, median and mode) and measures of variation (standard deviation, range, interquartile range) are two types of most commonly used statistics. For some specific purposes, such as threshold optimization (Yu, Wang, & Yang, 2017) and dead band setting (Afzal, Chen, Bandehkhoda, & Izadi, 2018) in alarm design, targeted statistics would be defined to solve different problems. For process variable monitoring, conventional statistical process control (SPC) charts are widely used for limit checking or trend checking of each individual variable (Harrou, Nounou, Nounou, & Madakyaru, 2015).

The key idea of multivariate statistical methods is projecting the high-dimensional process data onto a low-dimensional space represented by a smaller set of latent variables through dimensionality reduction. Hence, the low-dimensional data after dimensionality reduction can be provided to traditional risk assessment models, reducing the complexity of the model and making it more robust against false alarms (Adedigba, Khan, & Yang, 2017; Zadakbar, Imtiaz, & Khan, 2012). On the other hand, some statistical indexes like T^2, I^2 and the squared prediction error (SPE) constructed during dimensionality reduction can be used for risk assessment and process monitoring, which are called multivariate statistical process monitoring (MSPM) methods. The theoretical research and engineering applications of MSPM algorithms have been reviewed by Wang, Si, Huang, and Lou (2018). Four typical categories of MSPM, namely, principal component analysis (PCA), partial least-squares (PLS), independent component analysis (ICA) and Gaussian mixture model (GMM) are shown in Fig. 3.

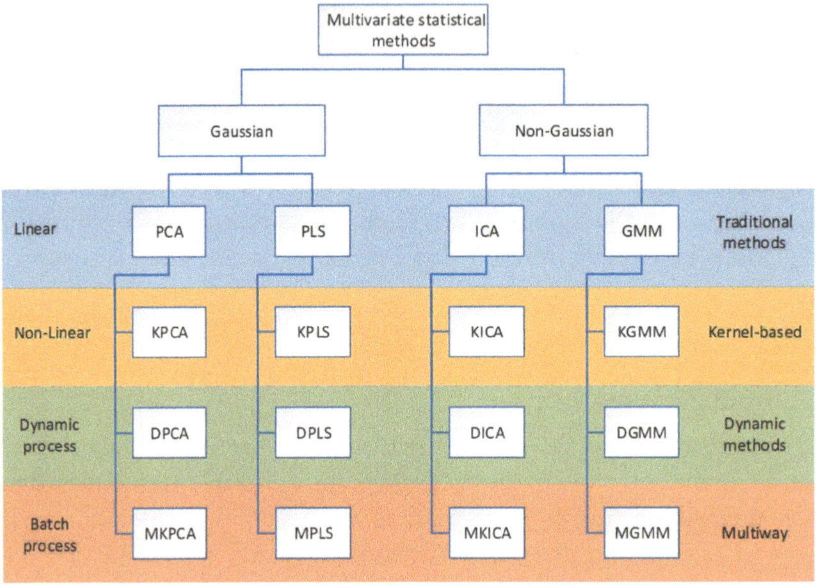

Fig. 3 Typical multivariate statistic methods and their modification.

The key idea of PCA is to decompose the high-dimensional data into a smaller set of principal components along directions that explain the maximum variation of the data (Krzanowski, 1992). Both input and output variables are collected in one dataset for a PCA model. Different from PCA, PLS separates variables into input and output variables and then finds the principal components to maximize the covariance between the two kinds of variables (Zhou, Li, & Qin, 2010).

Both PCA and PLS assume that the data obeys the Gaussian distribution. To solve non-Gaussian problems, independent component analysis (ICA) imposes independent components with high-order statistics; hence, it may reveal more meaningful information in non-Gaussian data (Li & Wang, 2002). Gaussian mixture model (GMM) provides another way to deal with non-Gaussian data by treating a complex process as a linear combination of several Gaussian models (Yu & Qin, 2008).

The above multivariate statistical methods assume that process variables are linearly correlated. Linear approximation and kernel function are two types of the most commonly used methods to handle the non-linearity issues. The core idea of linear approximation is to treat nonlinear problems as a combination of several local linear models, but it is hard to determine the number of local models. Different from linear approximation, kernel-based methods introduce a kernel function to transform the nonlinear variables in a higher-dimensional space where variables are linear. Many kernel-based modifications of conventional multivariate have been proposed, such as kernel PCA (KPCA) (Gharahbagheri, Imtiaz, & Khan, 2017) and kernel PLS (KPLS) (Wang, Yan, & Fei, 2015), kernel ICA (KICA) (Lee, Qin, & Lee, 2007) and nonlinear kernel GMM (NKGMM) (Yu, 2012)

The dynamic data from industrial chemical processes always contain process disturbances and data noises. In order to effectively identify the disturbances and noises, a dynamic PCA (DPCA) model was developed by Ku, Storer, and Georgakis (1995) for disturbance rejection in industrial process. The first step of this method is to augment the modeling dataset by incorporating several time-lagged data samples of each variable. By constructing the PCA model upon this augmented data matrix, the autocorrelation of the process data can be extracted. Similar dynamic methods by augmenting time-lagged data have used for improving traditional methods like PLS (Lee, Han, & Yoon, 2004), ICA (Stefatos & Ben Hamza, 2010) and GMM (Yu, 2012).

Batch processes are unique dynamic processes, the data of which have a special three-way structure (i.e., batches × variables × time). Nomikos and MacGregor (1994, 1995) proposed multiway model-based methods for batch process monitoring by unfolding of the three-way structure process data into a two-dimensional matrix and then using the traditional multivariate statistical methods to solve the two-dimensional matrix.

Data from complex chemical processes always show multi-features, such as non-Gaussianity, non-linearity, or multiple modes. Therefore, hybrid methods, such as MGMM, MKPCA, MKICA, DKPCA, or integrated approaches with both statistical methods and other methods have been proposed in recent years (Amin, Khan, Imtiaz, & Ahmed, 2019; Ge, 2018; Liu, Yang, Carlsson, Qin, & Yoo, 2019; Zhao & Huang, 2018; Zhu & Gao, 2018).

Another type of popular multivariate statistical methods is probabilistic model based on prior probabilities or prior distributions. The most typical one is Bayesian statistical methods (Roy, Srivastava, & Sinha, 2014), which have been discussed in Chapter 4 of this Volume by Nima Khakzad and Chapter 5 from Volume 1 (Seider et al., 2017). The n-gram model is another typical probabilistic model, which follows the Markov assumption and uses an n-1 order Markov model. It has been successfully applied in the area of sequence data analysis and risk prediction (Zhu, Wang, Li, Gao, & Zhao, 2016).

3.2 Artificial intelligence methods

A simple, visionary and useful definition Artificial Intelligence (AI) is "the study of how to make computers do things at which, at the moment, people are better" (Rich, 1985). The chemical engineers are excited about the potential applications of AI in the chemical industry (Venkatasubramanian, 2019).

As the most classic data-driven AI method, artificial neural network (ANN) was proposed to help people acquire knowledge automatically from historical data. A typical structure of a neural network is composed by an input layer, a hidden layer, and an output layer of neurons. The neurons between different layers are associated by transfer function with signals, weights and biases. Although there were many successful applications of ANN, the architecture of traditional neural network is too simple to solve problems coming from complex systems. Advanced AI methods with more complicated structures were needed to describe complex system.

With the development of data acquisition tools, computing power and AI algorithms, AI in chemical engineering entered a new era of data science or predictive analytics (Venkatasubramanian, 2019). Machine learning (ML) is an application of AI to give computers the ability to learn with data. It provides a given system with the ability to automatically learn and improve from experience without being explicitly programmed (Bagheri, Akbari, & Mirbagheri, 2019). Therefore, ML methods become popular in chemical engineering to mine the information behind massive amounts of data. As shown in Fig. 4, the ML methods can be divided into two categories: supervised learning methods and unsupervised learning methods. The supervised learning methods require a priori knowledge of what the result should be. The models of supervised learning methods are trained by labeled data. Supervised learning methods can be used for classification and regression. Unsupervised learning methods can be grouped into clustering and association.

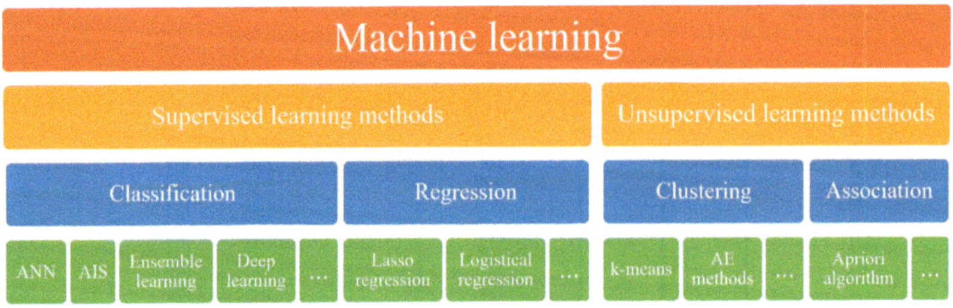

Fig. 4 Machine learning methods.

Classification methods are widely applied for industrial process when a large amount of labeled data is available. Deep learning (DL) methods can scan data to search for the features, which contributed a lot to the development of AI methods. As mentioned above, tools, computing power and methods are the bottleneck for training the deep neural network before. However, after 2005, significant progress has been made for AI applications. An important breakthrough is the developing of DL methods such as restricted Boltzmann machines (RBM) (Zhang & Zhao, 2017) and convolutional neural network (CNN) (Wu & Zhao, 2018a), which has been used for feature selection. Most of the DL methods are based on neural nets with more hidden layers. Up to now, advanced tools, higher computing power and new methods are developed to train the deep neural network.

Different from DL methods, ensemble learning (EL) attempts to construct multiple submodules to increase the accuracy of the total system (Zhou, Li, & Xiong, 2019). Multiple weak learners are used for building integrated models to reduce variance and prevent overfitting. Bagging and boosting are two typical strategies of EL methods. Bagging based methods train each learner on the sample data, where the output is average (regression) or voting (classification) of all learners. On the other hand, boosting based method iteratively train models, and each newly trained model tries to fix previous mistakes. Typical methods include random forest (RF) based on bagging strategy (Bahl et al., 2019) and gradient boosting decision tree (GBDT) based on boosting strategy (Cai, Gu, Ma, & Jin, 2019).

Artificial immune system (AIS) is another classic AI method. The key to design the AIS is to take full advantage of the immunology principles to generate accurate immune response against the invading antigens (Tan, 2016). Many efficient AIS models and algorithms have been explored, including negative selection algorithm, clonal selection algorithm, immune network model, Danger theory, and immune concentration.

All the above AI methods are supervised approaches based on labeled data. However, for industrial application, a large amount of unlabeled data reflecting process knowledge exists in IoTs. Unsupervised methods could be suitable for online application with less human intervention. Autoencoder (AE) neural network and k-means are the mostly common used unsupervised algorithms, which can make use of unlabeled data for feature extraction (Hinton & Salakhutdinov, 2006). However, the features extracted by unsupervised methods are unpredictable, which leads to the unstable performance of unsupervised methods.

3.2.1 AI based computer vision methods

In the past few years, with the rapid development of deep learning and artificial intelligence, computer vision has achieved spectacular success in many fields.

In computer vision domain, there are three main tasks: image classification, object detection and instance segmentation. Image classification aims to identify the classes of the images. Besides, object detection aims to detect and locate objects in images. At last, instance segmentation can frame object instances from one image with object detection method, and then uses a sematic segmentation method to mark each pixel in different instance areas.

Deep CNN was the most widely used method in computer vision. The general architecture of CNN is the stack of convolutional layers and pooling layers, which transforms the raw image into a representation at a higher level. In a convolutional layer, the representation of the output is composed of feature maps, within which each unit is connected to a local patch in the input feature map through a filter composed of a set of weights. All of the units in one output feature map share the same filter (Fig. 5). Assuming M input feature maps and N filters, the output feature maps of the lth layer are as follows (Wu & Zhao, 2018a):

$$x_j^l = f\left(\sum_{i=1}^{M} x_i^{l-1} * k_{ij}^l + b_j^l\right), \quad j = 1, \ldots, N \tag{1}$$

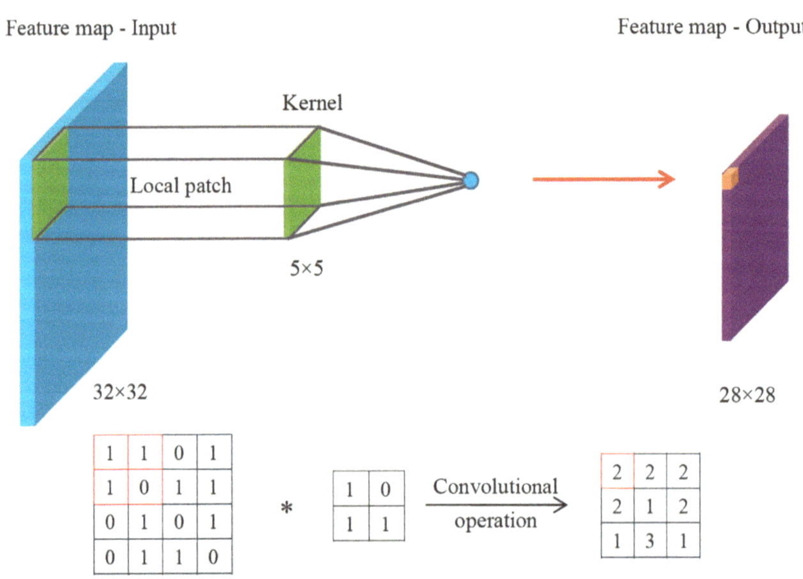

Fig. 5 Convolutional layer.

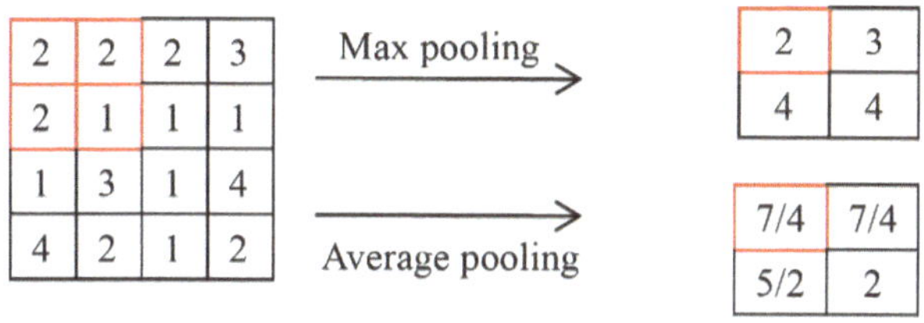

Fig. 6 Max pooling and average pooling operations.

where k^l_{ij} represents the kernel of the *j*th filter connected to the *i*th input map, x^{l-1}_i represents the *i*th input map and x^l_j represents the *j*th output map, b^l_j represents the bias corresponding to *j*th filter, f represents the activation function, and * represents the convolutional operation.

A pooling layer usually follows several convolutional layers and produces down-sampled versions of the input feature maps. The objective is to merge similar local features into one. There are two types of pooling operations: max pooling and average pooling (Fig. 6). Assuming M input feature maps, the output feature maps of the *l*th layer are as follows (Wu & Zhao, 2018a):

$$x^l_j = f\left(\beta^l_j down\left(x^{l-1}_i\right) + b^l_j\right), \quad j = 1, \ldots, N \tag{2}$$

where different from Eq. (1), β^l_j represent the additive bias and the multiplicative bias corresponding to the *j*th filter, respectively, and *down* represents the sub-sampling function.

In practical application, images come from video streams through cameras and internet/local area network (LAN). Most commonly used video recognition methods can be divided into two categories. The first is based on spatial-temporal two stream networks, where spatial stream network is for single frame and temporal stream network is for multi-frame optical flow (Simonyan & Zisserman, 2014). The other is based on a 3-D convolutional network, which means the convolutional operation is not only applied on spatial dimension but also on temporal dimension (Jiang, Rozgic, & Adali, 2017).

4. Application of data driven methods

As discussed in the first section of this chapter, risk assessment could be applied in the whole chain of chemical processes. Herein, we proposed a framework of data driven methods for risk assessment and management with IoTs shown in Fig. 7. Based on big data cloud, various types of data could

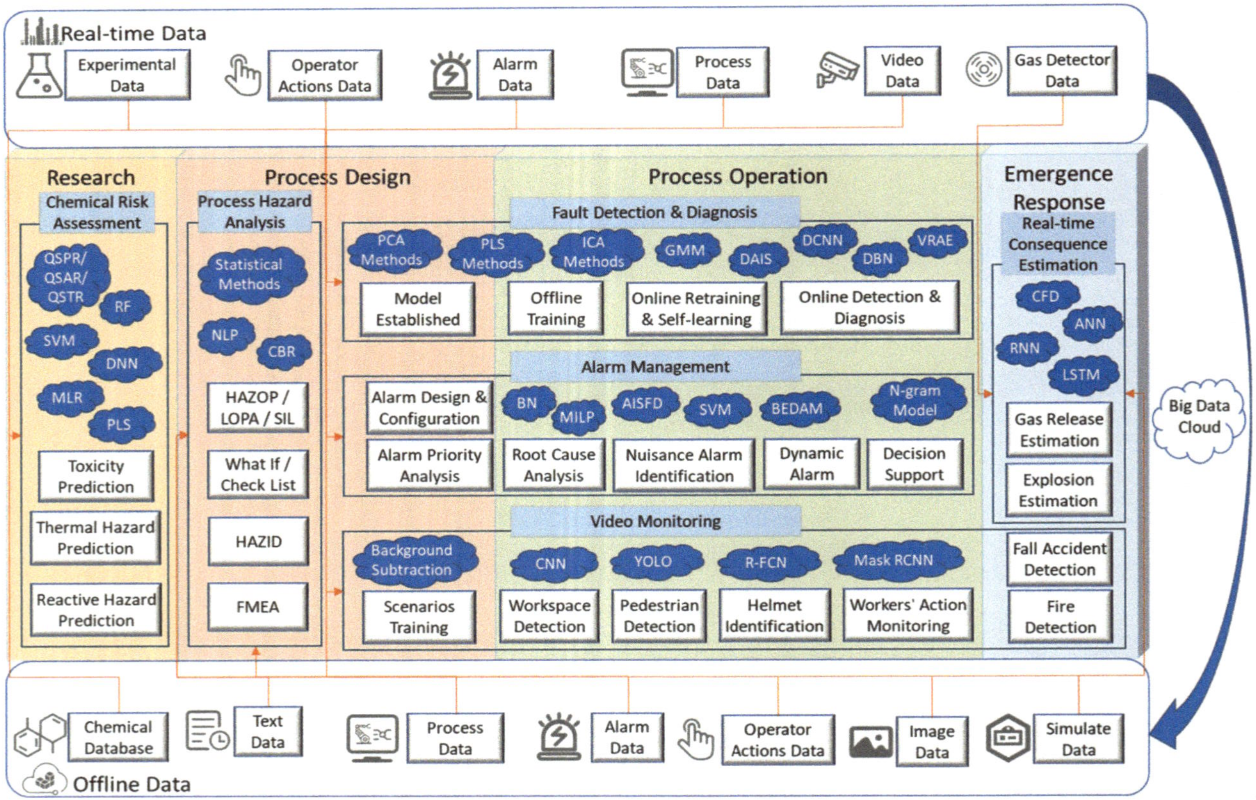

Fig. 7 Framework of data driven methods for risk assessment and management with IoTs in big data era.

support data-driven methods for risk assessment management in the whole four stages of a chemical process's life cycle, namely, research stage, process design stage, process operation stage and emergence response.

In the research stage, data driven methods could help predict chemical hazard like toxicity, thermal hazard and reactive hazard with less experiment and thus reducing laboratory risk. While in process design stage, traditional QRA methods have proven their reliabilities in numerous industrial applications. As an important approach for hazard identification of QRA, process hazard analysis (PHA) could use text data from historical cases. In process operation stage, process alarm, fault diagnosis and video monitoring would be suitable approaches of risk monitoring for DRA. Data driven methods could be used to establish prediction models with offline data, and then using real-time data for real-time risk assessment and management. At last, for emergence response, video monitoring could be used to detect the hazard scenarios, and the real-time consequences of them should be estimated. Data driven methods could help establishing fast calculation model to estimate gas release, fire or explosion with real-time data from industrial site. In the rest of the section, the application of data driven methods used in different stages will be discussed in a detail manner.

4.1 Risk assessment in research

Risk assessment of chemical product properties and reaction is the main task in research stage. Machine learning methods can be used to establish models to describe and predict the relationship between hazard properties and chemical structure. Multivariate statistical methods could be used to reduce the dimensions of input variables.

4.1.1 Chemical risk assessment

In research stage, new products and processes could bring more unknown risks. Since many of the hazardous properties of the chemical do not have a clear mechanism, data driven methods could use molecular structure and property database to help identifying chemical risk during the research stage.

In toxicology studies, chemical risk refers to a function of the chemical (toxicological) hazard potency and the degree of chemical exposure (Pease & Gentry, 2016). Usually, chemical risk analysis can be performed through conventional toxicity experiments with more risks and economic cost. With the advent of mathematical and statistical models, data driven methods are much faster and cheaper than experiment-based methods.

These data driven methods generally rely on a quantitative structure–property relationship (QSPR) model, whose input is the chemical structure and the output are chemical properties and potential risk. Wang, Su, et al. (2019) proposed a typical data driven chemical risk assessment method to predict the octanol-water partition coefficient (K_{OW}), which is an essential environmental property and could be further employed to predict toxicity. It consists of five steps:

Step 1: Data collection.

The experimentally chemical property data and simplified molecular-input line-entry system (SMILES) strings of compounds were collected.

Step 2: Feature extraction.

The SMILES strings of compounds were utilized to generate a list of numeric vectors to describe molecular structures and represent their features.

Step 3: Information processing.

The long short-term memory (LSTM) network was used in conjunction with signature descriptors for automatic feature selection of the SMILES strings.

Step 4: Model training.

After receiving the inputs from the LSTM networks, a QSPR model with a BPNN structure would be trained with collected data.

Step 5: Performance evaluation.

Based on the developed chemical risk assessment model, the generalization ability was assessed by the predictive performance of an external dataset.

Similar to QSPR model, quantitative structure–activity relationship (QSAR) model and quantitative structure–toxicity relationship (QSTR) model have also been proposed for chemical risk assessment. The models would be trained with historical data using AI based classification methods, and new structure of the product would be used to predict reactive hazard (Lu, Ng, & Mannan, 2011), thermal hazard (Pan et al., 2019), minimum ignition energy (Wang, Zhou, Xu, & Wang, 2017), and gas-liquid critical temperatures (Zhou, Wang, Jiang, Pan, & Wang, 2017).

4.2 Risk assessment in process design

In process design stage, the main purpose of risk assessment is to analyze the potential risk for future operation. Except for traditional QRA using statistical methods to estimate frequency, knowledge-based methods are mostly used for traditional PHA approaches. As in big data era, historical case stored as text data could be used for PHA.

4.2.1 Case based reasoning for PHA

HAZOP analysis is a widely recognized PHA technique used for hazard identification in the chemical process industry. HAZOP expert systems would support better analyzing. Large amounts of HAZOP analysis case have been accumulated which could be useful for new analysis. Zhao, Bhushan, and Venkatasubramanian (2005) developed a HAZOP analysis expert system called PHASuite using case-based reasoning (CBR). It aimed to facilitate modification of the existing models and creation of new models based on the knowledge in the existing models. However, PHASuite is still a model-based HAZOP expert system, while the "non-routine" HAZOP analysis relies on the human team. To solve this problem, Zhao, Cui, Zhao, Qiu, and Chen (2009) developed a learning HAZOP expert system called PetroHAZOP, based on the integration of CBR and ontology. In addition to HAZOP, layer of protection analysis (LOPA), safety requirements specification (SRS) and safety integrity level (SIL) validation are also widely used risk identification and risk assessment methods. It is necessary to ensure data consistency and management uniformity. Therefore, an intelligent software platform HASILT is developed on the basis of PetroHAZOP for facilitating the new PHA workflow, which integrates HAZOP, LOPA, SRS and SIL verification (Cui et al., 2012). Its architecture is shown in Fig. 8.

By using this intelligent PHA expert system, a HAZOP study in biofuel production has been performed. Through the analysis, the HAZOP expert team proposed up to 36 recommended action items, and the plant owner accepted all the recommendations and implemented all the action items. Based on the CBR principle, the exchange of recorded information

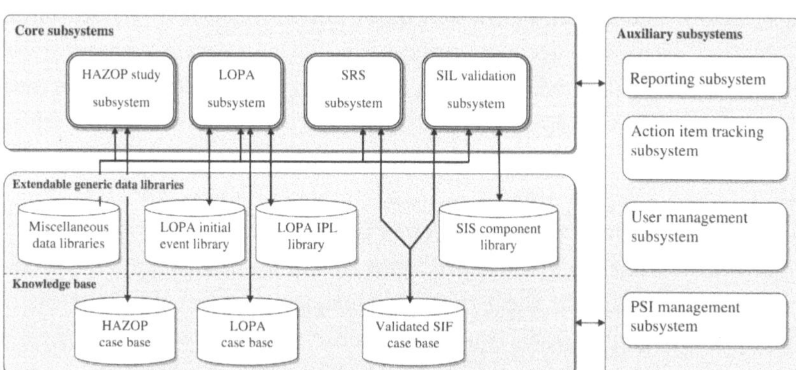

Fig. 8 System architecture of HASILT (Cui et al., 2012).

between industries and free access to a database can facilitate spreading experiences and/or lessons learned from other companies, which could help to identify the most probable safety tool to avoid accidents (Wu, Peñarrubia, Cui, & Zhao, 2017).

4.2.2 Natural language processing for PHA

Many PHA reports saved in text form could be an important data resource for risk assessment. Natural language processing (NLP) is an important artificial intelligence method for extracting key data information from unstructured languages. There are only a few reports on the application of NLP in PHA till now. Recently, Song and Suh (2019) proposed a text mining-based local outlier factor (LOF) algorithm approach to detect anomalous conditions using accident report documents, focusing on the text information. By extracting and prioritizing the anomaly conditions, the approach enables safety managers to monitor the natural language-based risk factors and reasons of infrequent, anomalous, and critical accidents. It is believed that NLP-based methods will play a more important role for improving risk assessment in the future.

4.3 Risk assessment and management in process operation

Traditional QRA is only applied during process design, and DRA could help to reduce risk during process operation. Paltrinieri, Khan, Amyotte, and Cozzani (2014) have proposed a dynamic risk management framework (DRMF), composed by two general stages, including risk understanding and decision. The first stage of learning and understanding including the phases of horizon screening and hazard identification. The decision stage, which refers to process of elaboration and judgment of information subsequent intervention, includes an assessment phase and a decision and action phase.

While for real-time decision support for operators, risk assessment should not only dynamically update risks with online data, but also real-time monitoring the risks. Once an unacceptable risk has been assessed with monitoring data, the targeted action should be done immediately to avoid hazardous consequences. A real-time risk management framework (RRMF) is proposed for online operation and emergency response, as shown in Fig. 9.

In fact, the risks of chemical processes are always caused by the deviation from normal condition. Real-time risk management should focus on the deviation with high risks. To understand the risks of chemical process, an offline identification and assessment should be taken. In this stage, an early warning system of risk monitoring could be designed to monitor the

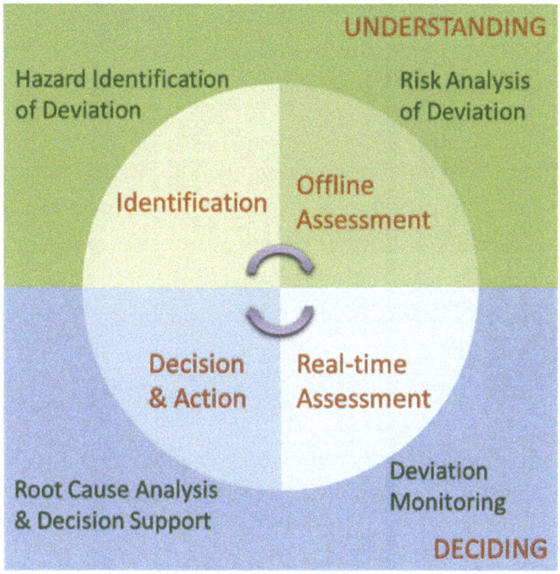

Fig. 9 Real-time risk management framework—RRMF.

deviation may cause hazard consequences. In practical application, some hazard scenarios with low frequencies are defined as acceptable risks in off-line risk assessment. However, in real-time risk assessment, once the deviation occurred, that means some protection layers failed, and the frequencies of hazardous consequences increased. In that case, the targeted actions should be taken. One process deviation from normal condition could be caused by different reasons, therefore, the root cause analysis is needed for decision support to reduce the risk. In the final step, the targeted decision could be given by the early warning system in understanding stage.

Among all the common approaches, industrial alarm and fault detection could be used for risk monitoring of processes and equipment. And, fault diagnosis could be used of root cause analysis of the process deviation.

4.3.1 Industrial alarm management

Alarm systems play important roles for risk monitoring in chemical plants, while the alarm data could be a key indicator to identify risks. This section will introduce the application of data driven methods on alarm management. Goel, Datta, and Mannan (2017) provided a literature review of latest research works on alarm design, reducing alarm flooding and effective alarm management, and highlight five focus areas in the area of alarm management. This section will introduce the up-to-date research outcomes with data driven methods in the five areas.

4.3.1.1 Configuration of an alarm and identifying incorrect alarm variables

The key point of this problem is to select proper process variables from alarm system to identify abnormal events in normal operation. In addition, the selection should be based on the associations between abnormal events and process variables. Wang, Yang, Chen, and Shah (2016) descirbed the relation mathematically as

$$X = g(Y, A) \tag{3}$$

where A stands for the set of abnormal events, Y is the set of process variables, and X is the set of selected process variables to be configured with alarms.

The existing association mining methods are mainly based on process knowledge (Hamaguchi, Takeda, Noda, & Kimura, 2012; Takeda, Hamaguchi, Kimura, & Noda, 2014; Yang, Xiao, & Shah, 2009) or historical data (Dalapatu, Ahmed, & Khan, 2013; Peng, Gu, Xu, & Zhu, 2015), while both types have their own limits. The knowledge-based methods have the problem of qualitative uncertainty, and the data-based methods are easily confined by the occurred phenomena in the historical data sets.

Hu, Chen, and Shah (2018b) proposed a data driven method to identify associations between alarms and states from A&E logs for state-based alarming in practice. Ariamuthu Venkidasalapathy, Mannan, and Kravaris (2018) proposed a systematic approach to treat alarm identification as a Mixed-Integer Linear Programming (MILP) problem, to decide which vairables would set as alarms for each fault.

4.3.1.2 Priority setting of an alarm

For industrial applications, when an abnormal situation occurs, hundreds of alarms would be shown to operators. Priorities of the alarms would support the operators to take corrective action. Currently, the assignment of alarm priorities is static with qualitative approaches, based on the understanding of severity in missing alarms and a rough estimation of safety time in handling alarms. Data driven methods could be alternatives or complements to the traditional approaches, such as to calculate the accurate estimation of the safety time with historical data sets (Chang, Khan, & Ahmed, 2011).

Besides risk-based approaches, event tree (ET) could also be used for half quantitative alarm priority analysis using process knowledge and historical data of failure probabilities (Dai & Qiu, 2019). Another focus of alarm priority research is dynamic alarm priority and limits analysis, which may be more reasonable for different process states and operational scenarios (Wang, Khan, Ahmed, & Imtiaz, 2016; Yu, Wang, & Ouyang, 2020; Zhu et al., 2016).

4.3.1.3 Handling nuisance alarms

Nuisance alarms are defined as the occurred alarms that do not affect the process, even if these alarms are ignored by operators.

Set delay time and dead band for an alarm could be useful in preventing nuisance alarms during process operation. Hollifield and Habibi (2010) poposed a recommendation for alarm dead bands and delays according to the signal types, such as temperature, flow rate, level, and pressure. Su et al. (2018) proposed a multi-setpoint delay-timer alarming strategy, which employs multiple setpoints and provides direct transitions from each of the delay-timer sub-status to the alarm or non-alarm status. Afzal et al. (2018) developed a mathematical model for time-deadband configurations for univariate alarm systems based on Markov processes. The design procedures are based on process data and alarm data with analytical expressions for the performance indices. Wang, Bai, Wang, and Yang (2019) formulated an index to determine whether deadbands are suitable for removing nuisance alarms of an analog signal, and proposes a method to design the deadband width based on historical process data.

Unreasonable alarm threshold is another main cause of nusiance alarms. Yu et al. (2017) proposes a method to design optimal alarm threshold for univariate alarm system based on so-called alarm probability plots, involving four statistics estimated from historical data. Tian, Zhang, Zhang, and Dong (2018) proposed an alarm threshold optimization method based on PCA weight and Johnson transformation. Traditional threshold optimization methods focus on univariate alarms, while many variables are interacting with each other. A multivariate thresholds optimization method was proposed by Han, Gao, Xu, and Zhu (2016). The causal relationship between process variables is constructed based on historical data. Then an objective funcion is established combined with false alarm probability (FAP), missed alarm probability (MAP), and the correlation between the alarm information and process information. At last, the optimal thresholds are obtained by the genetic algorithm. Various reasons may cause nuisance alarms, while some of the works to reduce the number of nuisance alarms could be found in the review of Goel et al. (2017).

4.3.1.4 Developing advanced alarming techniques

Goel et al. (2017) suggested the necessity of designing methodologies to reduce the alarm flooding in case of an abnormal condition. All the methods of alarm variable selection, dynamic alarm priority and handling nuisance alarm aim to solve the problem of alarm flooding. Researches on advanced alarming techniques should focus on the root cause anlaysis of mutiple alarms.

One direction of the research is to analyze the alarm variables' relationships when abnormal situation occurs. Wang, Khan, and Abimbola (2018) proposed a Bayesian network based method to analyze alarm system performance. Alarm variables relationships were analyzed with historical alarm, enabled alarm diagnostic to identify the root causes, and studied the variables dependence strength and finally improved the alarm system performance. Hu, Chen, and Shah (2017) proposed a method to analyze cause-effect relationships among binary-valued alarm variables, based on the estimated transfer entropies; as an unsupervised learning method, no labeled training data sets are required. Many new methods could be found for multiple alarm flood analysis, which are used to find out the root cause, or to locate the badly designed part in alarm systems and predict incoming alarm floods (Hu, Chen, & Shah, 2018a; Lai & Chen, 2017; Lai, Yang, Chen, & Cao, 2019). Considering the complex relationships between process variables of actual industrial systems, one scenario often corresponds to multiple univariate alarms. Multivariate alarm system (Chen & Wang, 2017) should be the futrue dirrection of alarm system design.

Another important direction of advanced alarm technologies is the dynamic alarm during transition state, such as start-up and shut-down. Based on the basis of the designed dynamic alarm limits and the calculation of an exact unsuppression delay time, Zhou, Hu, and Chen (2019) proposed an alarm monitoring method for equipment start-up operations. To solve the complex and dynmic problem during trasition states, fault diagnosis methods could be combined with alarm management methods for a hybrid strategy. Zhu, Shu, Zhao, and Yang (2014) proposed a dynamic alarm management strategy integrated with the artificial immune system-based fault diagnosis (AISFD) method and a Bayesian estimation based dynamic alarm management (BEDAM) method, shown in Fig. 10. During transitions,

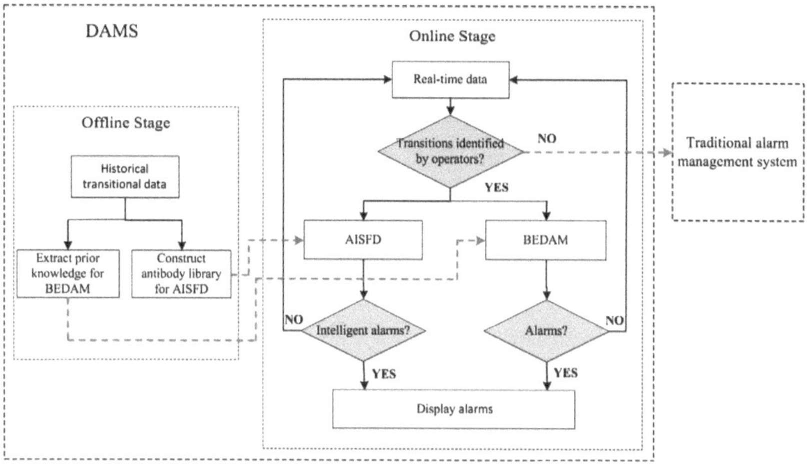

Fig. 10 Framework of the dynamic alarm management strategy (Zhu, Zhao, & Yang, 2014).

dynamic alarm limits obtained by the BEDAM method can control alarm floods. If a process fault occurs during transitions, a flood of alarms could still be yielded. To generate useful alarms in fault situations, DAIS is used for fault detection and diagnosis.

4.3.1.5 Assisting operator in decision making

In practice, human machine interface (HMI) acts as an interface between the operator and automated systems. Hu, Al-Dabbagh, Chen, and Shah (2018) presented timely work in the design studies on the development of visualization plots in three categories, including visualization toward overall performance, visualization toward pattern insights, and visualization toward real-time applications. Al-Dabbagh, Hu, Lai, Chen, and Shah (2018) proposed three types of visualization plots, which collectively compare the behavior of metrics, variables, and operation times, that can present information in an elegant, dense, and comprehensive fashion.

In fact, the above five focus areas are not isolated. To develop an outstanding alarm system, all the issues should be considered comprehensively. Hu, Shah, and Chen (2018) presented a framework for a smart data analytics platform integrating information from process and alarm databases complemented with process connectivity, which consists of four functional modules, including alarm data analysis, alarm configuration analysis, process data analysis, and connectivity & causality analysis.

4.3.2 Fault detection and diagnosis

Fault detection and diagnosis (FDD) focus on abnormal situations instead of univariate alarms, essential to maintain favorable operating conditions and predict risks of chemical processes. Venkatasubramanian, Rengaswamy, Yin, and Kavuri (2003) classified the FDD methods into two categories: model-based and process history based, with the latter also being referred as data-based. Model-based methods are usually developed based on some fundamental understanding of the physics of the process, which are not easily implementable due to the complexity of chemical process. Data driven methods have become more feasible for FDD. Many efforts have sorted out related work in the field of data driven FDD and conducted several reviews with analysis and comparison (Alauddin, Khan, Imtiaz, & Ahmed, 2018; Ge, 2017; Jiang, Yan, & Huang, 2019; Ming & Zhao, 2017). Multivariate statistical methods have been widely used for process monitoring introduced in Section 3.1, and more application of MSPM could be found in the review from Wang, Si, et al. (2018).

Benefits from the strong classification abilities, AI methods show exciting prospect in the field of FDD of chemical processes. Expert system is the earliest AI method applied to FDD of chemical process (Ramesh, Shum, & Davis, 1988; Rich & Venkatasubramanian, 1987). The main components of an expert system for FDD include: knowledge acquisition, choice of knowledge representation, the coding of knowledge in a knowledge base, the development of inference procedures for diagnostic reasoning and the development of input–output interfaces (Venkatasubramanian, Rengaswamy, Kavuri, & Yin, 2003). However, due to the system-specific knowledge for developing expert systems, the application scenarios of FDD models are quite limited, which has hindered its application in the chemical industry.

ANN is the most popular AI method for FDD, and back-propagation neural network (BPNN) is the most favorable type of ANN. However, limited by the capability of data acquisition in chemical processes and the poor performance of solving problems of complex systems, traditional ANN methods have been used for academic research but not industrial applications. With the popularization of information construction in chemical plants and the development of deep learning methods, the research of AI-based FDD has ushered in another climax, such as RBM, CNN and Deep Belief Network (DBN) (Gupta, 2018; Shu, Ming, Cheng, Zhang, & Zhao, 2016; Wu & Zhao, 2018a; Zhang & Zhao, 2017).

As a very successful method in object detection and recognition in the computer-vision domain, the advantage in feature extraction makes

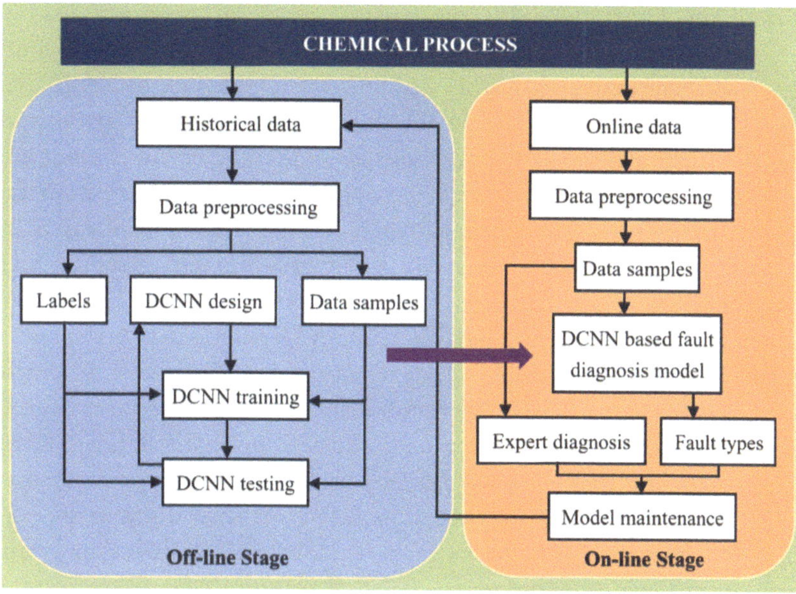

Fig. 11 The framework of the DCNN based fault diagnosis method (Wu & Zhao, 2018a).

CNN methods suitable for FDD of chemical processes. Wu and Zhao (2018a) proposed a Deep Convolutional Neural Network (DCNN) based fault diagnosis method for complex chemical processes, as shown in Fig. 11. Its diagnosis procedures including offline and online stages, while raw process data is transformed into a $m \times n$ matrix with time-varying features, where m represents the sample time length and n represents the number of variables.

Most of AI based FDD methods only monitor process faults in static condition. Considering the time-varying features of fault data and data in transition states, dynamic artificial immune system (DAIS) is proposed for FDD of chemical processes (Dai & Zhao, 2011). Different from traditional AIS methods, antibodies and antigens are represented by matrices of time-sampled data in DAIS. Benefits by its strong adaptation capability and independency on the number of training samples, DAIS based FDD could be used for full operating cycles of chemical processes (Zhao, Shu, Zhu, & Dai, 2014) In order to solve the problems from practical industrial applications, such as large-scale of process variables and few historical fault samples, feature selection algorithm (Ming & Zhao, 2018) and vaccine transplant method (Shu & Zhao, 2016) have been proposed to improve the DAIS method.

Both the DBN, DCNN, and DAIS belong to supervised learning methods. However, for various reasons, almost now records can be found after an incident or a near-miss. In addition, many historical data with great meaning which could be useful for FDD is unlabeled. As an learning algorithm, autoencoder based methods show a great advantage in industrial applications. Recently, variational autoencoder (VAE) was proposed as a deep generative modeling approach where a dataset drawn from any distribution can be generated by a set of latent variables and mapping them through a sufficiently complicated function which means that VAE can naturally handle nonlinearity in data (Kingma & Welling, 2013). Variational recurrent autoencoder (VRAE) uses recurrent neural network to extract the temporal feature and negative variational score (NVS) monitors in probability space, which is more suitable for chemical processes (Cheng, He, & Zhao, 2019).The basic idea of VRAE based methods is to first estimate the conditional probability density of the latent variables through the optimization framework of VRAE, then the estimated NVS of individual measurements are monitored for fault detection. The framework of VRAE based methods is shown in Fig. 12.

4.4 Risk assessment and management in emergency response

Risk assessment and management in emergency response should focused on two steps. The first one is hazard scenario detection, and the second is real-time consequence estimation of the hazard scenarios.

4.4.1 Hazard scenario detection

There are various of detectors in chemical plants to detect hazard scenarios like chemical leakage or fire. Early detection can effectively support decision making and minimize the damage. However, limited by the locations and functions of the detectors, some of the hazard scenarios cannot be detected immediately, such as outdoor fire or gas release and nonvolatile substance leakage. Therefore, cameras installed everywhere could help hazard scenario detection using computer vision methods.

Wu et al. developed an intelligent fire detection approach through cameras based on object detection and image classification methods (Wu, Wu, & Zhao, 2019). The approach uses YOLO to locate the fire coordinates when the flame is spotted on video screens. Due to the limitations of dataset and models, computer vision applications in reality may generate some false alarms. In order to meet the need of practical applications, an Xception

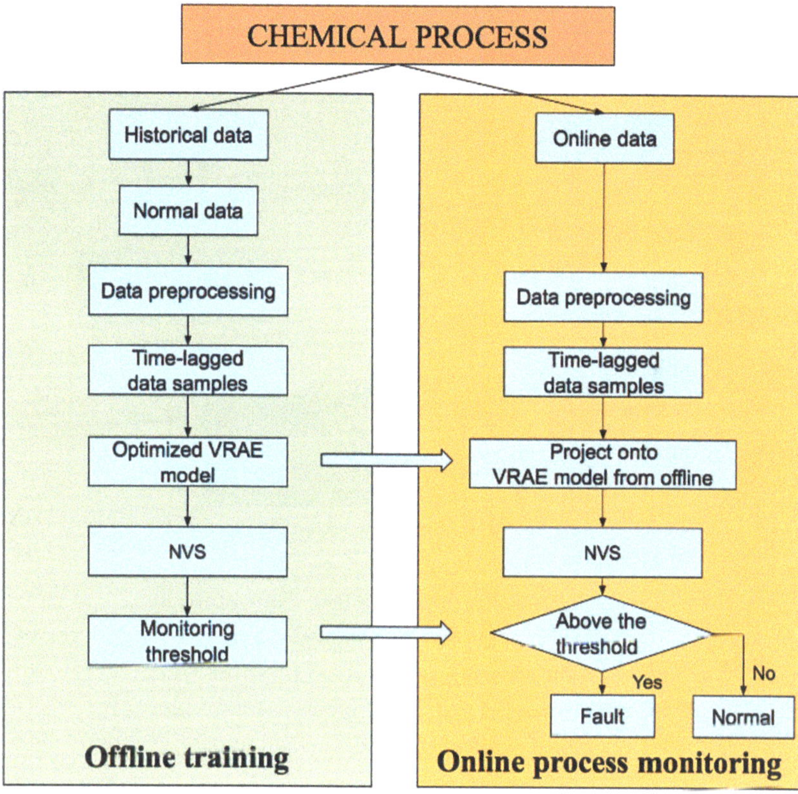

Fig. 12 Structure of VRAE-based process monitoring framework (Cheng et al., 2019).

network was used to distinguish the real fire and the false alarm. Its industrial deployment will help detect fire at the very early stage, which could facilitate the emergency management and significantly prevent loss (Fig. 13). Before flame appears, smoke is an early signal for abnormal events. However, smoke detection is more difficult than fire detection. Yuan et al. proposed a deep smoke segmentation network based on FCN for smoke detection and segmentation (Yuan et al., 2019). Smoke segmentation is to densely classify each pixel, which requires accurate separation of smoke components from background scenes at pixel levels.

As most of the chemical substances are not brightly colored or even colorless, conventional cameras can't separate them apart. Nowadays, many industrial monitoring cameras have the function of infrared radiation (IR) to record thermal information and visualize it in the form of an image. Since then, IR cameras have been used for substance leakage detection based on

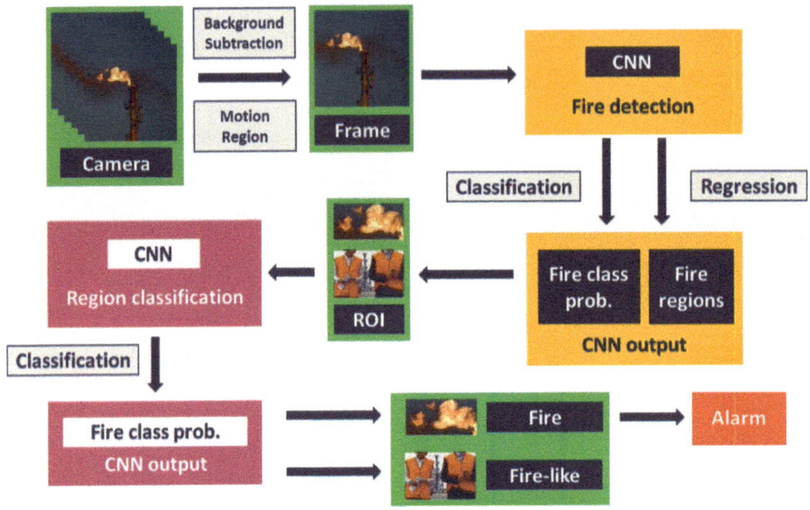

Fig. 13 Intelligent fire detection in factories (Wu et al., 2019).

temperature change caused by the leakage. Fahimipirehgalin, Trunzer, Odenweller, and Vogel-Heuser (2019) proposed a method based on PCA and K-Nearest Neighbor (KNN) classification methods to detect the liquid leakage from pipelines of a plant using thermographic video data.

4.4.2 Real-time consequence estimation for emergency response

A hazard scenario occurred does not necessarily cause a disastrous result. Once a hazard scenario is detected, real-time consequence estimation can support reliable decision making to reduce the severity level of consequence. The most common models for consequence estimation are Gaussian models, integrated models (PHAST, etc.), and CFD models. When an accident occurs, a real-time risk assessment must consider geographic and structural information to estimate the consequences. For this scenario, a CFD model is the only option because the influence of surface roughness is significant. However, hours of computation would be needed to get the detailed simulation results using CFD with a complex model configuration, which limits the application of CFD models in emergency response. In contrast, data driven methods are faster when applied online, which would be suitable for emergence response to estimate gas release, fire or explosion.

Wang, Chen, and Zhao (2015) developed a fast prediction approach integrated of gas detectors, artificial neural network (ANN) and gas dispersion model. The integration approach contains six steps, as shown in Fig. 14.

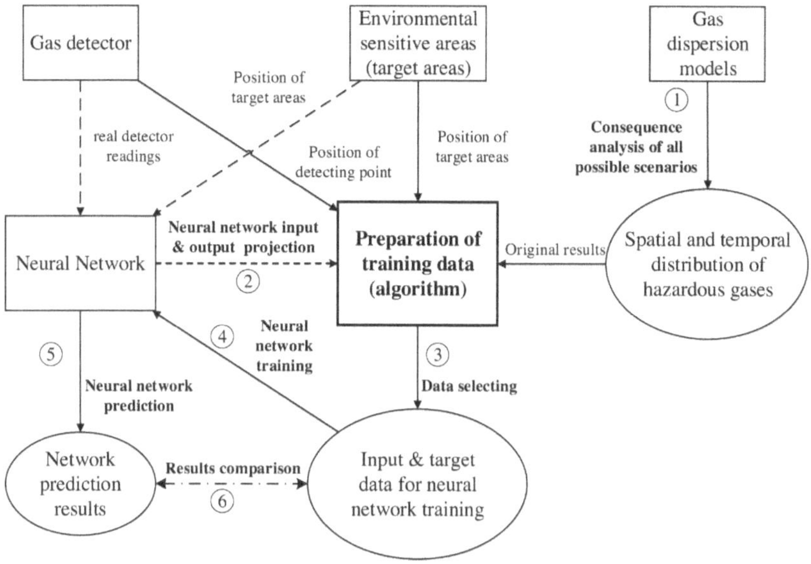

Fig. 14 The architecture of the integrated fast prediction model (Wang et al., 2015).

A CFD model would be used for consequence analysis of all possible scenarios to prepare training data for the fast prediction model. The structure of the prediction model is based on neural network whose input parameters are real-time data from gas detectors and weather stations. And the output parameters of the prediction model are the concentration of the hazardous gas in the specific target location and the time interval between the initial alarm and the time when the leaked gas reaches that target location.

In practice, chemicals released could be either in gas or liquid phase. Kim, Park, Kim, and Shin (2019) proposed a model that predicts the suspicious leak location(s) in real-time, using sensor data. Feed-forward neural network (FNN) and recurrent neural network (RNN) with long short-term memory (LSTM) that learned the data gathered from the installed sensors are proposed to predict the Top-5 possible points. Also, CFD simulations are used to train and verify the fast prediction model. Similarly, using AI methods and mechanism-based prediction model, data driven model could be established to fast estimate consequence of explosion (Li & Hao, 2018) and fire (Zhang et al., 2017; Zhang, Collin, Moireau, Trouvé, & Rochoux, 2019).

4.5 Risk management of human behaviors

Besides process and equipment error, accident caused by human behavior could happen in research, construction, operation, emergency management, transportation, and many other stages of a chemical process. Risk assessment and management of human behaviors should be taken into consideration seriously (Murphy & Paté-Cornell, 1996). However, human behaviors in chemical plant are personalized and independent, behaviors monitoring is an important factor of real-time risk management.

Wearing helmets is an effective and necessary safety precaution to reduce risks of head attack. However, due to discomfort, some workers would not always obey the regulations. It was reported that 47.3% victims did not have personal protective equipment or wear it incorrectly when danger arose. To ensure that workers follow the regulations and wear helmets on worksites, Wu and Zhao (2018b) developed an intelligent vision-based approach for helmet identification. The proposed approach can not only monitor whether workers are wearing helmets or not, but also be able to identify the helmet colors with high accuracy.

Workspace planning is a traditional management task, which is to align workers and groups with workspaces in advance to enable peak performance and ensure safety. Luo et al. (2019) described an initial effort on integrating the computer vision methods to implement automatic detection and visualization of dynamic workspaces of workers on foot. In this study, object detection, multiple object tracking, and action recognition are adopted to identify the actions of workers. These action recognition results are then visualized on the maps of the workspaces.

Accidental fall on work sites is a frequent accident and may cause workers injury and even more serious results. Because of unclear on-site status, supervisors cannot always respond appropriately to every emergency scenario where workers fall accidentally. Therefore, Tsai (2014) developed an image recognition system that enabled supervisors to rapidly recognize workers who are involved in accidental falls in work site. The system integrates accelerometer determination to monitoring the behavior of on-site workers and image recognition to highlight the falling people.

In construction, about 80–90% of accidents were caused by the unsafe actions and behaviors of employees. However, there is a lack of practical methods to measure workers' behaviors. Han, Lee, and Peña-Mora (2013) studied about ladder climbing, then developed a consequent motion-analysis technique to detect the unsafe actions of workers (i.e., backward-facing climbing, climbing with an object, and reaching far to a side). Similarly, there is an urgent need to study their unsafe actions using similar methods. In the future, we can develop an intelligent and automatic system for monitoring the safety of worker's actions. Furthermore, some machines and devices can also be monitored by computer vision applications. Chi and Caldas (2011) proposed an automated object identification for workers, loaders, and backhoes on construction sites. Then Kim and Chi (2019) proposed an action recognition application of earthmoving excavators based on sequential pattern analysis of visual features and operation cycles.

Besides on-site human behaviors, the performance of the control room operator is therefore critical to process safety. Eye tracking measures have the potential to monitor the human operator behavior in the process control room (Das, Iqbal, Bhavsar, Srinivasan, & Srinivasan, 2018). The gaze distribution of participants on different variables during abnormal situations is related to their performance in handling the situation (Sharma, Bhavsar, Srinivasan, & Srinivasan, 2016). Bhavsar, Srinivasan, and Srinivasan (2017) proposed gaze transition entropy to quantify the sequence of eye gaze transitions during the course of a task and dwell time entropy to quantifies the distribution of dwell time on the various AOIs on the HMI. Based on these two entropies, the insights into cognitive behavior of operators during process abnormalities could be extended.

5. Summary and perspective

In summary, risk assessment and management of chemical processes may be applied in the full life cycle. Chemical risk assessment, process hazard analysis, alarm management, FDD, substance leakage detection, fire detection, real-time consequence estimation and human behaviors monitoring could be used to assess and manage risks in different stages. As data driven methods need less prior knowledge and could deeply mine hidden information from large amounts of historical data, they have been widely used for risk assessment of chemical processes. Statistical methods and AI based methods are two most intensively studied data driven methods, introduced individually in this chapter. A framework of data driven methods for risk assessment in the full life cycle of chemical processes and the application have been discussed in detail. Also a real-time risk management framework is proposed in this chapter.

However, there is still a need to develop data driven methods further which can make deeper use of various data from IoTs. Some focus areas of future research in data driven methods for risk assessment and management are listed: (1) Various types of data can be gathered from IoTs in chemical industry, but most of the traditional data driven methods are still focus on only one data type individually. Thus advanced methods should be developed to deal with different types of data in a combined system. (2) Online data from production line of chemical plant always have uncertain quality problem, such as data missing, noise and drifting from actual value, therefore, data preprocessing based on the characteristics of chemical process should be considered to develop more robust data driven methods. (3) Various types of data are the foundation of risk assessment of chemical processes. However, due to the problems of confidentiality and differentiation between different companies, data sharing mechanisms are still missing. How to protect data privacy of different users while implementing data sharing for risk assessment with a shared database is an urgent problem to be studied.

References

Adedigba, S. A., Khan, F., & Yang, M. (2017). Dynamic failure analysis of process systems using principal component analysis and Bayesian network. *Industrial & Engineering Chemistry Research*, *56*(8), 2094–2106. https://doi.org/10.1021/acs.iecr.6b03356.

Afzal, M. S., Chen, T., Bandehkhoda, A., & Izadi, I. (2018). Analysis and design of time-deadbands for univariate alarm systems. *Control Engineering Practice*, *71*, 96–107. https://doi.org/10.1016/j.conengprac.2017.10.016.

Alauddin, M., Khan, F., Imtiaz, S., & Ahmed, S. (2018). A bibliometric review and analysis of data-driven fault detection and diagnosis methods for process systems. *Industrial & Engineering Chemistry Research*, *57*(32), 10719–10735. https://doi.org/10.1021/acs.iecr.8b00936.

Al-Dabbagh, A. W., Hu, W., Lai, S., Chen, T., & Shah, S. L. (2018). Toward the advancement of decision support tools for industrial facilities: Addressing operation metrics, visualization plots, and alarm floods. *IEEE Transactions on Automation Science and Engineering*, *15*(4), 1883–1896. https://doi.org/10.1109/TASE.2018.2827309.

Amin, M. T., Khan, F., Imtiaz, S., & Ahmed, S. (2019). Robust process monitoring methodology for detection and diagnosis of unobservable faults. *Industrial & Engineering Chemistry Research*, *58*(41), 19149–19165. https://doi.org/10.1021/acs.iecr.9b03406.

Ariamuthu Venkidasalapathy, J., Mannan, M. S., & Kravaris, C. (2018). A quantitative approach for optimal alarm identification. *Journal of Loss Prevention in the Process Industries*, *55*, 213–222. https://doi.org/10.1016/j.jlp.2018.05.012.

Bagheri, M., Akbari, A., & Mirbagheri, S. A. (2019). Advanced control of membrane fouling in filtration systems using artificial intelligence and machine learning techniques: A critical review. *Process Safety and Environmental Protection*, *123*, 229–252. https://doi.org/10.1016/j.psep.2019.01.013.

Bahl, A., Hellack, B., Balas, M., Dinischiotu, A., Wiemann, M., Brinkmann, J., et al. (2019). Recursive feature elimination in random forest classification supports nanomaterial grouping. *NanoImpact*, *15*, 100179. https://doi.org/10.1016/j.impact.2019.100179.

Bhavsar, P., Srinivasan, B., & Srinivasan, R. (2017). Quantifying situation awareness of control room operators using eye-gaze behavior. *Computers & Chemical Engineering*, *106*, 191–201. https://doi.org/10.1016/j.compchemeng.2017.06.004.

Cai, L., Gu, J., Ma, J., & Jin, Z. (2019). Probabilistic wind power forecasting approach via instance-based transfer learning embedded gradient boosting decision trees. *Energies*, *12*(1), 159. https://doi.org/10.3390/en12010159.

Carlos, M., Fatine, B., Nelly, O.-M., & Nadine, G. (2018). Deviation propagation analysis along a cumene process by using dynamic simulations. *Computers & Chemical Engineering*, *117*, 331–350. https://doi.org/10.1016/j.compchemeng.2018.06.010.

CCPS/AIChE. (2007). Process equipment reliability database. In *Guidelines for safe and reliable instrumented protective systems* (pp. 343–345). New Yotk, USA: John Wiley & Sons.

Chang, Y., Khan, F., & Ahmed, S. (2011). A risk-based approach to design warning system for processing facilities. *Process Safety and Environmental Protection*, *89*(5), 310–316. https://doi.org/10.1016/j.psep.2011.06.003.

Chen, K., & Wang, J. (2017). Design of multivariate alarm systems based on online calculation of variational directions. *Chemical Engineering Research and Design*, *122*, 11–21. https://doi.org/10.1016/j.cherd.2017.04.011.

Cheng, F., He, Q. P., & Zhao, J. (2019). A novel process monitoring approach based on variational recurrent autoencoder. *Computers & Chemical Engineering*, *129*, 106515. https://doi.org/10.1016/j.compchemeng.2019.106515.

Chi, S., & Caldas, C. H. (2011). Automated object identification using optical video cameras on construction sites. *Computer-Aided Civil and Infrastructure Engineering*, *26*(5), 368–380. https://doi.org/10.1111/j.1467-8667.2010.00690.x.

Cui, L., Shu, Y., Wang, Z., Zhao, J., Qiu, T., Sun, W., et al. (2012). HASILT: An intelligent software platform for HAZOP, LOPA, SRS and SIL verification. *Reliability Engineering & System Safety*, *108*, 56–64. https://doi.org/10.1016/j.ress.2012.06.014.

Dai, Y., & Qiu, Y. (2019). Risk matrix and event tree based half quantitative alarm priority analysis for alarm systems. *Process Integration and Optimization for Sustainability*, *3*(2), 159–166. https://doi.org/10.1007/s41660-017-0026-x.

Dai, Y., & Zhao, J. (2011). Fault diagnosis of batch chemical processes using a dynamic time warping (DTW)-based artificial immune system. *Industrial & Engineering Chemistry Research*, *50*(8), 4534–4544. https://doi.org/10.1021/ie101465b.

Dalapatu, P., Ahmed, S., & Khan, F. (2013). Alarm allocation for event-based process alarm systems. *IFAC Proceedings Volumes*, *46*(32), 815–820. https://doi.org/10.3182/20131218-3-IN-2045.00158.

Das, L., Iqbal, M. U., Bhavsar, P., Srinivasan, B., & Srinivasan, R. (2018). Toward preventing accidents in process industries by inferring the cognitive state of control room operators through eye tracking. *ACS Sustainable Chemistry & Engineering*, *6*(2), 2517–2528. https://doi.org/10.1021/acssuschemeng.7b03971.

exida. (2015). *Safety equipment reliability handbook* (4th ed.). Sellersville, USA: exdia.com L.L.C.

Fahimipirehgalin, M., Trunzer, E., Odenweller, M., & Vogel-Heuser, B. (2019). Automatic visual leakage inspection by using thermographic video and image analysis. In *Paper presented at the 2019 IEEE 15th international conference on automation science and engineering (CASE)*. 22–26 Aug. 2019.

Ge, Z. (2017). Review on data-driven modeling and monitoring for plant-wide industrial processes. *Chemometrics and Intelligent Laboratory Systems*, *171*, 16–25. https://doi.org/10.1016/j.chemolab.2017.09.021.

Ge, Z. (2018). Process data analytics via probabilistic latent variable models: A tutorial review. *Industrial & Engineering Chemistry Research*, *57*(38), 12646–12661. https://doi.org/10.1021/acs.iecr.8b02913.

Gharahbagheri, H., Imtiaz, S. A., & Khan, F. (2017). Root cause diagnosis of process fault using KPCA and Bayesian network. *Industrial & Engineering Chemistry Research*, *56*(8), 2054–2070. https://doi.org/10.1021/acs.iecr.6b01916.

Goel, P., Datta, A., & Mannan, M. S. (2017). Industrial alarm systems: Challenges and opportunities. *Journal of Loss Prevention in the Process Industries*, *50*, 23–36. https://doi.org/10.1016/j.jlp.2017.09.001.

Gupta, A. (2018). Introduction to deep learning: Part 1. *Chemical Engineering Progress, 114*(6), 22–29.
Håbrekke, S., Hauge, S., & Onshus, T. (2013). *PDS data handbook 2013: Reliability data for safety instrumented systems* (2013th ed.). Trondheim, Norway: SINTEF.
Hamaguchi, T., Takeda, K., Noda, M., & Kimura, N. (2012). A method of designing plant alarm systems with hierarchical cause-effect model. In I. A. Karimi & R. Srinivasan (Eds.), *Computer aided chemical engineering: Vol. 31.* (pp. 265–269). Elsevier.
Han, L., Gao, H., Xu, Y., & Zhu, Q. (2016). Combining FAP, MAP and correlation analysis for multivariate alarm thresholds optimization in industrial process. *Journal of Loss Prevention in the Process Industries, 40,* 471–478. https://doi.org/10.1016/j.jlp.2016.01.022.
Han, S., Lee, S., & Peña-Mora, F. (2013). Vision-based detection of unsafe actions of a construction worker: Case study of ladder climbing. *Journal of Computing in Civil Engineering, 27*(6), 635–644. https://doi.org/10.1061/(asce)cp.1943-5487.0000279.
Harrou, F., Nounou, M. N., Nounou, H. N., & Madakyaru, M. (2015). PLS-based EWMA fault detection strategy for process monitoring. *Journal of Loss Prevention in the Process Industries, 36,* 108–119. https://doi.org/10.1016/j.jlp.2015.05.017.
Hinton, G. E., & Salakhutdinov, R. R. (2006). Reducing the dimensionality of data with neural networks. *Science, 313*(5786), 504–507.
Hollifield, B., & Habibi, E. (2010). *The alarm management handbook: A comprehensive guide* (2nd ed.). Houston: PAS, Inc.
Hu, W., Al-Dabbagh, A. W., Chen, T., & Shah, S. L. (2018). Design of visualization plots of industrial alarm and event data for enhanced alarm management. *Control Engineering Practice, 79,* 50–64. https://doi.org/10.1016/j.conengprac.2018.07.005.
Hu, W., Chen, T., & Shah, S. L. (2017). An automated data-driven method to detect mode-based alarms. In *Paper presented at the 2017 American control conference (ACC).* 24–26 May 2017.
Hu, W., Chen, T., & Shah, S. L. (2018a). Detection of frequent alarm patterns in industrial alarm floods using Itemset mining methods. *IEEE Transactions on Industrial Electronics, 65*(9), 7290–7300. https://doi.org/10.1109/TIE.2018.2795573.
Hu, W., Chen, T., & Shah, S. L. (2018b). Discovering association rules of mode-dependent alarms from alarm and event logs. *IEEE Transactions on Control Systems Technology, 26*(3), 971–983. https://doi.org/10.1109/TCST.2017.2695169.
Hu, W., Shah, S. L., & Chen, T. (2018). Framework for a smart data analytics platform towards process monitoring and alarm management. *Computers & Chemical Engineering, 114,* 225–244. https://doi.org/10.1016/j.compchemeng.2017.10.010.
ISO. (2009). *ISO31000:2009 risk management—Principles and guidelines.* Geneva: International Organization for Standardization.
Jiang, Z., Rozgic, V., & Adali, S. (2017). Learning spatiotemporal features for infrared action recognition with 3D convolutional neural networks. In *Paper presented at the 2017 IEEE conference on computer vision and pattern recognition workshops (CVPRW).* 21–26 July 2017.
Jiang, Q., Yan, X., & Huang, B. (2019). Review and perspectives of data-driven distributed monitoring for industrial plant-wide processes. *Industrial & Engineering Chemistry Research, 58*(29), 12899–12912. https://doi.org/10.1021/acs.iecr.9b02391.
Kabir, S., & Papadopoulos, Y. (2019). Applications of Bayesian networks and petri nets in safety, reliability, and risk assessments: A review. *Safety Science, 115,* 154–175. https://doi.org/10.1016/j.ssci.2019.02.009.
Kim, J., & Chi, S. (2019). Action recognition of earthmoving excavators based on sequential pattern analysis of visual features and operation cycles. *Automation in Construction, 104,* 255–264. https://doi.org/10.1016/j.autcon.2019.03.025.

Kim, H., Park, M., Kim, C. W., & Shin, D. (2019). Source localization for hazardous material release in an outdoor chemical plant via a combination of LSTM-RNN and CFD simulation. *Computers & Chemical Engineering, 125*, 476–489. https://doi.org/10.1016/j.compchemeng.2019.03.012.

Kingma, D. P., & Welling, M. (2013). Auto-encoding variational bayes. *arXiv*. preprint arXiv:1312.6114, 1–14.

Krzanowski, W. J. (1992). A User's guide to principal components. *Journal of the Royal Statistical Society: Series A (Statistics in Society), 155*(1), 173–174. https://doi.org/10.2307/2982678.

Ku, W., Storer, R. H., & Georgakis, C. (1995). Disturbance detection and isolation by dynamic principal component analysis. *Chemometrics and Intelligent Laboratory Systems, 30*(1), 179–196. https://doi.org/10.1016/0169-7439(95)00076-3.

Lai, S., & Chen, T. (2017). A method for pattern mining in multiple alarm flood sequences. *Chemical Engineering Research and Design, 117*, 831–839. https://doi.org/10.1016/j.cherd.2015.06.019.

Lai, S., Yang, F., Chen, T., & Cao, L. (2019). Accelerated multiple alarm flood sequence alignment for abnormality pattern mining. *Journal of Process Control, 82*, 44–57. https://doi.org/10.1016/j.jprocont.2019.06.004.

Lee, G., Han, C., & Yoon, E. S. (2004). Multiple-fault diagnosis of the Tennessee Eastman process based on system decomposition and dynamic PLS. *Industrial & Engineering Chemistry Research, 43*(25), 8037–8048. https://doi.org/10.1021/ie049624u.

Lee, J.-M., Qin, S. J., & Lee, I.-B. (2007). Fault detection of non-linear processes using kernel independent component analysis. *The Canadian Journal of Chemical Engineering, 85*(4), 526–536. https://doi.org/10.1002/cjce.5450850414.

Li, J., & Hao, H. (2018). Far-field pressure prediction of a vented gas explosion from storage tanks by using new CFD simulation guidance. *Process Safety and Environmental Protection, 119*, 360–378. https://doi.org/10.1016/j.psep.2018.08.004.

Li, R. F., & Wang, X. Z. (2002). Dimension reduction of process dynamic trends using independent component analysis. *Computers & Chemical Engineering, 26*(3), 467–473. https://doi.org/10.1016/S0098-1354(01)00773-6.

Liu, H., Yang, C., Carlsson, B., Qin, S. J., & Yoo, C. (2019). Dynamic nonlinear partial least squares modeling using Gaussian process regression. *Industrial & Engineering Chemistry Research, 58*(36), 16676–16686. https://doi.org/10.1021/acs.iecr.9b00701.

Lu, Y., Ng, D., & Mannan, M. S. (2011). Prediction of the reactivity hazards for organic peroxides using the QSPR approach. *Industrial & Engineering Chemistry Research, 50*(3), 1515–1522. https://doi.org/10.1021/ie100833m.

Luo, X., Li, H., Wang, H., Wu, Z., Dai, F., & Cao, D. (2019). Vision-based detection and visualization of dynamic workspaces. *Automation in Construction, 104*, 1–13. https://doi.org/10.1016/j.autcon.2019.04.001.

Meel, A., & Seider, W. D. (2006). Plant-specific dynamic failure assessment using Bayesian theory. *Chemical Engineering Science, 61*(21), 7036–7056. https://doi.org/10.1016/j.ces.2006.07.007.

Meel, A., & Seider, W. D. (2008). Real-time risk analysis of safety systems. *Computers & Chemical Engineering, 32*(4), 827–840. https://doi.org/10.1016/j.compchemeng.2007.03.006.

Ming, L., & Zhao, J. (2017). Review on chemical process fault detection and diagnosis. In *Paper presented at the 2017 6th international symposium on advanced control of industrial processes (AdCONIP)*. 28–31 May 2017.

Ming, L., & Zhao, J. (2018). Feature selection for chemical process fault diagnosis by artificial immune systems. *Chinese Journal of Chemical Engineering, 26*(8), 1599–1604. https://doi.org/10.1016/j.cjche.2017.09.023.

Murphy, D. M., & Paté-Cornell, M. E. (1996). The SAM framework: Modeling the effects of management factors on human behavior in risk analysis. *Risk Analysis*, *16*(4), 501–515. https://doi.org/10.1111/j.1539-6924.1996.tb01096.x.

Necci, A., Antonioni, G., Bonvicini, S., & Cozzani, V. (2016). Quantitative assessment of risk due to major accidents triggered by lightning. *Reliability Engineering & System Safety*, *154*, 60–72. https://doi.org/10.1016/j.ress.2016.05.009.

Nomikos, P., & MacGregor, J. F. (1994). Monitoring batch processes using multiway principal component analysis. *AIChE Journal*, *40*(8), 1361–1375. https://doi.org/10.1002/aic.690400809.

Nomikos, P., & MacGregor, J. F. (1995). Multi-way partial least squares in monitoring batch processes. *Chemometrics and Intelligent Laboratory Systems*, *30*(1), 97–108. https://doi.org/10.1016/0169-7439(95)00043-7.

Paltrinieri, N., Khan, F., Amyotte, P., & Cozzani, V. (2014). Dynamic approach to risk management: Application to the Hoeganaes metal dust accidents. *Process Safety and Environmental Protection*, *92*(6), 669–679. https://doi.org/10.1016/j.psep.2013.11.008.

Pan, Y., Qi, R., He, P., Shen, R., Jiang, J., Ni, L., et al. (2019). Thermal hazard assessment and ranking for organic peroxides using quantitative structure–property relationship approaches. *Journal of Thermal Analysis and Calorimetry*. https://doi.org/10.1007/s10973-019-08966-3.

Pease, C. K., & Gentry, R. P. (2016). Systematic review in chemical risk assessment—A chemical industry perspective. *Environment International*, *92–93*, 574–577. https://doi.org/10.1016/j.envint.2015.12.007.

Peng, D., Gu, X., Xu, Y., & Zhu, Q. (2015). Integrating probabilistic signed digraph and reliability analysis for alarm signal optimization in chemical plant. *Journal of Loss Prevention in the Process Industries*, *33*, 279–288. https://doi.org/10.1016/j.jlp.2015.01.005.

Ramesh, T. S., Shum, S. K., & Davis, J. F. (1988). A structured framework for efficient problem solving in diagnostic expert systems. *Computers & Chemical Engineering*, *12*(9), 891–902. https://doi.org/10.1016/0098-1354(88)87016-9.

Rich, E. (1985). Artificial intelligence and the humanities. *Computers and the Humanities*, *19*, 117–122. https://doi.org/10.1007/BF02259633.

Rich, S. H., & Venkatasubramanian, V. (1987). Model-based reasoning in diagnostic expert systems for chemical process plants. *Computers & Chemical Engineering*, *11*(2), 111–122. https://doi.org/10.1016/0098-1354(87)80012-1.

Roy, A., Srivastava, P., & Sinha, S. (2014). Risk and reliability assessment in chemical process industries using Bayesian methods. *Reviews in Chemical Engineering*, *30*, 479.

Seider, W. D., Pariyani, A., Oktem, U. G., Moskowitz, I., Arbogast, J. E., & Soroush, M. (2017). Chapter five—Introduction to dynamic risk analyses. In F. Khan (Ed.), *Methods in chemical process safety: Vol. 1*. (pp. 201–254). Elsevier.

Seike, M., Kawabata, N., & Hasegawa, M. (2017). Quantitative assessment method for road tunnel fire safety: Development of an evacuation simulation method using CFD-derived smoke behavior. *Safety Science*, *94*, 116–127. https://doi.org/10.1016/j.ssci.2017.01.005.

Sharma, C., Bhavsar, P., Srinivasan, B., & Srinivasan, R. (2016). Eye gaze movement studies of control room operators: A novel approach to improve process safety. *Computers & Chemical Engineering*, *85*, 43–57. https://doi.org/10.1016/j.compchemeng.2015.09.012.

Shu, Y., Ming, L., Cheng, F., Zhang, Z., & Zhao, J. (2016). Abnormal situation management: Challenges and opportunities in the big data era. *Computers & Chemical Engineering*, *91*, 104–113. https://doi.org/10.1016/j.compchemeng.2016.04.011.

Shu, Y., & Zhao, J. (2016). Fault diagnosis of chemical processes using artificial immune system with vaccine transplant. *Industrial & Engineering Chemistry Research*, *55*(12), 3360–3371. https://doi.org/10.1021/acs.iecr.5b02646.

Simonyan, K., & Zisserman, A. (2014). Two-stream convolutional networks for action recognition in videos. *arXiv*. preprint arXiv:1406.2199, 1–11.

SINTEF, & NTNU. (2015). *OREDA handbook 2015: Offshore and onshore reliability data*. Trondheim, Norway: DNV—Det Norske Veritas.

Song, B., & Suh, Y. (2019). Narrative texts-based anomaly detection using accident report documents: The case of chemical process safety. *Journal of Loss Prevention in the Process Industries*, *57*, 47–54. https://doi.org/10.1016/j.jlp.2018.08.010.

Stefatos, G., & Ben Hamza, A. (2010). Dynamic independent component analysis approach for fault detection and diagnosis. *Expert Systems with Applications*, *37*(12), 8606–8617. https://doi.org/10.1016/j.eswa.2010.06.101.

Su, J., Guo, C., Zang, H., Yang, F., Huang, D., Gao, X., et al. (2018). A multi-setpoint delay-timer alarming strategy for industrial alarm monitoring. *Journal of Loss Prevention in the Process Industries*, *54*, 1–9. https://doi.org/10.1016/j.jlp.2018.02.004.

Takeda, K., Hamaguchi, T., Kimura, N., & Noda, M. (2014). A design method of a plant alarm system for first alarm alternative signals using a modularized CE model. *Process Safety and Environmental Protection*, *92*(5), 406–411. https://doi.org/10.1016/j.psep.2014.06.007.

Tan, Y. (2016). Artificial immune system. In Y. Tan (Ed.), *Artificial immune system applications in computersecurity* (pp. 1–25). Hoboken, NJ: John Wiley & Sons, Inc. https://doi.org/10.1002/9781119076582.ch1.

Tian, W., Zhang, G., Zhang, X., & Dong, Y. (2018). PCA weight and Johnson transformation based alarm threshold optimization in chemical processes. *Chinese Journal of Chemical Engineering*, *26*(8), 1653–1661. https://doi.org/10.1016/j.cjche.2017.10.027.

Tsai, M.-K. (2014). Automatically determining accidental falls in field surveying: A case study of integrating accelerometer determination and image recognition. *Safety Science*, *66*, 19–26. https://doi.org/10.1016/j.ssci.2014.01.012.

Venkatasubramanian, V. (2019). The promise of artificial intelligence in chemical engineering: Is it here, finally. *AIChE Journal*, *65*(2), 466–478.

Venkatasubramanian, V., Rengaswamy, R., Kavuri, S. N., & Yin, K. (2003). A review of process fault detection and diagnosis: Part III: Process history based methods. *Computers & Chemical Engineering*, *27*(3), 327–346. https://doi.org/10.1016/S0098-1354(02)00162-X.

Venkatasubramanian, V., Rengaswamy, R., Yin, K., & Kavuri, S. N. (2003). A review of process fault detection and diagnosis: Part I: Quantitative model-based methods. *Computers & Chemical Engineering*, *27*(3), 293–311. https://doi.org/10.1016/S0098-1354(02)00160-6.

Villa, V., Paltrinieri, N., Khan, F., & Cozzani, V. (2016). Towards dynamic risk analysis: A review of the risk assessment approach and its limitations in the chemical process industry. *Safety Science*, *89*, 77–93. https://doi.org/10.1016/j.ssci.2016.06.002.

Wang, Z., Bai, X., Wang, J., & Yang, Z. (2019). Indexing and designing deadbands for industrial alarm signals. *IEEE Transactions on Industrial Electronics*, *66*(10), 8093–8103. https://doi.org/10.1109/TIE.2018.2885718.

Wang, B., Chen, B., & Zhao, J. (2015). The real-time estimation of hazardous gas dispersion by the integration of gas detectors, neural network and gas dispersion models. *Journal of Hazardous Materials*, *300*, 433–442. https://doi.org/10.1016/j.jhazmat.2015.07.028.

Wang, H., Khan, F., & Abimbola, M. (2018). A new method to study the performance of safety alarm system in process operations. *Journal of Loss Prevention in the Process Industries*, *56*, 104–118. https://doi.org/10.1016/j.jlp.2018.08.017.

Wang, H., Khan, F., Ahmed, S., & Imtiaz, S. (2016). Dynamic quantitative operational risk assessment of chemical processes. *Chemical Engineering Science*, *142*, 62–78. https://doi.org/10.1016/j.ces.2015.11.034.

Wang, Y., Si, Y., Huang, B., & Lou, Z. (2018). Survey on the theoretical research and engineering applications of multivariate statistics process monitoring algorithms: 2008–2017. *The Canadian Journal of Chemical Engineering*, *96*(10), 2073–2085. https://doi.org/10.1002/cjce.23249.

Wang, Z., Su, Y., Shen, W., Jin, S., Clark, J. H., Ren, J., et al. (2019). Predictive deep learning models for environmental properties: The direct calculation of octanol–water partition coefficients from molecular graphs. *Green Chemistry*, *21*(16), 4555–4565. https://doi.org/10.1039/c9gc01968e.

Wang, M., Yan, G., & Fei, Z. (2015). Kernel PLS based prediction model construction and simulation on theoretical cases. *Neurocomputing*, *165*, 389–394. https://doi.org/10.1016/j.neucom.2015.03.028.

Wang, J., Yang, F., Chen, T., & Shah, S. L. (2016). An overview of industrial alarm systems: Main causes for alarm overloading, research status, and open problems. *IEEE Transactions on Automation Science and Engineering*, *13*(2), 1045–1061. https://doi.org/10.1109/tase.2015.2464234.

Wang, B., Zhou, L., Xu, K., & Wang, Q. (2017). Prediction of minimum ignition energy from molecular structure using quantitative structure–property relationship (QSPR) models. *Industrial & Engineering Chemistry Research*, *56*(1), 47–51. https://doi.org/10.1021/acs.iecr.6b04347.

Wu, H., Peñarrubia, I., Cui, L., & Zhao, J. (2017). Process safety management considerations for biofuel production. *Frontiers of Engineering Management*, *4*(3), 357–367. https://doi.org/10.15302/j-fem-2017025.

Wu, H., Wu, D., & Zhao, J. (2019). An intelligent fire detection approach through cameras based on computer vision methods. *Process Safety and Environmental Protection*, *127*, 245–256. https://doi.org/10.1016/j.psep.2019.05.016.

Wu, H., & Zhao, J. (2018a). Deep convolutional neural network model based chemical process fault diagnosis. *Computers & Chemical Engineering*, *115*, 185–197.

Wu, H., & Zhao, J. (2018b). An intelligent vision-based approach for helmet identification for work safety. *Computers in Industry*, *100*, 267–277. https://doi.org/10.1016/j.compind.2018.03.037.

Yang, F., Xiao, D., & Shah, S. L. (2009). Optimal sensor location design for reliable fault detection in presence of false alarms. *Sensors*, *9*(11), 8579–8592.

Yu, J. (2012). A nonlinear kernel Gaussian mixture model based inferential monitoring approach for fault detection and diagnosis of chemical processes. *Chemical Engineering Science*, *68*(1), 506–519. https://doi.org/10.1016/j.ces.2011.10.011.

Yu, J., & Qin, S. J. (2008). Multimode process monitoring with Bayesian inference-based finite Gaussian mixture models. *AIChE Journal*, *54*(7), 1811–1829. https://doi.org/10.1002/aic.11515.

Yu, Y., Wang, J., & Ouyang, Z. (2020). Designing dynamic alarm limits and adjusting manipulated variables for multivariate systems. *IEEE Transactions on Industrial Electronics*, *67*(3), 2314–2325. https://doi.org/10.1109/TIE.2019.2903757.

Yu, Y., Wang, J., & Yang, Z. (2017). Design of Alarm Trippoints for univariate analog process variables based on alarm probability plots. *IEEE Transactions on Industrial Electronics*, *64*(8), 6496–6505. https://doi.org/10.1109/TIE.2017.2682783.

Yuan, F., Zhang, L., Xia, X., Wan, B., Huang, Q., & Li, X. (2019). Deep smoke segmentation. *Neurocomputing*, *357*, 248–260. https://doi.org/10.1016/j.neucom.2019.05.011.

Zadakbar, O., Imtiaz, S., & Khan, F. (2012). Dynamic risk assessment and fault detection using principal component analysis. *Industrial & Engineering Chemistry Research*, *52*(2), 809–816. https://doi.org/10.1021/ie202880w.

Zeng, Z., & Zio, E. (2018). Dynamic risk assessment based on statistical failure data and condition-monitoring degradation data. *IEEE Transactions on Reliability*, *67*(2), 609–622. https://doi.org/10.1109/TR.2017.2778804.

Zhang, C., Collin, A., Moireau, P., Trouvé, A., & Rochoux, M. C. (2019). Front shape similarity measure for data-driven simulations of wildland fire spread based on state estimation: Application to the RxCADRE field-scale experiment. *Proceedings of the Combustion Institute*, *37*(3), 4201–4209. https://doi.org/10.1016/j.proci.2018.07.112.
Zhang, B., Liu, Y., & Qiao, S. (2019). A quantitative individual risk assessment method in process facilities with toxic gas release hazards: A combined scenario set and CFD approach. *Process Safety Progress*, *38*(1), 52–60. https://doi.org/10.1002/prs.11979.
Zhang, C., Rochoux, M., Tang, W., Gollner, M., Filippi, J.-B., & Trouvé, A. (2017). Evaluation of a data-driven wildland fire spread forecast model with spatially-distributed parameter estimation in simulations of the FireFlux I field-scale experiment. *Fire Safety Journal*, *91*, 758–767. https://doi.org/10.1016/j.firesaf.2017.03.057.
Zhang, Z., & Zhao, J. (2017). A deep belief network based fault diagnosis model for complex chemical processes. *Computers & Chemical Engineering*, *107*, 395–407. https://doi.org/10.1016/j.compchemeng.2017.02.041.
Zhao, C., Bhushan, M., & Venkatasubramanian, V. (2005). PHASuite: An automated HAZOP analysis tool for chemical processes: Part I: Knowledge engineering framework. *Process Safety and Environmental Protection*, *83*(6), 509–532. https://doi.org/10.1205/psep.04055.
Zhao, J., Cui, L., Zhao, L., Qiu, T., & Chen, B. (2009). Learning HAZOP expert system by case-based reasoning and ontology. *Computers & Chemical Engineering*, *33*(1), 371–378. https://doi.org/10.1016/j.compchemeng.2008.10.006.
Zhao, C., & Huang, B. (2018). A full-condition monitoring method for nonstationary dynamic chemical processes with cointegration and slow feature analysis. *AIChE Journal*, *64*(5), 1662–1681. https://doi.org/10.1002/aic.16048.
Zhao, J., Shu, Y., Zhu, J., & Dai, Y. (2014). An online fault diagnosis strategy for full operating cycles of chemical processes. *Industrial & Engineering Chemistry Research*, *53*(13), 5015–5027. https://doi.org/10.1021/ie400660e.
Zhou, B., Hu, W., & Chen, T. (2019). A new method for alarm monitoring of equipment start-up operations with applications to pumps. *Industrial & Engineering Chemistry Research*, *58*(26), 11251–11260. https://doi.org/10.1021/acs.iecr.8b05940.
Zhou, D., Li, G., & Qin, S. J. (2010). Total projection to latent structures for process monitoring. *AIChE Journal*, *56*(1), 168–178. https://doi.org/10.1002/aic.11977.
Zhou, Y., Li, S., & Xiong, N. (2019). Improved vine copula-based dependence description for multivariate process monitoring based on ensemble learning. *Industrial & Engineering Chemistry Research*, *58*(9), 3782–3796. https://doi.org/10.1021/acs.iecr.8b04081.
Zhou, L., Wang, B., Jiang, J., Pan, Y., & Wang, Q. (2017). Quantitative structure-property relationship (QSPR) study for predicting gas-liquid critical temperatures of organic compounds. *Thermochimica Acta*, *655*, 112–116. https://doi.org/10.1016/j.tca.2017.06.021.
Zhu, J., & Gao, F. (2018). Improved nonlinear quality estimation for multiphase batch processes based on relevance vector machine with neighborhood component variable selection. *Industrial & Engineering Chemistry Research*, *57*(2), 666–676. https://doi.org/10.1021/acs.iecr.7b03590.
Zhu, J., Shu, Y., Zhao, J., & Yang, F. (2014). A dynamic alarm management strategy for chemical process transitions. *Journal of Loss Prevention in the Process Industries*, *30*, 207–218. https://doi.org/10.1016/j.jlp.2013.07.008.
Zhu, J., Wang, C., Li, C., Gao, X., & Zhao, J. (2016). Dynamic alarm prediction for critical alarms using a probabilistic model. *Chinese Journal of Chemical Engineering*, *24*(7), 881–885. https://doi.org/10.1016/j.cjche.2016.04.017.
Zhu, J., Zhao, J., & Yang, F. (2014). Dynamic risk analysis with alarm data to improve process safety using Bayesian network. In *Paper presented at the proceeding of the 11th world congress on intelligent control and automation*. 29 June–4 July 2014.
Zio, E. (2018). The future of risk assessment. *Reliability Engineering & System Safety*, *177*, 176–190. https://doi.org/10.1016/j.ress.2018.04.020.

CHAPTER SIX

Rare event risk assessments

Terje Aven*
University of Stavanger, Stavanger, Norway
*Corresponding author: e-mail address: terje.aven@uis.no

Contents

1. Introduction: What is a rare event?

This chapter is about rare event risk assessments. A framework for the proper understanding, characterization, assessment and management of risk for such events will be presented and discussed. Section 2

Methods in Chemical Process Safety, Volume 4
ISSN 2468-6514
https://doi.org/10.1016/bs.mcps.2020.02.003

covers the definitions and interpretations of key concepts, like risk, uncertainties and probability, whereas Section 3 presents and discusses key features of the framework related to risk assessment and management. The framework is based on recent work conducted by the Society for Risk Analysis (SRA), documented in SRA (2015a, 2015b, 2017a, 2017b), as well as recent developments in risk science, including Aven (2014, 2016a, 2019a). Section 4 will provide some final remarks and conclusions.

The chapter addresses rare events, but what is a rare event? If we consult dictionaries, the answer is that the term refers to an event which does not occur often or whose frequency is low. Normally the focus is on events which have a potential for extreme consequences or impact. That is why these events are of special interest. It is also common to refer to "low probability–high consequence" types of events, basically reflecting the same idea.

In a process plant, an event leading to total loss of the plant is a rare event. There is a potential for such an event to occur, there are uncertainties, but the probability is low. Otherwise, it would be a poorly designed and operated plant, with a clearly unacceptable associated risk. However, if we consider all process plants in the world, and let us say for a period of 10 years, the probability of an event leading to total loss will be judged rather high. We can speak about frequency—some total loss events may occur, and the adequacy of the expression "rare event" can be questioned.

Referring to a "low probability–high consequence" type of event puts the focus on the possible occurrence of an extreme event, which is considered the scope of the present discussion; in the following, we use this understanding of the term "rare event." However, a comment is appropriate when it comes to the interpretation of probability. As we know, probability can have many different meanings, but for the present discussion it is sufficient to consider "frequentist probabilities" and knowledge-based (subjective) probabilities; see definitions in Section 2. In the present chapter, when not otherwise stated, probability refers to knowledge-based probability, expressing the uncertainty and degree of belief of the assessor. It is not a property of the process plant but a judgment made by the assessor on an aspect of the plant's performance. However, the term "low probability–high consequence" can also be understood in a frequentist context, but then the existence of this probability needs to be justified and estimated, as it is unknown. We will return to this issue in coming sections.

Assessing the risk of "low probability–high consequence" activities means that statistics cannot be the basis for the analysis, as few or no relevant data would be available. We are led to assessments based on modeling of the activity or system studied. Depending on knowledge of the activity (system), such risk assessments would have different forms and be given different weight in the risk management. Clearly, if we face a situation with large uncertainties about the phenomena studied, it will be difficult to build accurate models, and robust/resilience-based strategies would be needed to supplement the information obtained from the risk assessment. Section 3 will look more closely into this type of issue.

2. The framework—Definitions of key concepts

This section summarizes definitions of key concepts, starting with risk, uncertainty, probability and vulnerability. The concept of risk is highlighted, as it provides a central pillar for the framework. If we are to discuss risk assessment, we need to understand what risk is. Comments are provided for some of the terms. The definitions are in line with the SRA (2015a) glossary and the fundamental principles document (SRA, 2017b). Note that the glossary allows for alternative definitions.

2.1 Risk

In its most general form, risk can be seen as the two-dimensional combination of the consequences, C, of the activity considered (with respect to something that humans value) and associated uncertainties, U, about C.

The setting is this: We consider a future activity [interpreted in a wide sense to also cover, for example, natural phenomena], for example, the operation of a process plant, and define risk in relation to the consequences (effects, implications) of this activity with respect to something that humans value. The consequences are often seen in relation to some reference values (planned values, objectives, etc.), and the focus is often on negative, undesirable consequences. There is always at least one outcome that is considered negative or undesirable.

To illustrate, consider a tank with hydrocarbons in a process plant. This tank represents a risk source or hazard for people working at the plant. Hazardous events may occur, leading to explosions/fires, with loss of lives and injuries. These are the consequences—"nothing" happens, explosion/fire scenarios occur, with or without losses. Looking to the future, we do not know what events will occur and what the consequences will be; we face uncertainties and hence risk.

Often the consequences explicitly refer to events, A, having some effects. To highlight A, risk is denoted (A,C,U). Hence the term "consequences" is used for all effects of the activity considered but also for the effects given the occurrence of A.

Risk description: A qualitative and/or quantitative picture of the risk, i.e., a structured statement of risk usually containing the elements: risk sources, causes, events, consequences and uncertainty representations/measurements. Formally and in general terms, we write

$$\text{Risk description} = (C', Q, K)$$

where C' is the specified consequences of the activity considered, Q the measure (description) of the uncertainty, and K the background knowledge that Q is based on. Using the notation (A,C,U), we derive a general risk description (A',C',Q',K).

For Q, it is recommended to use probability (or imprecise probability), with associated judgments of the strength of the knowledge (SoK) supporting the probability judgments.

An example of a risk description would be a risk matrix for a set of events A', with axes capturing consequences C', probability P and a judgment of the strength of knowledge SoK supporting P. Another example would be a probability distribution of the number of fatalities as a result of accidents in a specific period of time, with an associated SoK judgment.

2.2 Uncertainty

Overall qualitative definitions:

- For a person or a group of persons, not knowing the true value of a quantity or the future consequences of an activity
- Imperfect or incomplete information/knowledge about a hypothesis, a quantity, or the occurrence of an event

Uncertainty metrics/descriptions (examples):

- A knowledge-based (subjective) probability
- The pair (Q,K), where Q is a measure (description) of uncertainty and K the background knowledge that supports Q

Epistemic uncertainty: As above for the overall qualitative definition of uncertainty and uncertainty metrics/descriptions (examples).

Aleatory (stochastic) uncertainty: Variation of quantities in a population of units (commonly represented/described by a probability model).

2.3 Probability

Frequentist probability of an event A: the limiting fraction of times A would occur if the situation considered were repeated (hypothetically) an infinite number of times under similar conditions.

Probability of an event A: the number, such that the uncertainty about (degree of belief in) the occurrence of A is considered equivalent, by the person assigning the probability, to the uncertainty about (degree of belief in) some standard event, for example, drawing at random a red ball from an urn that contains $P(A) \times 100\%$ red balls. Hence, a probability of 0.1 means that the assigner considers the uncertainty (degree of belief) of the event occurring to be the same as randomly drawing one particular ball out of an urn comprising 10 balls. The probability is conditional on the knowledge, K, of the assessor. We write $P(A|K)$.

Imprecise probability of an event A: A probability interval [a,b], expressing the assigner's unwillingness to be more precise than the interval specified, given his/her knowledge K. If, for example, an interval [0.1, 0.2] is specified, it means that the assigner's degree of belief that A will occur is higher than or equal to the degree of belief in randomly drawing one particular ball out of an urn comprising 10 balls and less than or equal to the degree of belief in randomly drawing one particular ball out of an urn comprising five balls. Alternatively, it means that the assigner has the same uncertainty or degree of belief as randomly drawing a red ball out of an urn comprising 100 balls, of which 10, 11, ..., or 20 are red. Given the knowledge K, he/she is not willing to be more precise.

2.4 Vulnerability

Overall qualitative definitions:

- The degree to which a system is affected by a risk source/event
- The degree to which a system is able to withstand specific loads
- Vulnerability is risk conditional on the occurrence of a risk source/agent

Using the above (A,C,U) risk notation, vulnerability can formally be expressed by (C,U|A).

Vulnerability metrics/descriptions (examples):

As for risk but conditional on the risk source or event (load). In it most general form, the vulnerability can be written

- (C′,Q,K | risk source/event A) (i.e., a risk description given the occurrence of a risk source or event).

A vulnerable system is a system whose level of vulnerability is judged to be high. A probability distribution for the number of fatalities, given a specific fire scenario, is an example of a concrete vulnerability metric. As for risk, a strength of knowledge (SoK) judgment should always be included.

Using the risk notation (A,C,U), we can formally write

$$\text{Risk} = (\text{A}, \text{U}) + (\text{C}, \text{U}|\text{A}) = \text{“event contribution”and vulnerability.}$$

The symbol "+" is not interpreted as a sum, as in mathematics, but as a symbol for combining the two elements. In the same way, we can write

$$\text{Risk description} = (\text{A}', \text{Q}, \text{K}) + (\text{C}', \text{Q}, \text{K}|\text{A}')$$

which expresses that risk is described by an "event risk contribution" and the vulnerability given the occurrence of the event A′.

2.5 Complex

- A system is complex if it is not possible to establish an accurate prediction model of the system based on knowing the specific functions and states of its individual components.

2.6 Event, consequences

Event:

- the occurrence or change of a particular set of circumstances, such as a system failure, an earthquake, an explosion or an outbreak of a pandemic
- a specified change in the state of the world/affairs

Consequences: The effects of the activity, with respect to the values defined (such as human life and health, the environment and economic assets), covering the totality of states, events, barriers and outcomes. The consequences are often seen in relation to some reference values (planned values, objectives, etc.), and the focus is often on negative, undesirable consequences. See also a comment in relation to the concept of risk.

2.7 Hazard

A risk source where the potential consequences relate to harm. Hazards could, for example, be associated with energy (e.g., explosion, fire), material (toxic or eco-toxic), biota (pathogens) and information (panic communication).

2.8 Knowledge

Two types of knowledge:
Know-how (skill) and know-that of propositional knowledge (justified beliefs).
Knowledge is gained through, for example, scientific methodology and peer review, experience and testing.

2.9 Model

A model of an object (e.g., activity, system) is a simplified representation of this object.

A probability model is a special type of model, based on frequentist probabilities (often referred to as chances in a Bayesian context).

2.10 Resilience

Overall qualitative definitions:
Resilience is the ability of the system to sustain or restore its basic functionality following a risk source or an event (even unknown).
Resilience metrics/descriptions (example):

- The probability that the system is able to sustain operation when exposed to some types of risk sources or events (which can be more or less accurately defined), with associated strength of knowledge judgments (SoK).

A resilient system is a system for which the resilience is judged to be high.

Using the above (A,C,U) notation introduced for risk, it is clear that (C,U|A)—the consequences and associated uncertainties given an event—captures an important aspect of risk (Aven, 2019a, 2019b). We may per definition refer to (C,U|A) as resilience; alternatively, it can be phrased as the "resilience-induced conditional risk" or "lack of resilience-induced conditional risk," given the occurrence of A (Aven & Thekdi, 2018). Alternatively, we could refer to (C,U|A) as vulnerability, as suggested above when introducing vulnerability. Some authors prefer to restrict the concept of vulnerability to situations where the event A is known. In the above set-up, both known and unknown types of events A are allowed.

2.11 Risk source or risk agent

An element (action, sub-activity, component, system, event, ...), which alone or in combination with other elements has the potential to give rise to some specified consequences (typically, undesirable consequences).

2.12 Robustness

The antonym of vulnerability.

2.13 Safe, safety

Safe: Without unacceptable risk.
Safety:

- Interpreted in the same way as safe (for example, when saying that safety is achieved)
- The antonym of risk (the safety level is linked to the risk level; a high safety level means a low risk level and vice versa)

Sometimes limited to risk related to non-intentional events (including accidents and continuous exposures).

2.14 Security, secure

Secure: Without unacceptable risk when restricting the concept of risk to intentional acts by intelligent actors.
Security:

- Interpreted in the same way as secure (for example, when saying that security is achieved)
- The antonym of risk when restricting the concept of risk to intentional acts by intelligent actors (the security level is linked to the risk level; a high security level means a low risk level and vice versa)

2.15 Threat

Risk source, commonly used in relation to security applications (but also in relation to other applications, for example, the threat of an earthquake).

Threat in relation to an attack: A stated or inferred intention to initiate an attack with the intention to inflict harm, fear, pain or misery.

2.16 Other related definitions

2.16.1 *Managerial review and judgment*

Process of summarizing, interpreting and deliberating over the results of risk assessments and other assessments, as well as of other relevant issues (not covered by the assessments), in order to make a decision (Aven, 2019a). This definition is not given in the SRA glossary.

2.16.2 *Model uncertainty*

Uncertainty about the model error, i.e., about the difference between the model output and the true value being modeled.

2.16.3 *Precautionary principle*

An ethical principle expressing that, if the consequences of an activity could be serious and subject to scientific uncertainties, precautionary measures should be taken, or the activity should not be carried out.

The cautionary principle is a more general version of this idea, with "scientific uncertainties" replaced by "uncertainties." Refer to Section 3.

2.16.4 *Risk acceptance*

An attitude expressing that the risk is judged acceptable by a particular individual or group.

2.16.5 *Risk analysis*

Systematic process to comprehend the nature of risk and to express the risk, with the available knowledge.

Risk analysis is often also understood in a broader way, in particular in the Society for Risk Analysis (SRA) community: risk analysis is defined as including risk assessment, risk characterization, risk communication, risk management, and policy relating to risk, in the context of risks of concern to individuals, to public and private sector organizations, and to society at a local, regional, national, or global level.

2.16.6 *Risk assessment*

Systematic process to comprehend the nature of risk, to express and evaluate risk, with the available knowledge.

2.16.7 *Risk evaluation*

Process of comparing the result of risk analysis (see Section 2.16.5) against risk (and often benefit) criteria, to determine the significance and acceptability of the risk.

2.16.8 *Risk management*

Activities for handling risk, such as prevention, mitigation, adaptation or sharing.

It often includes trade-offs between costs and benefits of risk reduction and choice of a level of tolerable risk.

2.16.9 *Risk perception*

A person's subjective judgment or appraisal of risk.

2.16.10 *Risk tolerance*

An attitude expressing that the risk is judged tolerable.

3. The framework—For assessing and managing risk

The framework is built on the following pillars

(1) A conceptual platform, as explained in Section 2

(2) A classification system for risk management strategies, based on three types of risk management strategies:

- **(a)** Risk-based requirements strategy
- **(b)** Risk assessments (risk-informed) strategy
- **(c)** Cautionary (use of robust/resilient arrangements and measures) strategy.

In addition to strategies (a)–(c), dialog and discursive strategies are also included but are not addressed in this chapter (refer to Aven and Kristensen, 2019; Renn, 2008).

This classification system, with its key elements, will be presented and discussed in the following. A main source is Aven and Kristensen (2019).

3.1 Basic ideas of the classification system

Consider the challenge of managing the risk related to the design and operation of a technical system, for example, a process plant. Two fundamental strategies are commonly adopted for this purpose: (i) the use of risk assessments (risk-informed strategy) and (ii) the use of robust/resilient arrangements and measures (founded on the cautionary/precautionary principles (Aven, 2019a; Renn, 2008; SRA, 2017b)). Simplified, we can say that the latter is justified as a result of the limitations in the former: the risk assessment's lack of ability to properly reflect all aspects of risk. These limitations are to a large extent about revealing uncertainties and lack of knowledge. We have some knowledge about the events and relevant systems and components; yet, there are always uncertainties. To meet these, we implement risk management systems. The aim of these is to control the risks, prevent the occurrence of incidents and accidents and/or to limit their consequences.

In practical (real life) situations, there is also a third, supplementary strategy to (i) and (ii), which is labeled "risk-based requirement strategy." This strategy is used when the knowledge is strong—the situations studied are well understood and there is nothing unusual or new about the activity or system considered. Experience, statistical analysis and traditional risk assessments are commonly used to establish the "risk-based requirement strategy." The idea is captured in standards, for example, the risk-related decision-making framework by ISO 17776 (ISO, 2016). In this standard, the "risk-based requirement strategy" is referred to as "good practice" and is added to the two other basic risk management strategies referred to above. We are led to a classification system based on (a)–(c).

Depending on the situation, these three management strategies are used to varying degrees. With strong knowledge, the "good practice" approach is justified, whereas, if the values at stake are high and the uncertainties are large, all strategies are needed. For other situations, with fewer uncertainties, the cautionary and precautionary strategy (ii) is not given the same weight.

For further analysis, we introduce the concepts of general knowledge (GK) and specific knowledge (SK). Consider a specific activity, for example, the operation of a process plant. The knowledge K about the consequences C of this activity is divided into GK and SK. The GK includes all knowledge available for related activities but not the specific activity. It also covers knowledge about the specific factors/conditions that are important for understanding and managing the risk related to the specific system considered. The specific knowledge includes knowledge concerning the specific system alone.

In addition, we need to clarify whose knowledge is referred to. We distinguish between the analysts' knowledge, the total knowledge available when adding other experts' knowledge, and the decision makers' knowledge.

In Fig. 1, a model for the selection of appropriate risk management strategies is presented, using the concepts of general and specific knowledge. It is based on Aven and Kristensen (2019) and can be seen as a version of the framework presented by ISO 17776 (ISO, 2016). A key idea of the classification system is that, when the general knowledge (GK), as well as the specific knowledge (SK), is strong, risk-based requirement strategy (a) can be implemented. If, however, the SK is weak (but the GK still strong), weight needs to be given to robustness/resilience, as the activity considered could have special features and there is a potential for surprises relative to the GK. If the SK is medium strong and the GK strong, risk assessment would play an important role. If the GK is weak, a robustness/resilience-based strategy is always needed, as the SK has a poor foundation. If the GK is medium strong, risk assessments will provide useful decision support, except in cases with weak SK. In the case of weak SK, the risk assessment will have rather limited information value.

When faced with a situation where the level of knowledge (GK or SK) is medium or weak, one should always consider the question: Should the level of knowledge be increased before a decision on the risk management strategies is made?

3.2 Risk assessments

In risk assessment, an aim is often formulated to assess the risk related to the occurrence of a specific event. As the event is rare in this case, we do not have enough data to use these directly to determine or describe the risk in a meaningful way. Alternative approaches are needed:

(1) We can develop a model of the system or activity considered, with elements for which we have more data and information

(2) We can use risk indicators

In the following, we will focus on (1), but first a few comments related to indicators.

Consider the task of assessing the risk related to major accidents in an industry, for example, the Norwegian petroleum industry. Such accidents are rare, and the use of risk indicators, such as the number of gas leakages and failure rates of safety barriers, are suggested, to monitor and evaluate the risk. Considering again risk as (A,C,U), the idea is to focus on A—or aspects of C such as the safety barrier failures—as these events occur rather frequently. Often indicators are developed on a more detailed system or activity level, to reflect technical, organizational and/or operational factors,

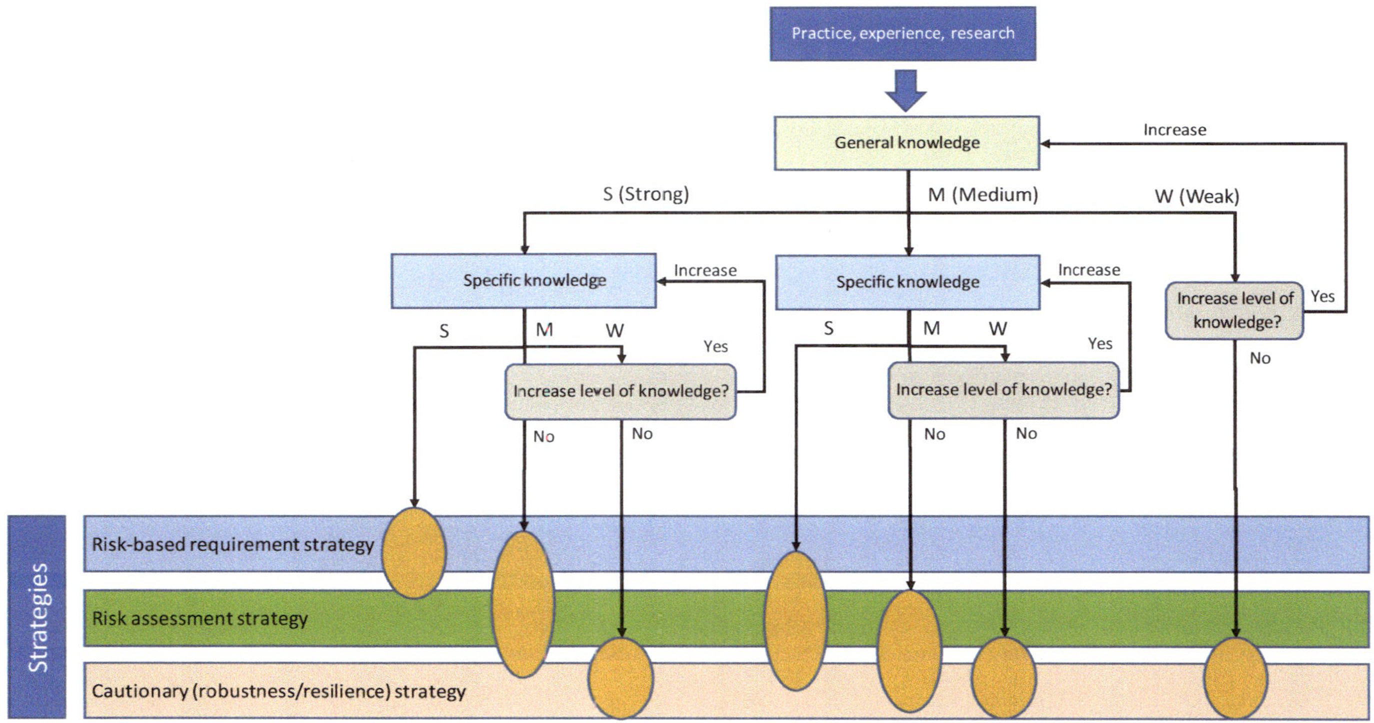

Fig. 1 A model for defining the main risk management strategy, based on categorizations of general and specific knowledge. *Based on Aven, T., & Kristensen, V. (2019). How the distinction between general knowledge and specific knowledge can improve the foundation and practice of risk assessment and risk-informed decision-making. Reliability Engineering and System Safety, 191, 106553.*

for example, related to maintenance work. Such indicators all provide knowledge about risk but need to be used with care, as there are many measurement challenges related to their use. A change in an indicator could, for example, be a result of an adjustment of the observation-collection regime. The relationship between changes in the indicator and risk could be difficult to establish. Focusing on events like relatively large leakages, the measurement problems can be avoided and still ensure a sufficient number of events. A comprehensive scheme for collecting and using such risk indicators has been developed for the Norwegian petroleum industry (see e.g., Abrahamsen, Selvik, Heide, & Vinnem, 2018; PSA-N, 2018; Vinnem, Aven, Husebø, Seljelid, & Tveit, 2006), and is considered to be an illustrative example of how to characterize risk with a basis in such indicators. The approach includes methods for identifying significant data trends.

Based on indicators, it is possible to derive risk metrics, for example, the probability of a major accident or an expected number of fatalities. This requires modeling, as will be discussed in the following. These models would typically be subject to considerable discussion and based on many assumptions. In the Norwegian petroleum industry example, it was decided not to give weight to such overall risk characterizations. Most attention is placed on the observations, for different categories of events, like large leakages. Clearly, recording such leakages is important for the proper management of the risks, and there are no associated measurement problems.

3.2.1 Use of models: Introduction

The models in (1) include as example events trees, fault trees and Bayesian networks. These models and related analysis are presented in textbooks in risk assessment and will not be further described here. Formally, we can formulate the problem like this. A quantity C is to be assessed. As C is unknown, we develop a model $C' = g(X)$ of C, where X is a vector of quantities on a more detailed level, for example, reflecting the states of the set of components of a fault tree. The difference between C and C' is the model error, and uncertainties about this error are referred to as model uncertainty; refer to Section 2.

Assuming that C represents the occurrence or not of a rare event, we are led to probability judgments related to C. Let p be a probability judgment related to C, for example, $p = P(C = 1)$, where "$C = 1$" refers to the occurrence of an accident. Then analogous to $C' = g(X)$ we establish a model $p' = f(q)$, where q is a set of probabilistic parameters on a more detailed level, for example, the probabilities of the basic events of a fault tree.

Instead of making a direct judgment p, we use the model f and derive the probability judgment p'.

Note that p'-p is not here a model error, as p is not a "true" value. If the point of departure had been frequentist probabilities, it makes sense to talk about a true value, but then the set-up is different. In this case true, underlying parameters and frequentist probabilities p and q are assumed to exist, and we can write $p=f(q)$. These quantities are unknown and need to be estimated; using the model f, we are led to an estimated frequentist probability p' of p, through $p'=f(q')$, with obvious interpretation. Now we can talk about an estimation error p'-p, and there is a need to address the uncertainties of this error.

The above discussion leads us to questioning the "quality" of the model. In cases of strong or medium knowledge, the model can be an accurate representation of the activity or system considered. However, in the case of weak knowledge, we will be faced with large model uncertainties. However, modeling is not only about accurate representations, it is also about simplicity, to allow the timely and efficient use of the model. Many approaches exist for how to make judgments about the quality of the models; see, for example, overview in Aven and Zio (2013) and Bjerga, Aven, and Zio (2014). As we are considering rare events, statistical approaches cannot be used. Argumentation and reasoning are the key instruments available. Often, models are based on the formulation of assumptions, and the quality issue can be reduced to a discussion about the reasonability of these assumptions; see, for example, discussions in Berner and Flage (2016), Bjerga and Aven (2016) and Khorsandi and Aven (2017).

This discussion can be generalized to one addressing the validity of the risk assessment; refer to Aven (2017a).

3.2.2 Validation

Clearly, validation of the model, and the risk assessment as such, is not easily conducted in cases as here considered. In its general form, the term "validity of an assessment" relates to the degree to which one is able to assess what one sets out to assess; alternatively, validity may just suggest that the assessment is solid in some sense, meeting some relevant criteria. But what should these criteria be?

Consider a risk assessment which produces output in the form of a probability P for an event A. We write $P(A|K)$. This probability expresses the analysts' degree of belief that the event A will occur, given some background knowledge K. This knowledge is essentially justified beliefs, founded on data, information, models, testing, argumentation, etc. The probability as such cannot meaningfully be validated, as it is a subjective or intersubjective tool to express uncertainties, and there is no true, objective reference against which to compare it. It is, however, possible to establish criteria for the process of making such judgments. These criteria can relate to K and the process of transforming K to P. For example, if the background knowledge K is weak, it can be argued that the probability assignment P does not have a solid foundation, and hence the assessment is not valid. Another example would be that the transformation process from K to P in some cases has strong limitations, as the use of probability forces the analysts to add information that is not really present in the data available (Flage, Aven, Baraldi, & Zio, 2014). To meet this challenge, the analyst may prefer to use imprecise probabilities, for example, expressing that P exceeds 0.90.

More generally, the issue of validity would question the choice of uncertainty measure, the events and quantities of interest, the strength of the knowledge supporting the measure used, as well as the process of transforming this knowledge into suitable risk metrics and characterizations. For this purpose, we can formulate criteria on a general basis from a risk science perspective, as indicated above, but equally important is the quality of the risk assessment as seen from the users' perspective (Aven, 2016c; SRA, 2015b): How credible are the results from the view of the different stakeholders? Are they trustworthy? And is the risk management trustworthy? The choice of metrics should obviously meet the needs of the decision makers. It is tempting to state that the quality of the risk assessment is mainly determined by its ability to meet the decision makers' expectations. However, such a perspective is easily refuted. The decision maker can be satisfied with a risk assessment expressing a low probability number for an extreme event, without giving weight to the knowledge supporting it or the process for deriving this number. The quality of K and this process are obviously essential elements of the assessment. As highlighted above, the scientific validity is the key to ensuring validity.

As another example to illustrate this point, consider an analyst team which produces risk numbers in the form of expected values. The decision makers and other stakeholders could find that the assessment meets their needs but, surely, such an approach would in most cases not be able to properly express risk from a risk science point of view, as risk is more than expected values (Aven, 2012, 2019a; Haimes, 2015; Paté-Cornell, 1999).

Note that, if risk assessment aims to estimate some underlying presumed true objective frequentist probabilities, as outlined above, the above discussion is still relevant. The frequentist probabilities are then the quantities of interest.

This leads us to the question of whether risk assessment can be used to accurately estimate risk. Surely, if accurate estimation—as well as accurate prediction—is a requirement, risk assessment is not in general valid (Aven, 2019a; Cumming, 1981; Weinberg, 1981). In cases of large uncertainty, such estimation and prediction cannot be ensured. Risk assessment in general does not meet the criteria for the traditional scientific method. However, current thinking on risk assessment considers it more a risk knowledge generation process; refer to definitions in Section 2 and, for example, SRA (2015a, 2015b, 2017b), where risk assessment is viewed as a systematic process to comprehend the nature of risk and to express and evaluate risk, with the available knowledge. The problems of risk assessment and quantified risk assessments (QRAs) in particular—for example, in dealing with uncertainties—have been thoroughly discussed in the literature; see, for example, Aven (2011b, 2016b), Rae, Alexander, and McDermid (2014), Goerlandt, Khakzad, and Reniers (2017) and Pasman, Rogers, and Mannan (2017). This discussion is also related to the concept of reliability in risk assessment, which basically expresses that the assessment yields the same result over repeated trials (Aven, 2011a, 2019a; Aven & Heide, 2009). See Goerlandt et al. (2017) for a review of works addressing the validity and validation of safety-related quantitative risk analysis. The review points to many important aspects in this respect, for example, peer review and quality assurance.

3.2.3 Scenarios and events addressed

A main task in risk assessment is to identify scenarios that may occur and assess the risk related to their occurrence. The number of scenarios could be very large, and not all are considered for further analysis. Broadly, we can distinguish between the following categories of scenarios:

- Not identified (an unknown unknown, i.e., a type of event that is not known; or an unknown known, i.e., an event type known by some but not by the analysts conducting this risk assessment).
- An identified scenario and included in the risk assessment—its likelihood and risk are assessed, the scenario is followed up, and measures to meet it are discussed.
- An identified scenario and included in the risk assessment—its likelihood and risk are assessed and found negligible. The scenario is not studied further.

This discussion, which is to a large extent based on Aven (2016b), is closely related to the concept of "completeness uncertainty." This concept is thoroughly discussed in the literature and can be viewed as an aspect of model uncertainty (Bjerga, Aven, & Flage, 2018). Assuming that risk is described according to (A′,C′,Q′,K), using the terminology introduced in Section 2, the concept of completeness uncertainty relates to the degree to which the set of events A′ is complete, i.e., covers the actual A occurring. There could be many reasons why A′ is not complete, for example, that some events are ignored because they are considered extremely unlikely, leading to a negligible risk contribution, or because they are unknown to the analysts. Clearly, this lack of completeness could contribute to model error, as an event could occur which is not reflected in the model used.

This discussion is also closely related to the concepts of common causes and special causes referred to in the quality discourse (Bergman, 2009; Deming, 2000; Shewhart, 1931, 1939). These two concepts refer, respectively, to variation that is predictable in the view of the historical experience base and to variation that is unpredictable and outside the historical experience base (it always comes as a surprise).

Now consider a risk metric computed by the formula $p' = f(q)$ and a set of events/scenarios A′ forming the basis of this metric. The actual event/scenario occurring is denoted A.

In relation to A′, we have two main challenges (refer to the different types of "black swans" studied in Aven (2014)):

(i) The analysis set-up excludes some events/scenarios B′ from the set A′ studied, as a result of unknown unknowns (a) and unknown knowns (b)
(ii) Some events/scenarios B′ are excluded from A′ and further analysis, due to judged low probability.

In theory, there is always a possibility of an unknown unknown, but if the phenomena are "well understood" (strong knowledge), such events would be rare. The unknown unknowns represent risk, but this risk contribution is not added to the risk metric used. Per definition, there is no way of explicitly incorporating this risk contribution at this point in time; hence the risk description needs to be understood as conditional on no such event occurring. Risk assessments are often conducted for complex systems, and then there is always the potential for surprises.

The category (b) of unknown knowns relates to the fact that the analysts may overlook some events from the analysis because they are not aware of them. If we add a category of "others," all types of events are of course covered.

Now let us move to the more detailed scenarios, as described by the models. If we think about scenarios linked to the branching events of an event tree and the outcomes, we will not find unknown knowns. However, if we look further into the details supporting these overall models, there could be such surprising scenarios. An illustrative example is the Heimdal incident in 2012 (Aven, 2016b; NOG, 2017; PSA-N, 2012).

Here, a hydrocarbon leak occurred on the Norwegian continental shelf, with a large leak rate in connection with the testing of two emergency shutdown valves (ESDVs). In preparation for testing of the valves, a pipeline was to be depressurized to flare. In the pipeline, a ball valve with a pressure of 16 bar was installed as the last barrier against the flare. This ball valve was in the closed position and was exposed to a pressure of 129 bar. The pressure caused the failure of the gasket flange to the valve and a consecutive gas leak estimated at 3500 kg with an initial leak rate of 16.9 kg/s. Gas was detected in a large area of the installation.

Prior to the incidents, the operating team was basing its risk scenarios on an understanding that the pipe was in accord with current design practice, as then all the relevant pipe sections, even past the valve to the flare, would have withstood the process pressure. Hence, the order in which the three valves were opened would not be safety critical. The relevant pipe section was, however, designed according to an older design practice, where spec break (change of pipeline specifications) to a lower pressure class is upstream of the last valve to the flare. With such a design, the order in which the three valves are operated is safety critical: If the last valve to the flare is opened last, the pipe will be subjected to higher pressure than designed for.

This knowledge was known to others in the organization, so the scenario that occurred did not come as a surprise for them—only for the operating team. Scenarios capturing this knowledge were not included in the analysis.

Now consider events ignored due to low probability (ii). Alternatively, the Heimdal event could be seen as excluded, due to arguments in line with category (ii), ignored as being so unlikely. Theoretically, a number of scenarios may occur, but in practice simplifications are made and some events are ignored. This practice is of special interest for the present discussion.

In the Heimdal case, the operating team was confident that the system was a standard system—and did not elaborate on related scenarios. The risk was concealed in the assumption made: The system is standard. Similarly, assumptions are made when developing models of the type f(q), for example, that the analysts exclude the possibilities that more than two fatalities could occur in relation to the scenarios described by an event tree model. A number of possible scenarios are obviously ignored by this type of modeling. The argument would be that the probability and risk are so small that we can ignore them. The question we will discuss below is: Can this be justified? And what are the "risks" involved in doing so?

To simplify, assume in the model that we allow for three as the maximum number of fatalities. Say the analysis has derived a probability of two fatalities associated with this tree equal to 2×10^{-5} and the analysts judge the probability of three fatalities to be a lower order of magnitude, hence about 2×10^{-6}. With a number of trees, the probability of three fatalities could be much larger but may still be considered so low in relation to two fatalities that it is not included in the analysis. To support the decision-making, the risk contribution from three fatalities is not considered essential.

Similar arguments could be used for 4,5, ..., fatalities. The probabilities become very low on the basis of the calculation of the model. The possibility of a large number of fatalities could be summarized by a probability of say 10 fatalities equal to 2×10^{-7}. For the purpose of the risk analysis, this contribution could be found to be of minor interest, when comparing the relevant decision alternatives or comparing the risk numbers with some predefined risk acceptance criteria (linked, for example, to the Fatal Accident Rate, the expected number of fatalities per 100 million exposed hours). It may not be ignored completely from the analysis, but as the probability is judged to be so low, this risk contribution is not given the same attention as the risk linked to one and two fatalities.

The probability judgments are based on background knowledge K, which captures assumptions, more or less tacitly formulated, as in the Heimdal case. The low probability numbers assigned for extreme scenarios are conditional on the belief that the system is a standard one. However, the system was a special one. The idea that it was special may not have been an issue at all, but, if it had been, the probability that the system was special could have been judged as so low that it would not have been believed to be a problem.

Let K_1 denote the background knowledge of the analysis team, which includes a belief (assumption H) that the system was a standard one in the Heimdal case. With this background knowledge, the operating teams do not consider scenarios with extreme outcomes to be in conflict with K_1 and H. Such scenarios are not thought of or they are ignored because the judged probabilities for the set of such scenarios are negligible: P(the set of such scenarios$|$ K_1, H)≈ 0.

In the following, we discuss how we can meet this challenge and see if there are possibilities to improve current risk analysis practice.

To formalize the discussion, we make a distinction between the outcomes (for example, expressed as the number of fatalities), hazardous events (for example, hydrocarbon leakages), and risk sources that can lead to these events and outcomes. If the number of fatalities is the quantity of interest, this is straightforward; the set must cover $\{0,1,2, \ldots\}$, or $\{0,1,2, \ldots, M\}$ if it is known that the maximum number of fatalities is bounded by M. In the event tree example referred to above, the number of fatalities was restricted to two. There may be reasons for this—there could be a maximum of two people exposed—but if this is not the case, it is important to have an initial model set-up that does not exclude unlikely events. It may be judged unlikely—at the design of the model—that more than two people are killed, but further analysis may challenge this judgment. The rule should, therefore, be to always frame the assessment in a way that includes all possible outcomes, using, if necessary, categories of "others" to stimulate thinking that goes beyond the defined outcome categories. This relates to a fundamental idea in risk analysis (the so-called backward approach—Aven (2009, p. 35)), where the point of departure is the outcome of interest (here, say at least three fatalities), and we ask what kind of events and risk sources can lead to such an outcome.

Using the event tree model, the key element of the analysis is the identification of the events, and to initially include a complete set of events (using a category "others" if needed), "complete" of course seen relative to the purpose of the analysis.

Then we come to the risk sources, the types of situations and sources for events to occur. This element is more difficult to define, as it covers so many different aspects. One example is a pipe section of hydrocarbons with flanges and valves, which has the potential to give rise to a leak. Another example is a maintenance activity on a valve, which also has the potential to cause a leak. In a fault tree, we develop sets of basic events which can lead to the top event, and we refer to this as a cause analysis—identifying sets of basic events that cause (lead to, explain) the top event. A fault tree is to be seen as a model of the real system and does not capture underlying factors linked to, for example, management and organizational issues, such as the importance of training. However, such factors are often modeled using influence diagrams—Bayesian belief networks; see, for example, Ayyub (2014) and Meyer and Reniers (2013). The models are crude representations of the many factors that can lead to a hazardous event, and often the justification of the models is rather poor, as there is no well-established theory supporting the modeling. The number of factors that can be combined to generate the event could in theory be huge but, to make a reasonable model of the system, only a limited number is selected. Yet the modeling can be useful, for example, in studying the effect of such factors on risk and seeing how various measures based on these factors influence the risk.

In a risk assessment, the degree to which the risk sources (causes, explanations) are in fact modeled varies, but for practical risk management, in particular in the operational phase, they are the key to ensuring safe activities, as the risk is "rooted" in these sources, and the risk-reducing measures to a large extent need to be based on these sources. However, the complex links between the sources, on the one hand, and the events and outcomes, on the other, represent a big challenge for the risk management, whether or not models of the phenomena are developed. Hypotheses and assumptions—more or less strongly justified—in reality often form the basis for the decision-making. The aforementioned Heimdal case provides an illustrating example; here, the thesis was that the system was a standard one.

A huge number of scenarios—linking risk sources with the events and outcomes—can be generated, but the hypotheses and assumptions exclude many or most of them. What is considered is often a limited set of thinkable scenarios, based on the risk assessment carried out and the basic beliefs that the relevant personnel have on the issue addressed.

Scenarios are here excluded from further analysis and decision-making, as they are not known to the operating team (unknown knowns) and because the likelihood is judged to be so small that it is not believed that the scenarios will occur. For the latter category, we again need to question whether neglecting scenarios due to low likelihood can be justified, as there are so many potential scenarios.

Seeing the link between sources, events and outcomes as a complex system, we would not be able to identify all scenarios of interest—the actual scenario occurring will come as a surprise (Turner & Pidgeon, 1997). The key for the analysis and management related to such scenarios is to acknowledge that not all scenarios will be identified—they can be viewed as unknown knowns, and some that are identified but not believed to occur due to judged low probability can in fact occur, as there are so many of them and the judgments can be based on more or less valid hypotheses and assumptions.

Methods for identifying and analyzing the risk sources and their link to events and outcomes (the key elements of the so-called bow-tie) can improve the insights about the system, but it is critical to acknowledge that the judgments made can never fully capture all risk aspects, for the reasons discussed above. Hence, we need to look for ways to think outside the normal box, challenging the hypotheses and assumptions on which the events and scenarios are built. It is beyond the scope of this chapter to describe methods that can be used for this purpose; see Aven (2014) and Bjerga and Aven (2016) for some selected approaches and methods, covering inter alia analysis methods to identify events/scenarios (such as the anticipatory failure determination (AFD) method of Kaplan, Visnepolschi, Zlotin, and Zusman (1999) and red teaming, which serves as a devil's advocate, offering alternative interpretations and challenging established thinking (Masys, 2012)). In addition, it is essential to be able to read signals and warnings—we must avoid missing or ignoring important early signals and precursors of serious events (but also of course avoid exaggerating them). The focus on signals and precursors of serious events is a common feature of most approaches that seek to meet surprises and the unforeseen ("black swans"). If we look at basic insights from organizational theory and learning, this feature is a main building block. A good example is the concept of collective mindfulness, linked to High Reliability Organizations (HROs) (Weich & Sutcliffe, 2007); one of its five principles is sensitivity to operations, meaning a focus on reading the landscape and taking adequate actions.

In the handling of unknown known scenarios, there is a vast body of literature on methodology covering *how to* involve expertise (those that may know) in the analysis and decision processes, which is relevant here. Special reference is made to the quality management area, with its focus groups, affinity diagrams, etc. See also Veland and Aven (2015), who present an adjusted job safety analysis, with the intention to better address this type of risk.

3.2.3.1 Summary of scenario discussion

To summarize and conclude this discussion: A key task of risk assessment is to identify events and scenarios that can lead to some specified outcomes. Depending on how these are defined, the list of such scenarios and events can be more or less complete, in the sense of capturing the actual one occurring. Hence, there could be risk contributors not captured by the analysis. To meet this risk, two main messages are drawn from the analysis in the previous sections. Firstly, it is essential to define a set-up for the analysis that is sufficiently broad on outcomes, so that all relevant types are covered. The use of a category "other" may be necessary in some cases to ensure this, stimulating questions about how to obtain such an outcome.

We should aim at similar thinking for the main events, where we need to have a special focus on unknown knowns. Unknown unknowns should also be covered but would normally be given less attention than the unknown knowns, as the latter category is much more likely to occur in most cases. Events at this level should not be excluded due to low probability.

For risk sources with links to events and outcomes, many scenarios will not be identified or will be ignored due to judged low probability. Care must be shown in this process, as the judgments on this part are so strongly based on hypotheses and assumptions, which are often difficult to justify. The actual occurrence of the scenarios will typically come as a surprise relative to the existing beliefs, but it should be noted that, from a broader perspective—having insights about the correctness of these hypothesis and assumptions, the probability for any such scenario could be quite large. This means incentives to scrutinize the probability judgments made, with their background knowledge, using methods as discussed above.

Any model and analysis of the source part needs to be seen in relation to the assumptions on which it is based. The risk assessments can improve on this area, but, for complex systems, they cannot predict with accuracy all scenarios that will occur. Hence, the appropriate management regime to meet the risk related to the operation of such a system must always combine risk analysis methods with measures of robustness and resilience, to cope with scenarios and events not identified or planned for. This is also the basic pillar for safety regulations in industry, but it is frequently challenged for being irrational and not placing sufficient weight on the real matters (Aven, 2011b).

3.2.4 Classifications and judgments of the severity of threats and risks

Based on the challenges of quantitative risk characterizations in cases of uncertainties, more qualitative approaches are developed. Here is an example (Aven & Kristensen, 2019). Threats and risks are classified according to three categories: High, Medium and Low. The category "High" is typically associated with an unacceptable risk, requiring additional measures in order to proceed, whereas "Low" is typically associated with an acceptable risk, where no additional measures are needed. The "Medium" category is associated with situations where the risk is acceptable but additional measures should be implemented if possible. Management considerations are needed to decide which measures to implement.

To classify a threat or the risk as High, the following criteria need to be considered (minimum one of these criteria applies):

(a) The risk is judged to be high when considering consequences and probability

(b) The risk is judged high, considering the potential for severe consequences and significant uncertainty (relatively weak knowledge)

(c) Lack of robustness/resilience

(d) Weak general knowledge about the risk sources, events, barriers, consequences (for short, denoted (A,C))
(e) Weak specific knowledge about (A,C)
(f) Strong general knowledge about undesirable features of (A,C)
(g) Strong specific knowledge about undesirable features of (A,C)

For example, the risk is classified as *High* if a critical failure on the blowout preventor (BOP) has been revealed through testing, by reference to (a), (c) and (g).

We can also classify the risk as *High* if the BOP has not been tested, with reference to (b), (c) and (e).

In the former example, we argued according to what we know and, in the latter, to what we did not know.

Analogously, we can define criteria for *Low* (all relevant criteria apply):

(h) No potential for serious consequences
(i) The risk is judged to be low when considering consequences and probability, and supporting knowledge is strong
(j) Solid robustness/resilience
(k) Strong general knowledge about (A,C)
(l) Strong specific knowledge about (A,C)

Medium applies to all other cases.

To evaluate the strength of the knowledge, we need to address issues such as (Aven, 2017b; Aven & Flage, 2018; Flage et al., 2014):

- The reasonability of assumptions made
- The amount and relevancy of data/information
- The degree of agreement among experts
- The degree to which the phenomena involved are understood and accurate models exist
- The degree to which the knowledge has been thoroughly examined (for example, with respect to unknown knowns; i.e., others, but not the analysis group, have the knowledge).

The last point relates to potential surprise issues like (Bjerga & Aven, 2016):

- The possibility of unknown knowns. Have special measures been implemented to check for this type of event (for example, the use of an independent review of the analysis)?
- The possibility that events are disregarded because of very low probabilities, although these probabilities are based on critical assumptions. Have special measures been implemented to check for this type of event (for example, signals and warnings influencing the existing knowledge basis)?

- Risk related to deviations from assumptions made
- Changes in knowledge over time

Based on judgments of such issues, qualitative score systems can be developed (see Aven, 2017c).

The distinction between general knowledge (GK) and specific knowledge (SK) for assessing risk leads to a new type of approach, as discussed in Aven and Kristensen (2019). We find, however, similar ideas in the Bayesian method, where a prior distribution and the underlying probability model are established on the basis of generic knowledge and combined with observations and measurements representing specific knowledge, to produce an integrated distribution, reflecting the total knowledge. The Bayesian method has a strong theoretical foundation and can be useful in many cases to systematically update knowledge. The above GK-SK approach differs, however, from the Bayesian approach in many ways. First, it is a qualitative approach, not a quantitative approach as the Bayesian one. The strengths of quantitative analysis are well known, as are the weaknesses of qualitative studies. Probabilistic quantification and the use of Bayes' formula make it possible to systematically combine knowledge about unknown quantities and observations (measurements). Using qualitative approaches, the stringency of the probability theory is lost, and it is not possible to ensure the same level of coherency in judgments. However, quantification also has strong limitations. To use a probability to represent uncertainty means that important aspects of risk are not reflected: aspects that are important for being able to properly evaluate the significance of the risk and make the right decisions. The argumentation for this assertion is well known (see, for example, Aven, 2014; Flage et al., 2014) but, to quickly recap, a probability representing or expressing uncertainty is conditional on some knowledge, and this knowledge could be more or less strong and even erroneous. For the proper use of the risk assessment, it is not enough to only report the probability numbers, as these reflect "conditional risk." The assessment may, for example, be based on a belief that the system studied has some specified properties, but in real life it could in fact have others, and this can lead to negative surprises (see examples in Aven, 2017c). The use of imprecise probability or interval probabilities (Flage et al., 2014) makes the transformation from knowledge to the probabilities more objective but does not eliminate the issue of the underlying knowledge being weak or wrong. Such intervals will also reduce the information value of the assessments, as the intervals seek to avoid incorporating analysts' judgments.

3.3 Robustness/resilience-based strategies

As discussed in Section 3.1, the risk-informed strategy alone has limitations in guiding the decision-making. The assessments have limitations in revealing and characterizing all aspects of risk, and there are value judgments and trade-offs that extend beyond risk assessments. We can agree on the risk but still come to different conclusions on how to treat it. For both reasons, uncertainty and value differences, a risk-informed approach is not sufficient and needs to be augmented with other principles, mainly the cautionary strategy in the case of high uncertainty and the discursive strategy in cases of different values (Aven & Renn, 2018; Klinke & Renn, 2002; Renn, 2008). The cautionary principle expresses that, if the consequences C of an activity could be serious and subject to uncertainties, cautionary measures should be taken, or the activity should not be carried out. The precautionary principle is a special case, applicable in the case of scientific uncertainties (Aven, 2019a). Robustness and resilience-based strategies can be seen as measures supporting this principle.

In risk management, there will always be a need to balance different principles and concerns. Yet, such principles can provide useful guidance, as they point to how to think and what aspects to consider and give weight to. As argued for in Aven (2019c), the cautionary principle is of special importance when risk is judged high in the following ways:

(1) There is a potential for C values that are extreme (highly negative).

(2) There is a potential for serious C values if the activity is not cautiously executed—C is very sensitive to how the activity is run.

(3) There is a potential for serious C values if something unlikely, surprising, or unforeseen should happen (for example, an event A not anticipated or a new type of event A).

(4) Weak knowledge about the consequences C of a specific type of activity, for example, about the effect of the use of a specific drug. There is a potential for serious C values.

(5) There is a potential for serious C values, and these are subject to scientific uncertainties.

Some of these criteria are closely linked and overlapping, as, for example, (4) and (5). What is a serious or extreme C is a judgment call. The consequences may relate to different types of values, such as life, health, the environment and economic quantities. Interpreting risk as above, the cautionary principle reflects a type of risk aversion, in the sense of disliking the risk (the negative consequences and the related uncertainties).

The cautionary principle states that, if risk is high in the sense of (1)–(5), caution is in place: Measures should be implemented, or the activity should not be realized. The principle provides guidance, it does not prescribe what to do. No risk management principle should prescribe what to do, as there is always a gap between principle and action. In the face of risk with the potential for serious consequences, there is no general formula or approach that can objectively produce the best decisions. Statistical analysis and traditional risk assessments can provide strong support for specific decisions in cases where "objective" probabilities can be established, but for situations characterized by uncertainties, the analytical approaches have severe limitations. Decision analysis theories reflecting the uncertainties exist, like the subjective expected utility theory, but they all have limitations in providing clear answers on what is the best decision. The subjective expected utility theory is suitable for individual decision makers but not for the organizational and societal levels (Lindley, 1985). The theory is difficult to use, with its demanding ways of assigning probabilities and utilities (Aven, 2012).

Consequently, the use of risk management principles is needed, to provide guidance on how to think and make good decisions. In relation to risk, there are two main types of concern: the need to create values on the one hand and protection on the other. The cautionary principle is of the latter type. It gives weight to the uncertainties. It has a role in notifying people and society in relation to protection against potential hazards and threats with serious consequences. A principle highlighting value creation is the use of traditional cost-benefit type of analysis, CBA (expected net present value calculations), as risk and uncertainties are here basically not given weight beyond expected values; refer to the discussion in Aven (2017b). Adopting one of these principles and ignoring the others would clearly lead to poor risk management. We need both categories of principles, as well as principles highlighting other concerns, in particular the need to obtain a balance between development and protection; refer to the discussion in Aven and Renn (2018).

In society, there is a continuous "battle" between development on the one side and protection on the other. This battle is rooted in differences in values and priorities but also in scientific and analytical argumentations. The cautionary principle has a role to play in relation to this type of consideration and management. A warning is in place—there is a potential for serious consequences and there are uncertainties. The development tools have spoken—now it is time for the protection side to highlight important aspects for the decision makers to adequately balance the various concerns.

The cautionary principle highlights the need for further checking of the assessments made: Is the knowledge supporting the judgment strong enough? Could there be a potential for surprises relative to current knowledge? The assessors may consider the situation to be characterized by rather strong knowledge, yet the cautionary principle stimulates both better analysis and robust/resilient measures. The key point made is that the analysis could have limitations in accurately reflecting the real world, and surprises can occur relative to current knowledge. This justifies robustness and resilience-based measures, for example, implementation of safety barriers, different layers of protection, "defense-in-depth," redundancy, diversification, the ALARP (as low as reasonably practicable) principle, etc. Current industry practice is based on this thinking and these types of measures—the cautionary principle is commonly adopted.

A remark is in place concerning the ALARP principle. It is a principle designed to support protection, as the basic idea is that a measure that will reduce risk should be implemented (Ale, Hartford, & Slater, 2015; HSE, 2001; Jones-Lee & Aven, 2011). Only in the case of it being demonstrated that the costs are grossly disproportionate to the benefits gained does the measure not need to be realized. Thus, it can be said to be highlighting cautionary thinking. However, in practice we often see that the gross disproportion criterion is verified by using cost-benefit types of analyses (Aven & Renn, 2018). These analyses are based on expected values and hence do not give much weight to risk and uncertainties. The result is that ALARP becomes a principle more highlighting development than protection. Alternative ways of verifying the disproportionate criterion are therefore suggested; see, for example, Aven (2017b). The essential point is that, if the cautionary thinking is to be given weight, measures should also be considered when the computed expected net present value is negative, provided these measure can contribute to a reduction in risks and uncertainties and a strengthening of the robustness and resilience of relevant systems (Aven & Renn, 2018).

See Jones-Lee (1989), Ayyub (2014) and Reniers and Noël Van Erp (2016) for thorough reviews of economic aspects of safety. These references also cover discussions of other economic approaches than CBAs, for example, the use of disproportion factors. As discussed by Aven and Flage (2009), to varying degrees, these approaches meet the criticism raised against CBAs and their expected value-based philosophy. Yet they all face the problem that they are based on probability or another measure of uncertainty, which is conditional on some knowledge, and this knowledge could be more or less strong and even wrong. Caution and precaution also mean highlighting the weakness and limitations of this knowledge.

Using the broader concept of the cautionary principle instead of the precautionary principle, we can avoid unnecessary discussion about what type of uncertainties we face, and focus can be placed on action and how to manage the risk. However, for some specific situations, it may be important to clarify whether the uncertainties are in fact scientific, so that the appropriate measures are taken. In relation to the approval of new products, the concept of scientific uncertainties is important, to ensure proper qualification processes. Clarifying when we face scientific uncertainties is also important in relation to other contexts, such as climate change, when more knowledge and science can reduce the scientific uncertainties and, in this way, clarify the issues and better distinguish between discussions about uncertainties and discussions about values (Aven, 2019c).

4. Final remarks and conclusions

Risk assessment is about comprehending the nature of risk, expressing risk and evaluating it, with the available knowledge. This is also the case for rare events. Accurate prediction is not the main issue. Either the event occurs or not. In some special cases, for example, natural hazards, we may have strong knowledge about a phenomenon and be able to produce rather objective probability distributions (based on frequentist probabilities) even for rare events. Refer to discussions about "perfect storms" in Paté-Cornell (2012) and Aven (2014). Then it may be argued that the event will occur if we just extend the period of time considered. However, the period may be so long that the conditions for its relevancy will change and make the analysis fail. More common, however, is the situation that there are more fundamental uncertainties, and such probabilities cannot be meaningfully defined. Probabilities are used to express uncertainties—they are based on the knowledge available. This knowledge can, however, be more or less strong and even wrong. An event that is judged to be very unlikely could occur. It may, for example, be the case that the judgment was based on an assumption that turns out to be erroneousness. Even though an event is judged to have a 10% probability of occurring, it does not need to occur. By proper risk management, the event may be avoided. The value of risk-reducing measures cannot therefore only be based on judgments expressed through expected value-based metrics, as also discussed in the previous section.

Risk assessment of rare events does not allow for direct high-level statistical analysis of data, but otherwise the risk assessments in such cases follow the same ideas and principles as for other situations. Some key aspects are discussed in the present chapters, using recent risk science and SRA documents (SRA, 2015a, 2017b).

References

Abrahamsen, E. B., Selvik, J. T., Heide, B., & Vinnem, J. E. (2018). A semi-quantitative approach for assessment of risk trends in the Norwegian oil and gas industry. In T. Aven & E. Zio (Eds.), *Knowledge in risk assessments* (pp. 297–312). New York: Wiley.

Ale, B. J. M., Hartford, D. N. D., & Slater, D. (2015). ALARP and CBA all in the same game. *Safety Science*, *76*, 90–100.

Aven, T. (2009). *Risk analysis*. Chichester: Wiley.

Aven, T. (2011a). *Quantitative risk assessment: The scientific platform*. Cambridge University Press.

Aven, T. (2011b). Selective critique of risk assessments with recommendations for improving methodology and practice. *Reliability Engineering and System Safety*, *96*, 509–514.

Aven, T. (2012). The risk concept—Historical and recent development trends. *Reliability Engineering and System Safety*, *99*, 33–44.

Aven, T. (2014). *Risk, surprises and black swans*. New York: Routledge.

Aven, T. (2016a). Risk assessment and risk management: Review of recent advances on their foundation. *European Journal of Operational Research*, *253*, 1–13. Open access.

Aven, T. (2016b). Ignoring scenarios in risk assessments: Understanding the issue and improving current practice. *Reliability Engineering and System Safety*, *145*, 215–220.

Aven, T. (2016c). Supplementing quantitative risk assessments with a stage addressing the risk understanding of the decision maker. *Reliability Engineering and System Safety*, *152*, 51–57.

Aven, T. (2017a). Risk analysis validation and trust in risk management: A postscript. *Safety Science*, *99*, 255–256.

Aven, T. (2017b). On some foundational issues related to cost-benefit and risk. *International Journal of Business Continuity and Risk Management*, 7(3), 182–191.

Aven, T. (2017c). Improving risk characterisations in practical situations by highlighting knowledge aspects, with applications to risk matrices. *Reliability Engineering and System Safety*, *167*, 42–48.

Aven, T. (2019a). *The science of risk analysis: Foundation and practice*. New York: Routledge.

Aven, T. (2019b). The call for a shift from risk to resilience: What does it mean? *Risk Analysis*, *39*, 1196–1203.

Aven, T. (2019c). The cautionary principle in risk management: Foundation and practical use. *Reliability Engineering and System Safety*, *191*, 106585.

Aven, T., & Flage, R. (2009). Use of decision criteria based on expected values to support decision-making in a production assurance and safety setting. *Reliability Engineering and System Safety*, *94*, 1491–1498.

Aven, T., & Flage, R. (2018). Risk assessment with broad uncertainty and knowledge characterisations: An illustrating case study. In T. Aven & E. Zio (Eds.), *Knowledge in risk assessments*. New York: Wiley.

Aven, T., & Heide, B. (2009). Reliability and validity of risk analysis. *Reliability Engineering and System Safety*, *94*, 1862–1868.

Aven, T., & Kristensen, V. (2019). How the distinction between general knowledge and specific knowledge can improve the foundation and practice of risk assessment and risk-informed decision-making. *Reliability Engineering and System Safety*, *191*, 106553.

Aven, T., & Renn, O. (2018). Improving government policy on risk: Eight key principles. *Reliability Engineering and System Safety, 176*, 230–241.
Aven, T., & Thekdi, S. (2018). The importance of resilience-based strategies in risk analysis, and vice versa. In B. D. Trump, M.-V. Florin, & I. Linkov (Eds.), *IRGC resource guide on resilience (Vol. 2): Domains of resilience for complex interconnected systems* (pp. 33–38). Lausanne, CH: EPFL International Risk Governance Center. Available at: irgc.epfl.ch and irgc.org.
Aven, T., & Zio, E. (2013). Model output uncertainty in risk assessment. *International Journal of Performability Engineering, 9*(5), 475–486.
Ayyub, B. M. (2014). *Risk analysis in engineering and economics* (2nd ed.). New York: Chapman & Hall/CRC.
Bergman, B. (2009). Conceptualistic pragmatism: A framework for Bayesian analysis? *IIE Transactions, 41*, 86–93.
Berner, C. L., & Flage, R. (2016). Strengthening quantitative risk assessments by systematic treatment of uncertain assumptions. *Reliability Engineering and System Safety, 151*, 46–59.
Bjerga, T., & Aven, T. (2016). Some perspectives on risk management—A security case study from the oil and gas industry. *Journal of Risk and Reliability, 230*(5), 512–520.
Bjerga, T., Aven, T., & Flage, R. (2018). Completeness uncertainty: Conceptual clarification and treatment. In T. Aven & E. Zio (Eds.), *Knowledge in risk assessment* (pp. 127–141). New York: Wiley.
Bjerga, T., Aven, T., & Zio, E. (2014). An illustration of the use of an approach for treating model uncertainties in risk assessment. *Reliability Engineering and System Safety, 125*, 46–53.
Cumming, R. B. (1981). Is risk assessment a science? *Risk Analysis, 1*, 1–3.
Deming, W. E. (2000). *The new economics* (2nd ed.). Cambridge, MA: MIT CAES.
Flage, R., Aven, T., Baraldi, P., & Zio, E. (2014). Concerns, challenges and directions of development for the issue of representing uncertainty in risk assessment. *Risk Analysis, 34*(7), 1196–1207.
Goerlandt, F., Khakzad, N., & Reniers, G. (2017). Validity and validation of safety-related quantitative risk analysis: A review. *Safety Science, 99*, 127–139.
Haimes, Y. Y. (2015). *Risk modelling, assessment, and management* (4th ed.). New York: Wiley.
HSE. (2001). *Reducing risk, protecting people*. London: Health and Safety Executive.
ISO. (2016). *17776. Petroleum and natural gas industries—Offshore production installations—Major accident hazard management during the design of new installations*. ISO.
Jones-Lee, M. W. (1989). *The economics of safety and physical risk*. Oxford: Blackwell.
Jones-Lee, M., & Aven, T. (2011). ALARP—What does it really mean? *Reliability Engineering and System Safety, 96*, 877–882.
Kaplan, S., Visnepolschi, S., Zlotin, B., & Zusman, A. (1999). *New tools for failure and risk analysis: Anticipatory Failure Determination (AFD) and the theory of scenario structuring*. Southfield, MI: Ideation International Inc.
Khorsandi, J., & Aven, T. (2017). Incorporating assumption deviation risk in quantitative risk assessments: A semi-quantitative approach. *Reliability Engineering and System Safety, 163*, 22–32.
Klinke, A., & Renn, O. (2002). A new approach to risk evaluation and management: Risk-based precaution-based and discourse-based strategies. *Risk Analysis, 22*(6), 1071–1094.
Lindley, D. V. (1985). *Making decisions*. New York: Wiley.
Masys, A. J. (2012). Black swans to grey swans: Revealing the uncertainty. *Disaster Prevention and Management, 21*(3), 320–335.
Meyer, T., & Reniers, G. (2013). *Engineering risk management*. Berlin: De Gruyter Graduate.
NOG. (2017). *Black swan*. Norwegian Oil and Gas Association. Retrieved August 3, 2019, from https://www.norskoljeoggass.no/en/page-not-found/migrering/publikasjoner/hse-and-operations/black-swans/.

Pasman, H. J., Rogers, W. J., & Mannan, M. S. (2017). Risk assessment: What is it worth? Shall we just do away with it, or can it do a better job? *Safety Science, 99*, 140–155.
Paté-Cornell, E. (1999). Conditional uncertainty analysis and implications for decision making: The case of WIPP. *Risk Analysis, 19*, 995–1003.
Paté-Cornell, E. (2012). On "black swans" and "perfect storms": Risk analysis and management when statistics are not enough. *Risk Analysis, 32*, 1823–1833.
PSA-N. (2012). *Report following the investigation of the hydrocarbon leakage on Heimdal 26.5.2012.* Stavanger: Petroleum Safety Authority Norway.
PSA-N. (2018). *Trends in risk level (RNNP).* Petroleum Safety Authority Norway. Retrieved August 3, 2019, from https://www.ptil.no/en/technical-competence/rnnp/.
Rae, A., Alexander, R., & McDermid, J. (2014). Fixing the cracks in the crystal ball: A maturity model for quantitative risk assessment. *Reliability Engineering and System Safety, 125*, 67–81.
Reniers, G. L. L., & Noël Van Erp, H. R. (2016). *Operational safety economics: A practical approach focused on the chemical and process industries.* Wiley. https://doi.org/10.1002/9781118871546.
Renn, O. (2008). *Risk governance: Coping with uncertainty in a complex world.* London: Earthscan.
Shewhart, W. A. (1931). *Economic control of quality of manufactured product.* New York: Van Nostrand.
Shewhart, W. A. (1939). *Statistical method from the viewpoint of quality control.* Washington DC: Dover Publications.
SRA. (2015a). *Glossary society for risk analysis. Retrieved August 3, 2019, from* www.sra.org/resources.
SRA. (2015b). *Foundations of risk analysis. Discussion note. Retrieved August 3, 2019, from* http://sra.org/sites/default/files/pdf/FoundationsMay7-2015-sent-x.pdf.
SRA. (2017a). *Core subjects of risk analysis. Retrieved August 3, 2019, from www.sra.org/resources.*
SRA. (2017b). *Risk analysis: Fundamental principles. Retrieved August 3, 2019, from www.sra.org/resources.*
Turner, B., & Pidgeon, N. (1997). *Man-made disasters* (2nd ed.). London: Butterworth-Heinemann.
Veland, H., & Aven, T. (2015). Improving the risk assessments of critical operations to better reflect uncertainties and the unforeseen. *Safety Science, 79*, 206–212.
Vinnem, J. E., Aven, T., Husebø, T., Seljelid, J., & Tveit, O. (2006). Major hazard risk indicators for monitoring of trends in the Norwegian offshore petroleum sector. *Reliability Engineering and System Safety, 91*, 778–791.
Weich, K. E., & Sutcliffe, K. M. (2007). *Managing the expected.* San Francisco: Wiley.
Weinberg, A. M. (1981). Reflections on risk assessment. *Risk Analysis, 1*, 5–7.

CHAPTER SEVEN

Decision making for risk management: A multi-criteria perspective

Hirushie Karunathilake[a,b], Ezzeddin Bakhtavar[a,c], Gyan Chhipi-Shrestha[a], Haroon R. Mian[a], Kasun Hewage[a], Rehan Sadiq[a,*]

[a]School of Engineering, University of British Columbia (Okanagan Campus), Kelowna, BC, Canada
[b]Department of Mechanical Engineering, University of Moratuwa, Moratuwa, Sri Lanka
[c]Department of Mining and Material Engineering, Urmia University of Technology, Urmia, Iran
[*]Corresponding author: e-mail address: rehan.sadiq@ubc.ca

Contents

Methods in Chemical Process Safety, Volume 4
ISSN 2468-6514
https://doi.org/10.1016/bs.mcps.2020.02.004

1. Introduction

Risk management is an integral part of engineering projects. This process is carried out in instances such as product development, process and system planning, and project implementation, and there may also be overlaps between the above-mentioned applications (Elahi, 2018; Lees, 2005; Szymański, 2017). In dealing with risks, it is necessary to both "assess" and "manage" (Aven, 2016). The work carried out in this field may focus on specific activities, to analyze and treat the risks. Alternatively, it may deal with more generic risk research and development, to understand and characterize the risks and the surrounding theories (Aven, 2016). The risk management process typically involves the following steps: (1) risk identification, (2) risk assessment and quantification, (3) implementing risk mitigation, and (4) monitoring and reviewing (Aven, 2016). Thus, merely identifying the risks and measuring their possibility of occurrence and potential impacts are not sufficient. It needs to be extended to identifying solutions and mitigation measures, implementing the said mitigation efforts, and constantly keeping track of the post-mitigation performance for changes and the necessity for remedial action.

All organizations face a wide variety of risks in carrying out their operations, and a range of different stakeholders are involved in these (Johansen, Eik-Andresen, & Ekambaram, 2014). In managing the risks, the perspectives and varying decision priorities of all stakeholder groups have to be considered. Risk management is also costly, and the expenses of risk assessment and mitigation have to be balanced with the benefits achieved by the elimination or reduction of the said risks. It needs to be assessed whether the costs incurred in identifying, quantifying, and mitigating risks is paid back by the averted costs due to the risks (Fatemi & Luft, 2002). Therefore, cost-benefit assessment is an important part of the risk management process.

Since risks can be defined as the uncertainty of outcomes, uncertainty analysis is important in risk management. Uncertainty concept can be defined as having limited knowledge, making it difficult to understand the current status or predict the future accurately (Avila, 2015). Uncertainties can be epistemic (due to lack of sufficient data) or aleatory (due to system stochasticity) (Ruparathna et al., 2017). Uncertainties coupled with adverse consequences thus lead to risks. The uncertainty created by the possibility of multiple alternative outcomes is a core part of risk, and both the probability and the consequence components of the risk function are regarded as uncertain

(Avila, 2015). Uncertainty integration in risk management can be done via sensitivity analysis, probabilistic analysis, or possibilistic modeling (Linares, 2002). Risk management needs to be dynamic, as the conditions and risk controlling strategies can change significantly with time and location. Risks are highly dependent on the macro environmental conditions, and may be influenced by economic or political factors (Fatemi & Luft, 2002).

With all of the above information, it is clear that many "*decisions*" have to be taken in the risk management process. These decisions are taken at different levels of the process and the organizational hierarchy, depending on the need and circumstances. For decision making to be effective, it needs a scientific and evidence-based approach that incorporates the available information, different priorities, and variability into the process. Many mathematical methods, models, and tools are in use for this purpose, especially in engineering applications. In addition, today, several software tools are in use to facilitate the computations, modeling, and simulations necessary for effective risk management.

It is clear that risk management is a multi-criteria decision making (MCDM) problem, as the conflicting priorities have to be accommodated while managing and mitigating the risks (Linkov et al., 2006). In MCDM, decisions are made to identify the best solution from a range of options, subject to several decision criteria and constraints (Polatidis, Haralambopoulos, Munda, & Vreeker, 2006). MCDM techniques can be used to rank and prioritize alternative solutions, or to identify an optimal system of solutions, based on the identified decision criteria, their relative importance, and the potential costs and benefits. These costs/benefits can be monetary, or alternatively, they can be in the form of environmental or social impacts both positive and negative. MCDM allows the decision makers attitudes toward risks to be reflected in risk management (Linares, 2002).

The objective of this chapter is to compile the knowledge and present the current advances in using multi-criteria decision making for risk management. Here, the principles and techniques of MCDM are introduced along with their applications in risk prioritization and management. The integration of uncertainties in risk assessment and management are also discussed with regards to practical applications of risk-based decision making. The goal is to familiarize the readers with the tools and techniques they can use for improved risk management in engineering applications. The knowledge can be utilized by the decision makers and professionals involved in a diverse range of applications such as product development, process planning, health and safety, and community development.

2. Principles of risk management

In order to explore the uses of risk-based multi-criteria decision making, it is first necessary to understand the concept of risk management itself. Risk Management process identifies, analyzes, and control the risk in any project and activity. It can be used to manage the opportunities and threats in projects (Jaafari, 2001). In short, risk management can be defined as the decision-making phase of risk assessment, where the necessary responses to identified and quantified risks are determined (Tzotzos, Head, & Hull, 2009). The objectives of project risk management are to increase the probability and impact of positive events and decrease the probability and impact of negative events in the project (PMI, 2004).

In recent years the term "risk" has been widely used in different sectors such as environmental, financial, public, occupational health, natural hazard, supply chain, and medicine. In addition, an increasing market competition and societal pressures have lead in establishing the risk management concept to evaluate, minimize, and mitigate the risk of any particular event or performed activity (de Almeida, Alencar, Garcez, & Ferreira, 2017). The risk management methodically considers management strategies to identify, assess, control, and reduce the undesirable consequences of a potential event and improve the reliability of the decision-making process (Shahriar & Bakhtavar, 2009). Risk response strategizing is a key step for the risk management process. These response strategies differ based on whether the risks are negative (threats) or positive (opportunities). Most organizations have a defined strategy on how they wish to handle the risks faced in their operations. Establishing this risk management strategy for a project or an organization (i.e., process steps, assessment method, responses and monitoring plan) is referred to as *risk framing* (Metheny, 2013). Risk framing can be in the form of regulations, policies, laws, contractual relationships, and other guidelines or process controls (Metheny, 2013).

The risk management concept has been standardized International Organization for Standardization (ISO) under ISO 31000 Risk Management (International Organization for Standardization, 2020). Under this, ISO 31000:2018, Risk management—Guidelines is a standard that provides a framework, process, and principles for risk management for organizations of all sizes, activities, and sectors. It increases the chances of realizing organizational goals by identifying opportunities and threats and allocating resources effectively for risk treatment. In addition, IEC 31010, Risk management—Risk assessment techniques can be used to help decision makers on understanding risks and measure the effectiveness of the controls (International Organization for Standardization, 2020). As per ISO 31000, risk management creates value as an integral part of the organizational processes. As such, it is a part of the decision making and needs to be structured, systematic, and timely. In addition, risk management needs to be timely, be based on the best information that is available, customized, transparent, and inclusive. The process is expected to be iterative, dynamics, and transparent, taking into account the socio-cultural factors and the need for continuous improvement (Olechowski, Oehmen, Seering, & Ben-Daya, 2016).

The following key steps have been identified in the PMBOK® Guide for the risk management process, which is defined as the "systematic process of identifying, analyzing, and responding to project risks" (Becker, 2004; PMI, 2004).

- Risk management planning
- Risk identification
- Qualitative risk analysis
- Quantitative risk analysis
- Risk response planning
- Risk monitoring and control

Thus, the major processes are risk identification, risk quantification, response development, and response control. Risk identification refers to the determination of the possible risks that can affect the project and their documentation. Quantification is done to evaluate the risks and their interactions to reveal their effect on the project outcome. Then, responses are developed to exploit opportunities and react to threats. Monitoring and control is done to identify and respond to the changes in risk and other conditions (PMI, 2004).

The tools used in risk treatment can be generally grouped into the following types: risk avoidance, risk transfer, risk abatement, risk spreading, and risk assumption (Purpura, 2013). The pathway used by each organization or individual to treat risk is determined based on their level of risk tolerance. The concept "risk appetite" is used to measure the amount and type of risk that an organization is willing to incur or bear in the attempt to achieve its objectives (Chase, 2015). It should be noted that the goal of incurring risk is to reap some reward ultimately, and the potential risk is evaluated and balanced against this reward in making the decisions regarding risk treatment. In general, a higher return translates to a higher risk. When multiple possible options are open to a decision maker, whether it is in a project or a single action, the preference is to select the option with the lowest risk and the highest return, and this is where robust decision making becomes necessary (Chase, 2015).

Several risk response strategies are available to address the negative risks that act as threats. They are briefly described as follows (Cantino, De Vincentiis, & Racca, 2011; PMI, 2004).

(a) Avoidance

In risk avoidance, once a potential risk has been identified, efforts are made to evade the risk, perhaps by not undertaking a particular activity. The best way to eliminate a risk is to avoid it completely as far as possible. This can be accomplished by several methods. One method is by changing the project scope. For example, groundwater using as source water can cause land subsidence in a city. The risk of land subsidence can completely be avoided by withdrawing water from rivers; however, it can be costly. Similarly, the project risks caused by site condition can be avoided by changing the construction site. In both cases, the project cost needs to be evaluated before the change of the scope.

(b) Transference

In risk transfer, the adverse consequences are passed on to a third party, usually through a monetary contract such as an insurance policy or a hedge. For example, the construction related risks can be transferred to the contractors through construction contracts. However, if the value of potential damages is large compared to the turnover of the contractors, they also transfer risk to the insurance or sureties. Furthermore, some risk can also be shared among different groups. For instance, a house is installing solar panels and there is a risk of unforeseen structural problems escalating the cost. Such risk can be shared by persuading the neighbors to install similar solar panels at the same time for a centralized solute PV facility.

(c) Mitigation

The project team reduces the level of the risk to an acceptable level to the project or organization, by reducing the probability and/or the impact of events (Kasap & Kaymak, 2007; PMI, 2004). For example, occupational health hazards to the workers during construction can be reduced by providing the safety devices and appropriate training. Risk abatement does not try to eliminate the occurrence, but tries to lessen the severity of the impact should the event occur. These are retroactive damage reduction measures. Risk spreading is done by distributing the possible risk over multiple locations or parties. For example, a high-value asset can be insured jointly by a number of insurers, who are then spearing the risk among themselves, or copies of a high-importance document can be stored in multiple geographic locations (Purpura, 2013).

(d) Acceptance

Finally, in risk assumption, the organization decides to assume the risk themselves and to move on with the venture despite the risk. This strategy is suitable for risks that are very small, very unlikely, or very difficult to respond by using one of the previous strategies. The project team can accept these risks with contingency planning (Kasap & Kaymak, 2007; PMI, 2004). This will decrease the vulnerability of the project toward the risk and if it happens, it can be managed by the contingency. So, such risk should accept with contingency planning. Moreover, they can also respond passively doing nothing except monitoring the status of the risk. The monitoring will help for immediate action if risk occurs that reduces the impacts of the risks.

In contrast, some risks are opportunities that need to be maximized. The risk response strategies that can be applied to opportunities are as follows (Cantino et al., 2011; PMI, 2004).

(a) Exploitation

Opportunities having strong potential benefits should be exploited. This can be done by improving the probability of occurrence of the opportunity or yet better ensuring that it will happen. However, this strategy will typically require some investment of time and money for implementation. If the benefit is higher than the cost, it is worthy to invest. For instance, municipalities are proactive to reduce water consumption in buildings and provide some funding if a building meets a specific water efficiency. This opportunity could be exploited by increasing water efficiency using water efficient water fixtures (e.g., toilet, laundry machine) and efficient lawn irrigation. The risk management team should compare the cost and potential benefit of the building retrofits and new lawn irrigation technique. If the potential benefits are higher than the cost, investment decision should be made.

(b) Share

Certain opportunities can be capitalized only if benefits are shared with others. This means the implementation of old saying "win-win" situation. Usually projects have many stakeholders with different preferences. In such a situation, a small collaboration can highly maximize opportunities for all of them. For instance, an owner and contractor can put a profit-sharing mechanism in their contract. This will encourage the contractor to develop cost saving modifications to the design. The saved cost is usually split between them with the contractor receiving a larger share. The US Army Corps of Engineers has been applying this strategy for decades, saving millions of dollars of cost.

(c) Enhancement

This strategy modifies the "size" of an opportunity by improving the probability of occurrence of the opportunity or the degree of the resulting benefits. Enhancement is not always certain, but it is a worthwhile approach. For instance, an opportunity is identified during risk analysis that a type of environmental document if prepared for a highway project could reduce 4 months of review duration for environmental assessment (EA) report. In addition, this opportunity could be enhanced by avoiding impacts in certain project areas, which may require modification in a project scope. In both events, the probability of the occurrence of the opportunity would be increased.

It is better to evaluate multiple strategies in dealing with risks, if the said risks have a high expected impact. A comprehensive method to analyze risk response strategies is to apply a value methodology. The value methodology can be used to optimize risk response strategies that will be more effective.

2.1 Risk in engineering applications

In order to understand the decision making needs that arise in engineering applications, it is first necessary to understand the risks and uncertainties that could arise in such ventures. From community development and construction to product development and market entry, projects are fraught with various types of risk. As previously mentioned, risk arises due to the uncertainty of outcomes, and risk management needs to clearly consider the potential sources of this uncertainty.

In planning for risk management, it is necessary to keep in mind that the problem is affected by both internal and external factors and stakeholders. For instance, consider the example of launching a new car into the market. It is clear that this business decision is affected by a variety of risks and uncertainties, both within and outside of the control of the organization. There may be weaknesses in the product design and the manufacturing, or the consumers may not embrace the new model. Alternatively, a rival company may release a competing product with a similar design, leading to unexpected market competition. In this example, it can be seen that each of these risks involve a different set of stakeholders. The design and manufacturing involves internal stakeholders such as designers, engineers, and other technical personnel. Sourcing the necessary materials and components to form an efficient supply chain involves external suppliers. The market acceptance, at the first glance, seems to be outside the control of the organization involving an external set of stakeholders. Yet, in mitigating this risk, market analysts and other internal experts can play a role in pre-project planning.

As individuals and organizations have limited resources that need to be utilized in the most effective manner, risk management needs to be well-planned. Due to the limitation in resources, it is not possible to treat each and every potential risk, leading to the need for prioritization and selection of risks (Linares, 2002). Risk is essentially quantified through a combined function of impact and likelihood of occurrence (Talabis & Martin, 2012). Therefore, when risks need to be prioritized in for risk treatment, the severity of the impact and the likelihood are both considered together.

When treating any given risk, there can be a number of potential treatment options, with different costs and benefits. Then it becomes necessary to evaluate the treatment options or "solutions" with regards to multiple factors. Looking back to the previous example, it can be seen that the easiest option to *avoid* the risk is to stop the product from being launched. However, there are other possible options for risk mitigation, such as doing a market analysis beforehand prior to design finalization, conducting a targeted marketing campaign, and changing the product design. The best treatment strategy could also be a mix of the above-mentioned treatment modes. In order to identify the best solution, it is necessary to do a thorough analysis. While economics are an important consideration, direct costs and revenue are not the only factors that need to be considered. The brand reputation of the company, the environmental and social impacts of the product, the match with the current portfolio, the trends in the macro-environment are all important factors in this decision. While ad hoc decision making based on the intuition and the preferences of the decision makers is sometimes used in making such decisions, a more comprehensive and scientific method is necessary to make the best choices. This is where the science of decision comes into play.

The steps of decision making can be summarized as (1) goal setting, (2) compiling options for achieving that goal, (3) rank-ordering options, and (4) selecting the highest ranked alternative (Byrnes, 2002). However, in reality, the decisionmaking effort may not be such a clearly sequential step-by-step process. Instead, it is a dynamic process with possible repetitive iterations to arrive at a final conclusion (Byrnes, 2002). Making decisions based on scientific evidence and methods makes the above process more effective, by improving predictability, confidence, transparency, and administrative discretion (Allio, Ballantine, & Meads, 2006). Mathematical techniques are useful in this endeavor, not only in processing the data inputs, but also in devising decision algorithms and incorporating uncertainty.

As indicated previously, many engineering problems have to consider multiple points of information and judgment parameters in decision making, making multi-criteria decision assessment a necessity. The next sections summarize the use of MCDM techniques in engineering problem solving and particularly risk assessment. These techniques can be used by all types of organizations in managing their project risk effectively.

2.2 Decision making for risk management

Multi-criteria decision-making in general follows six steps including, (1) problem formulation, (2) identify the requirements, (3) set goals, (4) identify various alternatives, (5) develop criteria, and (6) identify and apply decision-making technique (Sabaei, Erkoyuncu, & Roy, 2015). Various mathematical techniques can be used for this process, and the choice of techniques is made based on the nature of the problem and the level of complexity assigned to the decision-making process. All methods have their own pros and cons.

In risk management, risk should be prioritized and analyzed through the decision-making concepts (Ayyub, 2014). Risk prioritization aids managers in identifying and assessing high-risk sources, essential control strategies, and adjustment techniques for the control strategies (Behraftar, Hossaini, & Bakhtavar, 2017). Risk prioritization and assessment should be performed based on the triple main risk factors concept including, occurrence probability (O), severity (S), and detection probability (D) as addressed in the concept of failure mode and effects analysis (FMEA) method (Bakhtavar & Yousefi, 2019). FMEA is a method which can be used to identify the causes and consequences of failure events (Sharif et al., 2017). FMEA was first introduced within an aerospace agency in the 1960s. Later, the technique was adopted in many sectors particularly addressing the issues of quality and safety. FMEA is a powerful and one of the most widely-applied techniques for identifying and assessing risks (Liu, Chen, Duan, & Wang, 2019). It considers the triple risk factors of occurrence, severity, and detectability simultaneously in ranking the risks by assigning a risk priority number (RPN) (Rah, Manger, Yock, & Kim, 2016).

In this type of analysis, the failure modes (what can go wrong?), the causes of failure (the factors that contribute to the failure), and the detections or control mechanisms (what are the mechanisms to prevent or identify the failures?) are investigated for a given system. Numerical values are assigned to the severity of the impacts of failure, the probability (or frequency) of occurrence, and the ease of detection. Based on the above value, the RPN is calculated for each combination of failure modes.

$$RPN = severity \times occurence \times detection$$

The FMEA technique holds an important position in the operations research sector. Moreover, the term "risk" is very common in operations research. Hence, it is important to discuss the risk management from the context of FMEA.

FMEA is unable to assign different importance weights to the triple risk factors, separating failure modes with different severities but having similar risk priority number (RPN) and fully prioritizing (Bakhtavar & Yousefi, 2019). The literature on FMEA is steadily growing and covering many sectors, along with some modifications and applications. Determining the risk ranking of failure modes in FMEA is a multifaceted challenge that needs a multi-criteria decision making (MCDM) analysis (Liu, Chen, et al., 2019). Therefore, FMEA can be viewed as a MCDM problem because of the involvement of multiple risk factors, which includes prioritization and assessing the failure modes based on the triple risk factors. Several comprehensive studies have overviewed the application of MCDM techniques in different fields including but not limited to; energy, environment, and sustainability; supply chain management; materials; quality management; construction and project management; safety and risk management; operation research and soft computing, etc. However, it was observed that operation research and soft computing has a relatively higher application of MCDM (Mardani et al., 2015).

MCDM considers the importance of the risk factors, deconstructs the risk assessment process into distinct phases, and prioritizes failure modes through mathematical models (Liu, Chen, et al., 2019). According to a recent literature review by Liu, Chen, et al. (2019), there were more than 150 research articles published in the last two decades showing the application of MCDM in the context of FMEA under different scenarios. At a broader level, the common MCDM techniques applied in FMEA includes but not limited to distance-based techniques, outranking techniques, compromised techniques, pairwise comparison techniques etc. In addition, various hybrid and multiple factor-based techniques have been developed to address the FMEA.

In a MCDM problem, the basic ingredients are the criteria and alternatives. Different alternatives evaluated against set criteria to formulate a comparison of alternatives. The results can be improved further by assigning weights to different criteria, as the importance can vary extremely from one decision-maker to another. Hence, for selected criteria, there can be a different level of importance from the perspective of different decision-makers (Sabaei et al., 2015). It important to evaluate the assign weights to each criterion from different decision-makers to ensure the reliability of results.

The selection of the MCDM technique for a solving a particular problem can vary depending on the context, which emphasizes the need to understand decision-making classifications. The MCDM techniques are categorized based on (1) compensatory and non-compensatory, (2) discrete and continuous, and (3) individual and group decision-making. Classification of MCDM based on discrete and continuous data is the most commonly applied ones (Sabaei et al., 2015). From the perspective of discrete and continuous data, the MCDM is divided into multi-attribute decision making (MADM) and multi-objective decision making (MODM) (Zavadskas, Antucheviciene, & Kar, 2019). MADM considers the problems under an inherent discrete decision space (Triantaphyllou, 2000). MODM is based on mathematical theory and deals with the problems under continuous decision space.

3. Multi-attribute decision making

MADM is the subdiscipline of MCDM for making preference decisions, such as prioritization, evaluation, and selection over different options, which could be characterized by usually conflicting, multiple, and attributes (Hwang & Yoon, 1981). In a MADM technique, alternatives are specified and evaluated based on effectual attributes, attributes' weights are determined, and scoring alternatives with respect to attributes (Amaral & Costa, 2014).

3.1 MADM risk-based techniques

MADM techniques have been extensively used to aid FMEA in risk prioritization and analysis by enhancing the performance of FMEA. Liu, Chen, et al. (2019) also highlighted the main MCDM techniques in the case of risk prioritization of failure modes as mentioned. Among these techniques, some specifically come under MADM-based techniques. Moreover, modified MADM techniques are developed to deal with the uncertainties in the decision process; they are referred as fuzzy-based MADM techniques and have been extensively used by researchers. Fig. 1 shows a summary of the main

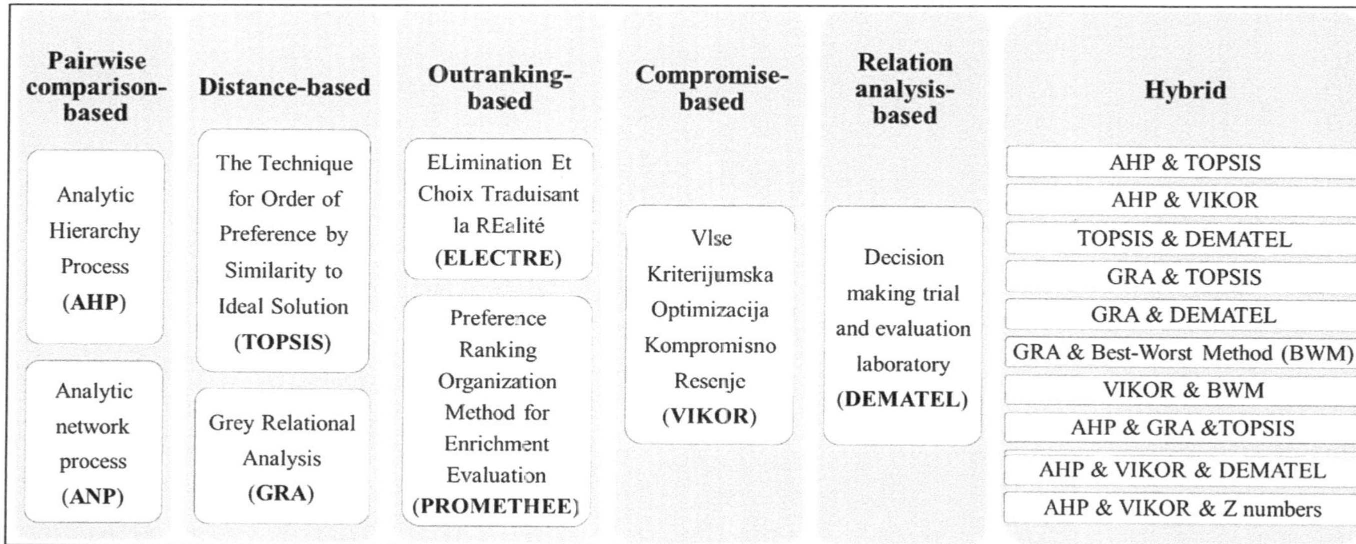

Fig. 1 The most applicable MADM techniques for risk prioritization and evaluation.

MADM techniques that have been largely applied by researchers for risk prioritization and evaluation of failure modes particularly in the context of FMEA.

Among pairwise comparison-based techniques, the analytic hierarchy process (AHP) is applied for the risk evaluation process in FMEA based on the triple risk factors (Braglia, 2000). Moreover, previous studies have used a fuzzy version of AHP in FMEA to prioritize risks (Ilangkumaran, Shanmugam, Sakthivel, & Visagavel, 2014). The analytic network process (ANP) has been proposed as another main pairwise comparison-based technique for the risk prioritization of failure modes through FMEA (Zammori & Gabbrielli, 2012). AHP has been applied for decision making during the planning phase of several sectors, such as forest planning, a risk factor for spruce beetle outbreaks, farmland preservation, incorporating risk attributes in this phase (Ananda & Herath, 2009). In AHP, the criteria weights are set based on the decision-maker, which makes this a subjective process (Sabaei et al., 2015). In addition, the values are varied depending on the shape of the hierarchy structure and simultaneously ensuring the consistency of weights (Song & Kang, 2016). The technique for order of preference by similarity to ideal solution (TOPSIS) is a distance-based technique applied to improve the traditional FMEA in risk prioritization and evaluation of failure modes (Sachdeva, Kumar & Kumar, 2009). TOPSIS also involves assigning weights and calculating the scores of alternatives. TOPSIS has been effective in producing results without requiring the decision makers or participants to define criteria scale (Cegan, Filion, Keisler, & Linkov, 2017). Fuzzy TOPSIS, which is based on the entropy technique, can be used for risk prioritization and evaluation in FMEA (Song, Ming, Wu, & Zhu, 2013). Grey relational analysis (GRA) with fuzzy linguistic variables is also used for risk prioritization and evaluation of failure modes in FMEA (Chang, Wei, & Lee, 1999). A fuzzy-based GRA technique has been developed for FMEA-based evaluation of risks (Li, Li, Sun, & Wang, 2018). In addition, a combined AHP and fuzzy TOPSIS has been proposed to optimize maintenance activities (Carpitella, Certa, Izquierdo, & La Fata, 2018). Besides fuzzy TOPSIS, other extensions have been developed to evaluate risks and rank potential failures such as intuitionistic fuzzy hybrid TOPSIS (Shi et al., 2016). Similarly, risk priority models based on the D numbers and TOPSIS have been developed to generate a more rational ranking method for failure modes (Bian, Zheng, Yin, & Deng, 2018).

In the category of outranking-based techniques, ELimination Et Choix Traduisant la REalité (ELECTRE) is applied to prioritize and evaluate failure modes in FMEA along with expert judgment and entropy-based technique for weighing risk factors (Liu, You, Li, & Su, 2016). The application of ELECTRE technique is very diverse, as this technique has been applied for various purposes such as evaluating transportation investments for urban planning and selecting the location of water treatment plants. ELECTRE is even used by shipping companies as a tool, and several software packages have been developed for some ELECTRE techniques, including ELECTRE IS, ELECTRE III-IV, and ELECTR TRI (Govindan & Jepsen, 2016). An FMEA based on preference ranking organization technique for enrichment evaluation (PROMETHEE) was developed to prioritize risk-based failure modes. The application of PROMETHEE technique is also very diverse; one study categorized the application of PROMETHEE in environment management, water hydrology, business and financial management, chemistry, transportation, manufacturing, energy management, and social sector (Behzadian, Kazemzadeh, Albadvi, & Aghdasi, 2010).

In compromise-based techniques, the fuzzy version of Vlse Kriterijumska Optimizacija Kompromisno Resenje (VIKOR) has been proposed to evaluate the risk of failure modes in FMEA (Liu, Liu, Liu, & Mao, 2012). Decision-making trial and evaluation laboratory (DEMATEL), as a relation analysis-based technique, DEMATEL-based FMEA is used for the risk prioritization purpose (Shin, Lee, & Yoon, 2018).

In previous literature, it can be noted that a wide variety of hybrid MADM techniques have been applied for the risk prioritization and evaluation of failure modes in FMEA. As indicated in Fig. 1, the main hybrid techniques are an integration of AHP and fuzzy TOPSIS (Carpitella et al., 2018), fuzzy AHP and fuzzy VIKOR (Liu, You, You, & Shan, 2015), TOPSIS and DEMATEL (Chang, Chang, & Lee, 2014), GRA and TOPSIS (Liu, Wang, You, & Wu, 2019), GRA and DEMATEL (Chang, Chang, & Tsai, 2013), GRA and best-worst technique (BWM) (Lo & Liou, 2018), fuzzy VIKOR and fuzzy BWM (Tian, Wang, & Zhang, 2018), AHP, GRA, and fuzzy TOPSIS (Sakthivel & Ikua, 2017), fuzzy AHP, fuzzy TOPSIS, and fuzzy VIKOR (Sakthivel, Saravanakumar, & Muthuramalingam, 2018), AHP, VIKOR, and DEMATEL (Liu, You, Ding, & Su, 2015), AHP, VIKOR, and Z numbers (Mohsen & Fereshteh, 2017).

In FMEA, assigning relative importance to risk factors can be subjective and leads to inaccurate results. However, MADM techniques can be applied to calculate the weights of risk factors (criteria). Broadly, the weighting of risk factors can be performed using subjective weighting, objective weighting, combination weighing, and incomplete weighting. MADM techniques are more applicable in performing subjective weighing (Liu, Chen, et al., 2019). The opinion of experts is often used in determining the weights of the factors in any MADM problem (Karunathilake, Hewage, Mérida, & Sadiq, 2019).

3.2 Risk prioritization by MADM

In this section, a general methodology framework (Fig. 2) based on MADM techniques under fuzzy environment is provided to improve FMEA in the determination of the importance weights of the triple risk factors and risk prioritization and evaluation of failure modes. This method is important in solving risk prioritization problems in an uncertain and stochastic environment.

Let there be a risk prioritization and evaluation problem consisting of m failure modes (U_i, $i = (1, 2, \ldots, m)$) that should be evaluated with respect to n risk factors (Y_j, $j = (1, 2, \ldots, n)$) based on the judgments of l FMEA experts (E_k, $k = (1, 2, \ldots, l)$). In this case, FMEA experts could have different roles;

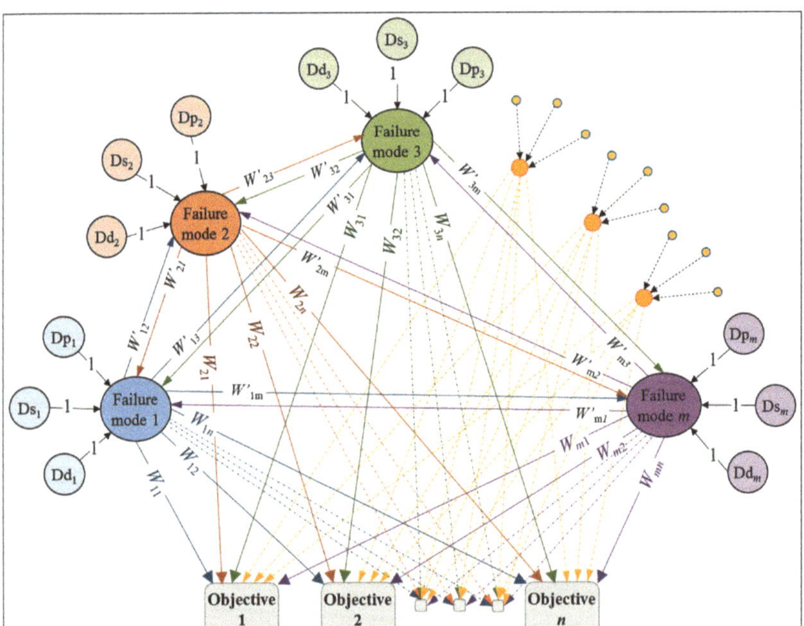

Fig. 2 Multi-objective risk-based cognitive map (updated based on Bakhtavar & Yousefi, 2018).

therefore, different judgment weights (ψ_k, $\psi = (\psi_1, \psi_2, \ldots, \psi_l)$, where $\psi_k > 0$ and $\sum_{k=1}^{l} \psi_k = 1$) could be assigned to the experts. Fuzzy risk evaluation matrix that should be completed for the failure modes (U_i) with respect to the considered risk factors (Y_j) according to the judgment of the kth FMEA expert is given by $\tilde{X}^k = \left(\tilde{x}_{ij}^k\right)_{m \times n}$, where $\tilde{x}_{ij}^k$ is the fuzzy value examined for failure mode i with respect to risk factor j based on the judgment by expert k. Based on the notations mentioned above, the methodology could be described in the following main steps:

- *Step 1*: Failure modes (U_i) as risk alternatives and risk factors (Y_j) as evaluation criteria are identified according to the FMEA concept through FMEA experts.
- *Step 2*: Fuzzy risk evaluation matrix ($\tilde{X}^k = \left(\tilde{x}_{ij}^k\right)_{m \times n}$) is formed based on the judgments by kth expert. To this end, a fuzzy rating scale or the standard tables (such as Table 1, which could be updated based on any problem) could be used.
- *Step 3*: The fuzzy risk evaluation matrices given by all experts are aggregated (if there is more than one expert). One of the common formulas to determine the aggregated fuzzy evaluation values ($\tilde{x}_{ij}^{\prime}$) of the aggregated matrix is as given in Eq. (1). If no weight does not consider the FMEA experts for their roles and judgments, Eq. (2) is applicable.

$$\tilde{x}_{ij}^{\prime} = \left\{ \prod_{k=1}^{l} \left(\tilde{x}_{ij}^k\right)^{\psi_k} \right\}^{\left(1 / \sum_{k=1}^{l} \psi_k\right)} \tag{1}$$

$$\tilde{x}_{ij}^{\prime} = \left\{ \prod_{k=1}^{l} \tilde{x}_{ij}^k \right\}^{(1/l)} \tag{2}$$

- *Step 4*: Importance degree weights are determined for the risk factors (Y_j) using the well-known weighting approach, such as fuzzy pairwise comparison (in fuzzy AHP and fuzzy ANP), gray pairwise comparison (Bakhtavar, Lotfian, & Mikaeil, 2019), fuzzy BWM, Entropy, or a combination of them. In the case of several experts, their judgments are aggregated through the updates of Eqs. (1) and (2) by considering fuzzy pairwise values given by experts for each risk factor with respect to one other risk factor instead of the fuzzy values examined for failure mode i

Table 1 Fuzzy rating scale for the triple risk factors (O, S, and D).

Occurrence probability (O)	Severity (S)	Detection probability (D)	Fuzzy number (degree)
More than one event per day	Complete destruction	Undetectable	(9,10,10)
More than one event per week	Serious damage and long-time stoppage	Very unexpected detection	(8,9,10)
An event per week	Serious damage and temporary stoppage	Unlikely event detection	(7,8,9)
More than one event per month	Medium damage and temporary stoppage	Likely to detect too little	(6,7,8)
One event per month	Minor damage and temporary stoppage	Low detection probability	(5,6,7)
More than one event per 6 months	No damage but temporary stoppage	Medium detection probability	(4,5,6)
More than one event per year	No damage and continuing after major modifications	Almost high detection probability	(3,4,5)
More than one event per 2 years	No damage and continuing after minor modifications	High detection probability	(2,3,4)
One event over 2–5 years	No damage and continuing with the minimum limitations	Very high detection probability	(1,2,3)
One event in more than 5 years	No impact	It can be definitely detected	(1,1,2)

Adapted and updated based on Bakhtavar E., & Yousefi S. (2019). Analysis of ground vibration risk on mine infrastructures: Integrating fuzzy slack-based measure model and failure effects analysis. *International Journal of Environmental Science and Technology*, 16, 6065–6076, https://doi.org/10.1007/s13762-018-2008-0.

with respect to risk factor j based on expert judgments. Using fuzzy pairwise comparison, a decision hierarchy could be formed. Notably, the resulted weights of risk factors could be valid only if the pairwise comparison judgments provided by the experts are consistent through checking the consistency degree that must be less than 0.1.

- *Step 5*: The aggregated fuzzy risk evaluation matrix ($\tilde{X}' = \left(\tilde{x}'_{ij}\right)_{m\times n}$) is defuzzified through one of the defuzzification methods, such as the center of gravity, the center of mass, centroid weighted-average method, mean of maxima, the center of the largest area, etc.
- *Step 6*: The most appropriate single or hybrid MADM techniques (Fig. 1) must be applied to prioritize the identified failure modes (U_i). In this case, GRA-TOPSIS could be one of the appropriate hybrid MADM techniques.

3.3 New developed techniques of MADM

Alireza and Javad (2019) classified more than 20 MADM techniques under the "newly developed" category. However, in this chapter, those MADM techniques developed in the last two decades are discussed broadly. These MADM-based techniques could be also used in the sixth step of the general methodology presented in this section.

Superiority and inferiority ranking: Superiority and Inferiority Ranking (SIR) was developed in 2001. In this technique the concept of superiority and inferiority scores are proposed via matrices (S-matrix and I-matrix). These matrices describe the attitude of the decision-maker in terms of intensities of superiority and inferiority of each alternative. SIR can be combined with TOPSIS and SAW. Moreover, SIR-TOPSIS is considered a special case for PROMETHEE (Xu, 2001). The technique has been applied in the energy sector and planning sector in selecting the concrete pump or in contractor selection (Alireza & Javad, 2019).

Additive ratio ASessment: Additive Ratio ASessment (ARAS) was developed in 2010. This technique is known for selecting the alternatives considering number of attributes, and the priorities of alternatives can be determined according to the utility function value. The degree of alternative utility is estimated by comparing it with the ideal solution (Zavadskas & Turskis, 2010). This technique has been used in recruitment and selection of personnel and ranking of factories. ARAS has been recently used in fuzzy and gray environments in combination with other techniques (Alireza & Javad, 2019).

Weighted aggregates sum product assessment: The weighted aggregates sum product assessment (WASPAS) was first introduced in 2012. This is a new and relatively accurate MCDM approach, which can increase the ranking accuracy of alternatives. This technique is a combination of weight sum and weight product method. Some studies indicated the applicability of

WASPAS technique in many fields and sectors. The relative importance of each attribute is determined and then the alternatives are evaluated and prioritized. The application of this technique includes material selection and process analysis (Badalpur & Nurbakhsh, 2019). A recent application of WASPAS is for a construction project in the context of qualitative risk analysis.

Stepwise weight assessment ratio analysis: The stepwise weight assessment ratio analysis (SWARA) was developed in 2010. This technique is different from other MADM approaches as the ratio of multiple criteria is the key element of the SWARA technique. The relative importance and initial prioritization of alternatives with respect to each attributes are determined by decision-maker's opinion (Panahi, Khakzad, & Afzal, 2017). This technique has been used for supplier selection and in evaluating and choosing project alternatives. SWARA can be used with other techniques due to this feature this technique in recent years has been widely used (Alireza & Javad, 2019).

Integrated determination of objective CRIteria weights: The Integrated Determination of Objective CRIteria Weights (IDOCRIW) technique was introduced in 2016. This technique determines a relative impact loss and the weight of attributes in a combination with Entropy and Criterion Impact LOSs (CILOS). IDOCRIW is applied for analysis of rotor systems, assessing the performance of the construction sectors. IDOCRIW is a first technique that combines two different techniques for the attributes evaluation. This technique is more reliable and has relatively higher accuracy (Alireza & Javad, 2019).

Evaluation based on distance from average solution: The Evaluation based on Distance from Average Solution (EDAS) technique was developed in 2015. This technique can consider the conflicting attributes by calculating the average solution. The model can describe the difference between all selected alternatives based on the positive distance from average and negative difference from average. The alternative with a higher positive different and lower negative distance from the average is selected as the most suitable one. The EDAS technique is applied in evaluating the services and personnel selection. It is suggested to apply the fuzzy EDAS model for risk analysis and in many other uncertain and fuzzy environments (Zhang, Gao, Wei, Wei, & Wei, 2019).

Multi-attributive border approximation area comparison: The Multi-Attributive Border Approximation area Comparison (MABAC) was developed in 2015. This technique has gained attention due to its simplicity in terms of application and stable solution. In this technique, the performance of each criterion is taken from the context of the upper approximation area having ideal alternatives and lower approximation area having non-ideal alternatives. This technique has been used in the evaluation of railway stations, selecting the location of wind farms, and site energy generation technology (Roy, Ranjan, Debnath, & Kar, 2016).

4. Multi-objective decision making for optimization

In many decision problems, more than one objective is followed by decision-makers. This concept is based on MODM problems, which could be formulated by multi-objective mathematical programming (MOMP) models in continuous decision space. In a multi-objective optimization problem, conflicting criteria are optimized together to provide the best solution to a given problem, in applications such as planning and design (San Cristóbal, 2011; Yoon & Hwang, 1995). A MOMP model includes two or more objective functions subjected to several constraints. Fuzzy numbers could be considered in these optimization models for referring to uncertainties in judgments and ambiguities in data (Lu, Zhang, & Ruan, 2008). To this end, triangular fuzzy numbers could be used in both objective functions and constraints. In addition to the mathematical programming-based MODM techniques, some other MCDM techniques, such as multi-objective optimization by ratio analysis (MOORA) and data envelopment analysis (DEA) could apply as MODM techniques due to their nature.

Goal programming is the main technique to model and solve the mathematical programming-based MODM problems based on mathematical programming (Charnes & Cooper, 1977). In goal programming, the optimal solution could be obtained when deviations from the goals that considered by decision-makers are minimized (Bakhtavar, Aghayarloo, Yousefi, Hewage, & Sadiq, 2019). To provide precise achievement for each objective function, fuzzy set theory could be integrated to goal programming. Therefore, a problem with several alternatives and goals based on fuzzy sets could be considered in the form of fuzzy goal programming (Ramík, 2000).

4.1 MODM risk-based techniques

MODM risk-based research reported in literature is limited compared to MADM risk-based research. In this section, three main techniques that apply multi-objective concept for risk prioritization and evaluation are presented. The techniques are MULTIMOORA, multi-objective fuzzy cognitive map with MADM (MOFCM-MADM), and alpha-level fuzzy-based DEA (AL-FDEA).

The base of MULTIMOORA is MOORA technique introduced by Brauers and Zavadskas (2006). This technique has been effectually employed in many decision-making problems. The main advantage of MOORA is the ability to select the best possible (satisfactory) option through simultaneously optimizing numerous contrary criteria subject to essential constraints. Other advantages of MOORA are its simplicity, minimum mathematical calculations, stability and robustness, and comprehensiveness (Bakhtavar, Yousefi, & Jafarpour, 2019). MULTIMOORA was developed for the risk evaluation of failure modes in FMEA under fuzzy environment (Liu, Fan, Li, & Chen, 2014). MULTIMOORA technique was then further developed by using continuous weighted entropy and fuzzy sets based on interval-valued intuitionistic to improve the capabilities of FMEA (Zhao, You, & Liu, 2017). This version of MULTIMOORA is considered in Section 4.1.1.

Bakhtavar and Yousefi (2018) developed MOFCM-MADM based on the fuzzy cognitive map (FCM) introduced by Kosko (1986) and weighting scenarios-based TOPSIS for the risk prioritization and evaluation of failure modes with respect to risk factors (O, S, and D) in FMEA. FCM is a graph-based knowledge representation technique to represent the causal relationships among a set of concepts. It could proficiently analyze and solve problems with the concepts of decision support, classification, time series prediction, and system control (Bakhtavar & Shirvand, 2019). Also, the causality weights of components in a decision-based system were suitably resulted from FCM and joined to MCDM techniques such as FANP (Bakhtavar, Aghayarloo, et al., 2019). In general, FCM determines causal relationships among different components based on a value in the interval $[-1.1]$ under a mental process of individuals based on a specific FCM structure (Kosko, 1986). Expert-based knowledge is required to draw a specific FCM model consisting of nodes, interconnecting nodes with arcs, and signs on the arcs. Criteria (such as risk factors and objectives) as the concepts of a system are indicated with nodes, while causal relationships among the concepts (nodes) are described through arcs and their signs based on weights (W_{ij}) between concepts i and j. Positive, negative, and no causal relationships are presented with $W_{ij}>0$, $W_{ij}<0$, and $W_{ij}=0$, respectively (Bakhtavar, Aghayarloo, et al., 2019).

AL-FDEA was developed by Bakhtavar and Yousefi (2019) based on DEA (Charnes & Cooper, 1977), which is a non-parametric technique with the mathematical modeling concept to prioritize failure modes with respect to the triple risk factors. DEA, which consists of objective function and constraints, calculates the efficiency of decision-making units (DMUs). Fuzzy concept based on alpha-level cut and the slack-based model (SBM) of DEA were applied in the AL-FDEA technique to consider uncertainties in the risk-based decision-making problem. The fuzzy DEA part of AL-FDEA calculates the importance weights of the risk factors through mathematical programming and scores for failure modes. The multiple objectives could be defined as the minimization of occurrence probability and severity and maximization of detection probability (the triple risk factors), which come together in a single objective function. There are two single objective functions for the lower and upper bounds of the efficiency of DMUs (Bakhtavar & Yousefi, 2019).

4.1.1 Fuzzy MULTIMOORA

A risk-based fuzzy MULTIMOORA technique proposed by Liu et al. (2014) is studied for the risk prioritization and evaluation of potential failure modes under the fuzzy environment in FMEA. The technique initiates with a decision matrix consists of the values of alternatives with respect to objectives (criteria). The ratio system, reference point method, and full multiplicative form are three main concepts involved in the MULTIMOORA technique. Every coordinate of the reference point method, which is defined based on the ratio system, could be described as the minimization or maximization of a certain objective. The full multiplicative form could be expressed as the minimization and maximization of the purely multiplicative utility function (Liu et al., 2014).

A risk problem is assumed to be consisting of m failure modes (U_i, $i=(1, 2, \ldots, m)$) that are evaluated with respect to n risk factors (Y_j, $j=(1, 2, \ldots, n)$) based on the judgments by l FMEA experts (E_k, $k=(1, 2, \ldots, l)$). FMEA experts can have different judgment weights ($\psi_k, \psi=(\psi_1, \psi_2, \ldots, \psi_l)$, where $\psi_k>0$ and $\sum_{k=1}^{l}\psi_k=1$). Fuzzy risk evaluation matrix for the failure modes (U_i) with respect to the considered risk factors (Y_j) based on the judgment of the kth FMEA expert is represented by $\tilde{X}^k=\left(\tilde{x}_{ij}^k\right)_{m\times n}$, where $\tilde{x}_{ij}^k$ is the fuzzy value examined for failure mode i with respect to risk factor j based on the judgment by expert k. Based on the notations mentioned above, the technique is described through the following main steps:

- *Step 1*: Failure modes (U_i) and the triple risk factors (Y_j) are defined as risk alternatives and objectives, respectively.
- *Step 2*: Fuzzy risk evaluation matrix ($\tilde{X}^k = \left(\tilde{x}_{ij}^k\right)_{m\times n}$) based on the judgments by kth expert is formed. A fuzzy rating scale such as that given in Table 1 or any updates of that could be applied for this purpose.
- *Step 3*: All fuzzy risk evaluation matrices provided based on different experts are aggregated (if there are more than one expert) through Eq. (1). If the importance of the judgments by experts is the same, Eq. (2) is applicable instead of Eq. (1). $\tilde{X}' = \left(\tilde{x}'_{ij}\right)_{m\times n}$ denotes the aggregated fuzzy risk evaluation matrix. Moreover, the fuzzy importance weights of the risk factors (objectives) could be determined through one of the weighting methods (such as fuzzy pairwise comparison). Judgments could be aggregated based on Eqs. (1) and (2) if more than one expert may provide pairwise comparisons.
- *Step 4*: The aggregated fuzzy risk evaluation matrix is normalized as $\tilde{R} = \left(\tilde{r}_{ij}\right)_{m\times n}$ through Eq. (3). $\tilde{r}_{ij}$ notifies a normalized triangular fuzzy number. Then, the weighted normalized fuzzy risk evaluation matrix ($\tilde{R}' = \left(\tilde{r}'_{ij}\right)_{m\times n}$) is formed by multiplication of the weights of risk factors and the normalized triangular fuzzy numbers through Eq. (4).

$$\tilde{r}_{ij} = \left(r_{ij1}, r_{ij2}, r_{ij3}\right) = \left(\frac{x'_{ij1}}{x^*_{ij3}}, \frac{x'_{ij2}}{x^*_{ij3}}, \frac{x'_{ij3}}{x^*_{ij3}}\right)$$
$$x^*_{ij3} = \sqrt{\sum_{i=1}^{m}\left(x'_{ij3}\right)^2} \tag{3}$$

$$\tilde{r}'_{ij} = \left(r'_{ij1}, r'_{ij2}, r'_{ij3}\right) = w_j \otimes \tilde{r}_{ij} = \left(w_j r_{ij1}, w_j r_{ij2}, w_j r_{ij3}\right) \tag{4}$$

- *Step 5*: Based on the concept of the fuzzy ratio system, the evaluations adopted by decision-makers are summed in the maximization (positive) case and subtracted in the minimization (negative) case to calculate the normalized performance values or ratios ($\tilde{\eta}_i$) for each failure mode through Eq. (5). Positive and negative objectives are indicated by g and n-g, respectively (Liu et al., 2014). Then, the fuzzy performance ratios for each failure modes with respect to all risk factors as the objectives ($\tilde{\eta}_i$) are defuzzified by one of the defuzzification methods, such as the centroid method (Eq. 6). The symbols $\oplus$ and $\ominus$ are the fuzzy summation and subtracting, respectively.

$$\tilde{\eta}_i = (\eta_{i1}, \eta_{i2}, \eta_{i3}) = \left(\overset{g}{\underset{j=1}{\oplus}} \tilde{r}'_{ij} \right) \ominus \left(\overset{n}{\underset{j=g+1}{\oplus}} \tilde{r}'_{ij} \right) \tag{5}$$

$$\eta_i = \frac{\eta_{i1} + \eta_{i2} + \eta_{i3}}{3} \tag{6}$$

- *Step 6*: Based on the fuzzy reference point method, the fuzzy maximum objective reference point vector is calculated based on the weighted normalized fuzzy evaluation matrix $\tilde{R}' = \left(\tilde{r}'_{ij}\right)_{m \times n}$ obtained by Eq. (4). The jth fuzzy component of the reference point is denoted by $\tilde{r}^*_j$ and is determined in the form of maximum or minimum of the jth risk factor (positive or negative) through Eq. (7). Accordingly, the distance of each failure mode from the fuzzy maximum objective reference point is calculated through Eq. (8). Then, failure modes could be prioritized according to the deviation from the reference point and the min-max Metric of Tchebycheff (Liu et al., 2014).

$$\tilde{r}^*_j = \left\{ \begin{array}{l} \left(\max_i r'_{ij1}, \max_i r'_{ij2}, \max_i r'_{ij3} \right), j \le g \\ \left(\min_i r'_{ij1}, \min_i r'_{ij2}, \min_i r'_{ij3} \right), j > g \end{array} \right\} \tag{7}$$

$$d_i = \max_j d\left(\tilde{r}'_{ij}, \tilde{r}^*_j\right) = \max_j \left\{ \sqrt{\frac{1}{3}\left[\left(r'_{ij1}, r^*_{j1}\right)^2 + \left(r'_{ij2}, r^*_{j2}\right)^2 + \left(r'_{ij3}, r^*_{j3}\right)^2 \right]} \right\} \tag{8}$$

- *Step 7*: Based on the fuzzy full multiplicative form through Eq. (9), the overall utility of ith failure mode could be stated to rank failure modes. $\tilde{A}_i$ denotes the product of the risk factors of ith failure mode that must be minimized and $\tilde{B}_i$ denotes the product of the risk factors of ith failure mode that must be maximized (Liu et al., 2014). Then, the overall utility of failure modes are deffuzified through Eq. (6) to rank the failure modes. The higher overall utility indicates a failure mode with the higher rank among other failure modes.

$$\tilde{U}_i = \tilde{A}_i \oslash \tilde{B}_i; \quad \tilde{A}_i = \prod_{j=1}^{g} \tilde{r}'_{ij}; \quad \tilde{B}_i = \prod_{j=g+1}^{n} \tilde{r}'_{ij} \tag{9}$$

- *Step 8*: As given in the dominance theory (Brauers & Zavadskas, 2011), final prioritization of failure modes is done based on the three ranks obtained from Steps 5 to 7.

4.1.2 Multi-objective fuzzy cognitive map with MADM

MOFCM-MADM technique integrates FCM with multiple risk-oriented objectives and different weighting scenarios in TOPSIS. This technique prioritizes and evaluates failure modes with respect to the triple risk factors in three main steps as follows (Bakhtavar & Yousefi, 2018).

- *Step 1*: Failure modes are recognized and considered as risk alternatives (R_i), and the fuzzy degrees of occurrence probability (D_p), severity (D_s), and detection (D_d) are determined for each failure mode through the scale given in Table 1. Besides, the main objectives (goals) of the system (such as maximizing safety, minimizing stoppage in operation, and minimizing operational and capital costs) are defined.
- *Step 2*: A specific FCM is designed to investigate causal relationships among the failure modes in the presence of the risk factors and considered multiple objectives (Fig. 2). The figure indicates a cognitive map that considers the effects of risk factors on each failure mode and the failure modes on the management objectives (goals) during the risk evaluation process. Accordingly, the multi-objective FCM model consists of the following steps:
  - *Step 2.1*: Concepts (failure modes, risk factors, and objectives) are considered as the map nodes in the system.
  - *Step 2.2*: The initial matrix of causal relationship weights is considered by determining the causal relationship between every two concepts based on expert opinions. Accordingly, the specific cognitive map is

designed. Fig. 2 indicates a sample cognitive map representing three different sets of required nodes (failure modes, risk factors, and objectives). In this figure, the risk factors' nodes are linked to each failure mode by a weight of 1. Because the amounts of the risk factors' concepts should be entirely given to any failure mode. Other causal relationships and weights between other every two concepts (W_{ij} for $i=(1, 2, \ldots, m)$ and $j=(1, 2, \ldots, n)$) should be specified by expert opinions.

- *Step 2.3*: Scenarios are made, and the values of the nodes are calculated through obtaining a sustainable structure of the cognitive map. In each scenario, one failure mode is assumed to occur, and a value of 1 is assigned to its related concept accordingly. Therefore, the values of the triple risk factors are normalized between 0 and 1. For this purpose, after determining the values of a certain node, the values for other nodes related to that node can be calculated by Eq. (10); where, $A_i^{(k+1)}$ is the value of concept i in iteration $k+1$, f is transformation function, and $A_i^{(k)}$ is the value of concept i in iteration k. A hybrid learning algorithm of nonlinear Hebbian and differential evolution could be used to improve the precision of the final weights of concepts when the initial causal relationship matrix is completed based on expert opinions. Using the learning algorithm, the dependency on expert opinions could be decreased, and the structure of FCM could be improved. The output of the three concepts resulted by running the hybrid algorithm in each scenario indicates the effect degree of each failure mode on the system objectives.

$$A_i^{(k+1)} = f\left(A_i^{(k)} + \sum_{\substack{j=1 \\ j\neq i}}^{n} W_{ij} A_j^{(k)}\right) \tag{10}$$

- *Step 3*: A single or hybrid MADM technique (Fig. 1) is applied to prioritize failure modes as alternatives with respect to the system's multiple objectives (as criteria). Notably, the designed FCM in Step 2 results in the weights of the failure modes with respect to objectives, which is considered as a decision matrix. Before doing any process by a MADM technique on the FCM-based decision matrix, the matrix is normalized. TOPSIS or GRA-TOPSIS, for example, could be applied for risk prioritization. As a result of the prioritization, the failure mode with the highest value represents the most critical failure mode (risk), which needs high consideration by the system management. As pointed out by Bakhtavar and Yousefi (2018), different weighting scenarios could be regarded for the objectives with different importance to cover the different opinions of any system management policy.

4.1.3 Alpha-level fuzzy-based DEA

AL-FDEA techniques applied the SBM of DEA under uncertainty using the alpha-level cut concept for the risk prioritization and evaluation of failure modes based on the triple risk factors in FMEA. This technique reduces dependency on expert opinions and could consider a multi-objective function based on the minimization of occurrence probability (O) and severity (S) and maximization of detection probability (D) subject to the required constraints. These concepts are defined in two single objective functions, which address the lower and upper bounds of the efficiency of DMUs. In this technique, the determination of scoring weights for failure modes is based on mathematical programming (Bakhtavar & Yousefi, 2019). The main steps of AL-FDEA are summarized as follows:

- *Step 1*: Failure modes are defined as DMUs with risk factors as inputs to extract outputs.
- *Step 2*: Triangular fuzzy numbers based on (L, M, U), as described in Table 1 are assigned to the triple risk factors (evaluation objectives) with respect to each failure mode.
- *Step 3*: The triangular fuzzy values obtained from Step 2 are converted into interval values by (L^*, U^*) based on α-level cut concept for importing as inputs to the DEA models through Eqs. (11) and (12). To this end, cuts in different α levels (for $\alpha = [0,1]$) could be considered, and the interval values for the triple risk factors are determined through $L^* = \alpha M + (1-\alpha)L$ and $U^* = \alpha M + (1-\alpha)U$ (Bakhtavar & Yousefi, 2019). The minimization of the two first risk factors and maximization of the third one could be followed as multiple objectives to remove the criticality of failure modes.
- *Step 4*: The lower and upper bounds of the efficiency scores based on different α-level per each failure mode (DMU) are determined by running the mathematical models in Eqs. (11) and (12), respectively. In these equations, $(E_k)_\alpha^U$ and $(E_k)_\alpha^L$ are the upper and lower bounds of the

efficiency of DMU k at different α levels, respectively; $(x_{ik})_\alpha^L$, $(x_{ik})_\alpha^U$, $(y_{rk})_\alpha^L$, and $(y_{rk})_\alpha^U$ are the lower and upper bounds of inputs and outputs for DMU k at different α levels. Notably, an SBM without outputs (a constant output equal to one in Eqs. (11) and (12)) is applied for the efficiency evaluation problem due to numerous DMUs (Bakhtavar & Yousefi, 2019).

$$
\begin{aligned}
&\operatorname{Min}(E_k)_\alpha^L = q - \frac{1}{m}\sum_{i=1}^{m}\left(S_i^-\right)^U/(x_{ik})_\alpha^U \\
&\text{s.t.} \\
&\quad 1 = q - \frac{1}{s}\sum_{r=1}^{s}\left(S_r^+\right)^L/(y_{rk})_\alpha^L \\
&\quad q(x_{ik})_\alpha^U = \sum_{j=1,\neq k}^{n}\left(x_{ij}\right)_\alpha^L\lambda_j' + (x_{ik})_\alpha^U\lambda_j' + \left(S_i^-\right)^U \quad i=1,\ldots,m \\
&\quad q(y_{rk})_\alpha^L = \sum_{j=1,\neq k}^{n}\left(y_{rj}\right)_\alpha^U\lambda_j' + (y_{rk})_\alpha^L\lambda_j' - \left(S_r^+\right)^L \quad r=1,\ldots,s \\
&\quad \sum_{j=1}^{n}\lambda_j' = q \\
&\quad \lambda_j' \geq 0, j=1,\ldots,n, \quad \left(S_i^-\right)^U \geq 0,\ i=1,\ldots,m, \quad \left(S_r^+\right)^L \geq 0,\ r=1,\ldots,s, \quad q>0
\end{aligned} \tag{11}
$$

$$
\begin{aligned}
&\operatorname{Min}(E_k)_\alpha^U = q - \frac{1}{m}\sum_{i=1}^{m}\left(S_i^-\right)^L/(x_{ik})_\alpha^L \\
&\text{s.t.} \\
&\quad 1 = q - \frac{1}{s}\sum_{r=1}^{s}\left(S_r^+\right)^U/(y_{rk})_\alpha^U \\
&\quad q(x_{ik})_\alpha^L = \sum_{j=1,\neq k}^{n}\left(x_{ij}\right)_\alpha^U\lambda_j' + (x_{ik})_\alpha^L\lambda_j' + \left(S_i^-\right)^L \quad i=1,\ldots,m \\
&\quad q(y_{rk})_\alpha^U = \sum_{j=1,\neq k}^{n}\left(y_{rj}\right)_\alpha^L\lambda_j' + (y_{rk})_\alpha^U\lambda_j' - \left(S_r^+\right)^U \quad r=1,\ldots,s \\
&\quad \sum_{j=1}^{n}\lambda_j' = q \\
&\quad \lambda_j' \geq 0, j=1,\ldots,n, \quad \left(S_i^-\right)^L \geq 0,\ i=1,\ldots,m, \quad \left(S_r^+\right)^U \geq 0,\ r=1,\ldots,s, \quad q>0
\end{aligned} \tag{12}
$$

- *Step 5*: Failure modes are prioritized based on $1 - I(\widetilde{E}_{\tilde{k}}, R)$ through Eq. (13). Where, $I(\widetilde{E}_{\tilde{k}}, R)$ represents a prioritization score for *k*th DMU (failure mode), $(E_k)^L_{\alpha_i}$ and $(E_k)^U_{\alpha_i}$ are the lower and lower bounds of the efficiency of *k*th DMU for α_i level (*i* is the index for different α levels by $i=0,1, \ldots, m$). Notably, in the case of risk prioritization problems and the absence of outputs in DEA, the highest value of $1 - I(\widetilde{E}_{\tilde{k}}, R)$ represents a failure mode with the highest priority among other failure modes. This failure mode should be considered as the highest critical risk in the system. It means that a failure mode with the lowest $I(\widetilde{E}_{\tilde{k}}, R)$ has the highest critical priority.

$$I(\widetilde{E}_{\tilde{k}}, R)_{m\to\infty} = \frac{\sum_{i=0}^{m}\left[(E_k)^U_{\alpha_i} - \left(\min_{i,k}\left\{(E_k)_{\alpha_i}\right\}\right)\right]}{\sum_{i=0}^{m}\left[(E_k)^U_{\alpha_i} - \left(\min_{i,k}\left\{(E_k)_{\alpha_i}\right\}\right)\right] - \sum_{i=0}^{m}\left[(E_k)^L_{\alpha_i} - \left(\max_{i,k}\left\{(E_k)_{\alpha_i}\right\}\right)\right]} \tag{13}$$

4.2 Risk management strategy optimization

When the risk management strategy is optimized for a given problem, community, or an organization, it is necessary to consider various parameters. One such parameter is the stakeholders.

Certain environmental decisions such as risk management, sustainability enhancement, carbon pricing, etc., involve multiple stakeholders. These stakeholders may have their own preferences that affect an ultimate goal. Such preferences can be conflicting to each other. The preferences of all the stakeholders should be considered and optimized while making decisions in this context. Such a problem can be solved by using game theory. Game theory can provide the optimal solution for multiple stakeholders or agents. It is applicable to resolve socio-economic aspects of conflicts and planning, design, and policy problem even using qualitative information (Madani, 2010). Game theory was developed by von Neumann and Morgenstern with the publication of "Theory of Games and Economic Behavior" book in 1944 (Madani, 2010; Nash, 1951).

Game theory investigates the outcomes of interaction among self-interested agents. "Self-interest" means each agent has his own preference and acts in an attempt to achieve the preferences (Leyton-Brown & Shovan, 2008). The interest of an agent can be modeled employing utility theory that quantifies an agent's level of preference over a set of available alternatives. In addition, game theory demonstrates a change in preferences because of the uncertainty present in alternative (Leyton-Brown & Shovan, 2008). Typically, games are of two types: non-cooperative and cooperative. In a non-cooperative game the basic modeling unit is the individual who plays individually concerning only his/her benefits and does not make binding commitments with other players to coordinate strategies (Leyton-Brown & Shovan, 2008; Ma, Li, Lin, Zhu, & Yuan, 2015). In a cooperative game, also known as coalitional game, the basic modeling unit is a group in which players form a coalition to improve their collective payoff (Leyton-Brown & Shovan, 2008; Ma et al., 2015).

The strategic interactions in game theory are commonly represented in a normal form, also called the strategic or matrix form. A (finite, n-person) normal-form game is a tuple (N, A, u), where N is a finite set of n players, indexed by i (Leyton-Brown & Shovan, 2008); $A = A_i \times \cdots \times A_n$, where A_i is a finite set of actions available to player i.

Each vector $a = (a_i, \ldots, a_n) \in A$ is called an action profile; $u = (u_i, \ldots, u_n)$ where u_i: $A \rightarrow R$ is a real-valued utility (or payoff) function for player i.

The most fundamental solution concepts in a game theory are Pareto optimality and Nash equilibrium (Leyton-Brown & Shovan, 2008). In any non-cooperative game Nash equilibrium is a solution from which no player can unilaterally deviate to improve his payoff (Nash, 1950, 1951). In a cooperative game, Pareto optimal is the best solution in which no one can be made better off without making someone worse off (Chew, Aya, Tan, Chwan, & Foo, 2011). Usually, the Pareto optimal solution in a cooperative game does not coincide with the Nash equilibrium solution in a non-cooperative game (Chew et al., 2009). Some case studies are briefly described below.

Chhipi-Shrestha, Rodriguez, and Sadiq (2019) proposed a framework by combining multi-criteria decision analysis (MCDA) and game theory for a selection of a sustainable water reuse application, incorporating human health risk (Chhipi-Shrestha et al., 2019). The framework was applied to a City of Penticton (BC, Canada). Several criteria were considered, such as social: human health risk, government policy, and public perception; economic: life cycle cost; and environmental: freshwater saving, energy use, and carbon emissions were considered for three stakeholders: municipality, citizens, and farm operators. Altogether eight water reuse options were included. The analysis shows that lawn, golf course and public park irrigation and toilet flushing with an equal sharing of municipal benefits between the municipality and citizens is the optimal solution.

Another study proposed a decision framework that can model the stakeholder's conflicting priorities over the sustainability criteria using game theory for the selection of a sustainable municipal solid waste treatment option without significant environmental risks (Soltani, Sadiq, & Hewage, 2016). The proposed framework compares life cycle sustainability impacts of the selected options incorporating stakeholders' preferences. Then, game theory was applied for the fair share of the costs and benefits among stakeholders. This helps stakeholders to reach an agreement on a mutually sustainable solution. The framework was applied for the selection of waste-to-energy technology for Vancouver (BC, Canada) for two stakeholders: cement industry and municipality. The analysis shows that the cement industry and the municipality both can take benefit from the refuse-derived fuel, if the industry pays a tipping fee to access the required amount of solid waste from the municipality.

Other instances such as Wang, Cai, Tan, and Zeng (2017) used a game theory to assist in selecting the best groundwater remediation technologies to reduce the risk. Aviso, Tan, and Culaba (2010) applied game theory to analyze the responses of participating firms to the strategies of other firms for the maximization of their own respective payoffs, including human health risks. Similarly, Romero-Gomez, Choi, Woods, and Kang (2011) used game theory approach to analyze the impact of decentralized reclaimed water distribution networks in Tucson, Arizona, USA. These instances show that game theory can be applied to optimize the payoffs and reduce risks for all the multiple stakeholders as a way of risk management.

5. Practical application of decision making principles in risk management

The practical applications of risk management can be seen in a diverse array of fields, ranging from health sciences and national security to basic factory floor management. In real life context, information rarely has 100% accuracy, and conditions do not remain static. This creates the need for uncertainty analysis-based decision making in risk management.

5.1 Stochastic process-based risk management

Certain processes affecting risk can be stochastic. Stochastic processes involve random variables and characterized by variability in space or time. These processes as opposed to deterministic processes result in uncertainty in the risk estimate. The stochastic uncertainty can be incorporated in decision analysis using Monte Carlo Simulations (MCS). *Monte Carlo Simulation* is a technique for characterizing the uncertainty and variability in risk estimates by repeatedly sampling the probability distributions of the risk equation inputs and using these inputs to calculate a range of risk values (US EPA, 2001). MCSs are widely used to incorporate uncertainty in decisions since its introduction in the 1940s, e.g., in probabilistic risk estimate (Martani, 2015). Some examples are given below.

Martani (2015) used MCSs in the analysis of microbial risks to include variability and uncertainty in input data (e.g., microbial concentration, exposure parameters) for developing guidelines for microbial quality of reclaimed water (Chhipi-Shrestha, Hewage, & Sadiq, 2017a). Such simulations provide the risk estimates in a range rather than single value. A specific value can be selected based on the benchmarks, targets, regulations, or capability to handle the risk. They also used sensitivity analysis to show the effects of the variability of input parameters on the final risk estimate. The most sensitive input parameters in all water reuse types were found to be pathogenic *E. coli* ratio and morbidity. The effects of pathogenic *E. coli* ratio on the final output can range from −44% to 91% across all reuse types.

Borges (2017) applied MCSs to incorporate uncertainty in lifecycle cost analysis in the oil and gas industry. The author run MCSs in different input variables by selecting a range and probability distribution for each input variable affecting production efficiency and capital expenditure (Borges, 2017). In addition, a tornado chart was used to show sensitivity analysis results for final outputs: production efficiency and net present value of the revenues. The analysis shows that ±20% variation in input parameters: mean time to repair and mean time to failure can result the production efficiency below the minimum acceptable target of 97%. Similarly, the sensitivity analysis shows that variation in oil price and the mean time to failure play a significant role in the variance of the net present value of the revenues.

Leontaris, Morales-Nápoles, and Wolfert (2016) used MCSs to incorporate uncertainty in cable installation activities for offshore wind farms (Leontaris et al., 2016). In particular, they applied MCSs to take into account the weather risk affecting cable installation duration. The research presented two scenarios: (1) infield cable installation and (2) cable (un)availability. Scenario 1 did not account the unavailability of cable. Scenario 2, i.e., the cable unavailability had longer installation duration as expected. Although 90th percentile of installation durations for Scenarios 1 and 2 have a small difference, it is noteworthy that the day rates of the vessels that are required in the installation are in a scale of hundreds of thousands of euros. When 80th percentile, which is usually preferred by planners, is considered, the infield cable installation duration may require up to additional 2 days when unavailability of cable is accounted. Such analysis demonstrates the applicability of MCSs in incorporating uncertainty in risk management decisions.

5.2 Decision makers' attitudes

Decision makers can have varying attitude about a subject. Due to which the selection of the optimal alternative is uncertain. The available decision approaches are as follows (Engemann, Miller, & Yager, 1993).

(a) *Pessimistic*: The decision maker selects the worst possible outcome in each alternative under this strategy. Then the best-worst alternative is selected.

(b) *Optimistic*: The decision maker selects the best possible outcome in each alternative in this approach. Then, the best of the best alternative is selected.

(c) *Hurwicz*: Under this strategy, the decision maker selects a level of optimism, α, between 0 and 1. The decision maker applies a weighted average of the best, C^{+}_{i}, and worst, C^{-}_{i}, payoffs in each alternative as:

$$H = \alpha C^{+}_{i} + (1 - \alpha) C^{-}_{i}$$

Then the decision maker selects the alternative that has the highest H value.

(d) *Normative:* Under this strategy, the decision maker takes the mean of the payoffs across all n outcomes for each alternative. Then the alternative with the highest mean value is selected.

Typically, the decision process for the decision under uncertainty is as follows:

1. For each alternative A_i, calculate $X_i = f(C_{i1}, C_{i2}, \ldots, C_{in})$, where f is an aggregation function which depends on the decision maker's attitude.
2. Select alterative A^* such that $X^* = \text{Max}[X_i]$.

In this approach, any aggregation operator can be used such as weighted arithmetic mean, harmonic mean, geometric mean, ordered weighted average, min operator, max operator, etc. For example, Dinius (1987) and Brown, McClelland, Deininger, and Tozer (1970) used the weighted arithmetic mean method for aggregation and Dojlido, Raniszewski, and Woyciechowska (1994) proposed the square root harmonic mean for aggregation. Similarly, some studies have used Ordered Weighted Averaging (OWA) (Sadiq & Tesfamariam, 2007; Yager, 1988), which is a class of generalized mean operator developed by Yager. Moreover, Smith (1990) used minimum operator as an aggregation method for the development of environmental indices (Smith, 1990).

5.3 Dynamic risk management

A dynamic risk assessment (DRA) refers to a method that updates the estimated risk of a dynamic process. A process may change with time due to varying environmental conditions, operational and maintenance activities, human factor, safety barriers, regulations, procedures, etc. Compared to traditional risk assessment, dynamic risk assessment includes an additional phase of monitoring the process and update the risk estimate (Khan et al., 2016). Dynamic risk management includes the key approach–dynamic risk assessment. Khan et al. (2016) proposed a dynamic risk management (DRM) framework by integrating Plan-Do-Check-Adjust (PDCA) management method and DRA approach as shown in Fig. 3. The steps of DRM are briefly mentioned below.

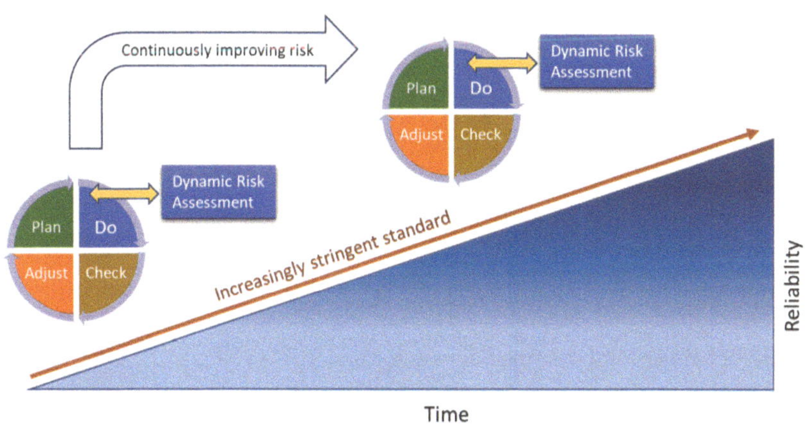

Fig. 3 Dynamic risk management framework. *Adapted from Khan, F., Hashemi S. J., Paltrinieri N., Amyotte P., Cozzani V., & Reniers G. (2016). Dynamic risk management: A contemporary approach to process safety management.* Current Opinion in Chemical Engineering, *14, 9–17.*

(a) *Plan*: Identify and recognize new potential hazards; plan a change.
(b) *Do*: Implement and test the change; apply dynamic risk assessment and update the estimated risks; show residual risks.
(c) *Check*: Examine the performance of the change; compare the residual risk to the expected risk resulted from the changed plan; ensure procedures/audits are in place.
(d) *Adjust*: Identify additional control measures required; prepare/revise performance standards; apply changes in the control measures.

Then, the PDCA cycle is repeated based on the new information. The application of dynamic risk management framework will continuously improve the process reducing the risk and increasing the reliability. Although it seems that the DRM framework is complex, resource-demanding and requires iterations and numerical work, in fact these are the strengths of dynamic methods aiming to ensure continuous improvement (Khan et al., 2016). Moreover, by incorporating the economic evaluation the dynamic risk management framework will assist decision makers in investing on appropriate measures to reduce risk.

Many studies have shown that accidents in industries, e.g., the BP Texas City refinery accident in 2005, the BP Deepwater Horizon disaster in 2010, and other major industrial accidents might have been prevented if a dynamic risk management was implemented (Paltrinieri, Khan, Amyotte, & Cozzani, 2014; Paltrinieri, Khan, & Cozzani, 2015; USCSB, 2010). The cost of the application of DRM would definitely be lesser than the damage costs of those accidents, indicating the importance of DRM.

5.3.1 System dynamics for multi-period planning

System dynamics is a well-established methodology to quantify complex feedbacks in system interactions (Forrester, 1961; Forrester, 1968). A system refers to "a collection of elements that continually interact over time to form a unified whole." Dynamics means change over time, where variables and parameters change their values over time. Therefore, system dynamics is a method that is used to study how a system changes over time (Martin, 1997). The system dynamics model is commonly used to measure system behaviors with feedback loops for effective projections (Qi & Chang, 2011). The model allows for the effective trade-off analysis of multiple scenarios and of multiple attributes in multiple periods (Sehlke & Jacobson, 2005). It consists of the construction of "stock and flow diagrams" to represent a dynamic system.

System dynamics approach can be used to model the process to estimate and manage risks. For an example, one study proposed a system dynamics-based chemical supply chain risk mitigation framework to predict the risks present in the transportation stage of the chemical supply chain in different risk mitigation decisions over time (Li, Ren, & Wang, 2016). Thirteen risk scenarios were established, and experts were asked to assign input values to each risk scenario regarding the probability and consequence severity. After the input of the collected data into the developed model, it is simulated to determine the performance of time-dependent chemical supply chain model. Based on the results, the recommended risk mitigation methods were to increase the transportation equipment capacity and transportation equipment number. In addition, the effect of the different increase of transportation equipment capacity from 5% to 20% and that of transportation equipment number from 5% to 20% of current situation were analyzed. Furthermore, economic evaluation can be integrated into a developed system dynamics model to continuously improve the system performance incorporating economic aspect and optimize risk mitigation decisions in chemical supply chain risk mitigation.

Other examples are as follows. Nawarathna, George, and Malano (2009) applied system dynamics to predict future water supply and demand under changing land use and climate by considering irrigation, urban and environmental water demand. Similarly, system dynamics has been used to predict the water resources carrying capacity by considering water resource, industrial, agricultural and residential water and wastewater treatment and reuse (Zhang, Feng, Mao, Zheng, & Wang, 2009). Qi and Chang (2011) applied system dynamics to estimate domestic water demand under changing macro-economy by considering socio-economy, population, and water. Karamouz, Goharian, and Nazif (2012) assessed the reliability of water quantity and quality employing system dynamics by considering daily time step. Willuweit and O'Sullivan (2013) used system dynamics to model the effects of urban development and climate change on urban water cycle by considering social, environmental, economic, and functional indicators. Chhipi-Shrestha et al. proposed a system dynamics-based water-energy-carbon nexus model for sustainable planning and management of urban water systems over multiple periods (Chhipi-Shrestha, Hewage, & Sadiq, 2017b). Applied system dynamics has been used for the long-term planning of net-zero water development in a city (Chhipi-Shrestha, Hewage, & Sadiq, 2018). Furthermore, some studies have used applied system dynamics to identify an optimal plan by considering population, industry, agriculture, and water resource (Zhang, Zhang, Chen, Chen, & Zhao, 2008; Zhang, Zhang, & Zhang, 2009). All these studies illustrate the applicability of system dynamics for planning in multi-periods, which assist in risk management.

5.4 Bayesian networks for risk assessment

Bayesian networks (BNs) have been used in a variety of engineering applications in the past, including medical diagnosis and clinical decision support, risk assessment and management, and spatial dynamics modeling (Pourret, Naim, & Marcot, 2008). BNs can model complex networks composed of multiple variables, and can compute their probability distributions based on observations and prior knowledge (Weber, Medina-Oliva, Simon, & Iung, 2012). In a BN, the "nodes" are used to indicate the variables, and the arcs connecting them are used to signify the causal relationships between the variables (Weber et al., 2012).

Risk assessment through BN has been a widely studied area in literature, in applications such as decision support in clinical activities, health risk assessment, crime or terrorism risk analysis, and road accident risk modeling (Gregoriades & Mouskos, 2013; Liu & Lu, 2009; Pourret et al., 2008). Identification of risk factors or predictor variables, and their values is initially done to define BN-based risk models (Gregoriades & Mouskos, 2013). This is followed by defining the relationships between the variables, and finally parameterizing the network by defining the conditional probability tables (CPT) for each node in the BN (Pourret et al., 2008). Continuous variables can be parameterized with distributions or otherwise defined equations. It is important to note that in order to develop a BN, even continuous variables need to be discretized, as the network outputs are in the form of CPTs (Pourret et al., 2008). Accordingly, several different "states" can be defined for a variable.

Bayesian networks are a highly effective in contexts such as energy system planning, spatial dynamics assessment, and multi-period industrial operations management, due to their robustness in interpreting uncertain knowledge and assumptions, and ability to represent the vague or linguistic qualities of human decision making related to policy choices (Liu & Lu, 2009). This technique can be used to not only to develop models based on exact and certain data on the risk factors, but also for analyses based on a decision makers' preferences and beliefs. Previous studies have discussed the possibility of taking either a data based approach (where the possession of a significant amount of data enables the model to be constructed through a data miming strategy), or a knowledge-based approach (where the causal knowledge of domain experts is used in the network construction) (Daim, Kayakutlu, Suharto, & Bayram, 2014).

Netica is a software which can be used to develop Bayesian belief networks. It has been widely used in similar studies focusing on risk assessment and decision making (Daim et al., 2014; Liu & Lu, 2009; Pourret et al., 2008). The software has the capacity of computing conditional probabilities of decision variables based on the input data which can be tabulated into the developed network model (Daim et al., 2014). Three different main node types are used in Netica in constructing the networks.

Nature nodes: Nature nodes can be used to represent variables, often external parameters which are outside of the decision maker's direct control. A nature node can be deterministic or probabilistic.

Decision nodes: A decision node is used to represent a choice which can be controlled by the decision maker.

Utility nodes: Utility nodes can be used in decision nets. In these, the expected value of the decision is maximized when computing the best decision under the network parameters.

After the network has been constructed and trained with the conditional probability tables (CPT), the network is compiled. Findings (evidence) can be entered to the nodes, based on known or estimated information. Netica uses this evidence information to then find the conditional probabilities for the other nodes. This belief updating process is known as probabilistic interference. Model validation can be done using actual statistics, by comparing true-positive and false-negative predictions from the model (Pourret et al., 2008). Alternatively, domain expert consensus can be used in validating models (Daim et al., 2014).

5.5 Emerging applications in risk management

Risk management has been widely used in many fields such as engineering (construction, operation and maintenance of buildings, etc.), business, traffic management, medical, etc. In addition, new techniques like real-time monitoring and control and artificial intelligence have been applied in risk management. Real-time sensing and artificial intelligence such as neural networks, genetic algorithms, Fuzzy inference system, and support vector machine can be applied for performance monitoring and control. Few case studies are illustrated below.

Risk management on building performance, for example, can be done by real-time monitoring through environmental sensing. The Senseable City Laboratory of the Massachusetts Institute of Technology implemented the projects "Future ENEL" and "Enernet" at the urban scale. The project "Future ENEL," aimed at investigating futuristic sensor technologies applied to the safe construction and operation life cycle management of energy plants. The second project is "Enernet," which proposed a new method to measure human activity, using WiFi connections as an indicator of occupancy. A comparison was made between the number of WiFi connections and energy consumption for two buildings in the campus. The results show that the operation of the heating, ventilation and air conditioning (HVAC) systems had a higher correlation with external temperature rather than occupancy (Martani, 2015). The common aspect of these two projects is that all of them use process monitoring as opposed to condition monitoring on components, to understand the state of systems. Process monitoring observes operating characteristics of an object or group of objects to see whether they are running efficiently and safely. However, condition monitoring checks objects periodically and is less efficient compared to process monitoring. Real-time monitoring is particularly interesting for the process of risk management on building performance (Martani, 2015).

Other studies are as follows. Valderrama and Useche (2003) used modified genetic algorithm to detect leaks in water distribution systems, which will help to reduce to the risk of large-scale supply disruption. Similarly, Mounce, Boxall, and Machell (2010) applied artificial neural network (ANN) and fuzzy inference system for detection of leaks and bursts in water distribution systems. Peleato, Legge, and Andrews (2018) applied ANN coupled with fluorescence data to improve the predictability of drinking water disinfection by-products (DBPs) in order to reduce human health risks. Kisi, Shiri, and Nikoofar (2012) used ANN, adaptive-neuro-fuzzy inference system (ANFIS), and gene expression programming (GEP) to forecast daily lake-level variations up to 3-day ahead time intervals for source water management. Similarly, (Eaton & Barker, 2006) used ANN and heuristic algorithms to assess water demand and flow pressure. These case studies show the applicability of artificial intelligence for future predictions and control for risk management.

6. Conclusions

Risk management is a multi-period and multi-stakeholder effort that is integral in handling projects and various engineering applications.

The contents of this chapter indicate that there is a variety of different techniques and tools that can be used in risk assessment and management. The choice of these techniques to remain with the decision makers, based on their particular needs, preferences, and situation. Using mathematical programing, multi-criteria decision making, and dynamic analysis methods such as system dynamics and Bayesian networks can assist decision makers to arrive at better judgments regarding their risk management strategy.

This chapter summarizes the key principles of project risk management and a few of the possible techniques that can be used. Interestingly, a collaborative approach among all the stakeholders is the most effective manner to identify, categorize, and rate the risks facing a project or an organization. The emerging trends in risk management point toward increased use of IT and data sciences for data-driven risk management, which can assist greatly in the decision making with better information and reduced uncertainty. Moreover, it has now been identified that converging risk management with strategic planning can greatly increase the effectiveness of the outcomes for an organization. For this, a risk culture needs to be cultivated within an organization, and the social and behavioral aspects of the stakeholders need to be taken into account. It is hoped that the readers of this chapter have gained an insight to the possibilities of evidence-based and scientific decision making for risk management, which can be used for dynamic multi-period planning and operations.

References

Alireza, A., & Javad, K. (2019). *New methods and applications in multiple attribute decision making (MADM)*. Switzerland: Springer.

Allio, L., Ballantine, B., & Meads, R. (2006). Enhancing the role of science in the decision-making of the European Union. *Regulatory Toxicology and Pharmacology*, *44*(1), 4–13.

Amaral, T. M., & Costa, A. P. C. (2014). Improving decision-making and management of hospital resources: An application of the PROMETHEE II method in an Emergency Department. *Operations Research for Health Care*, *3*, 1–6.

Ananda, J., & Herath, G. (2009). A critical review of multi-criteria decision making methods with special reference to forest management and planning. *Ecological Economics*, *68*(10), 2535–2548.

Aven, T. (Aug. 2016). Risk assessment and risk management: Review of recent advances on their foundation. *European Journal of Operational Research*, *253*(1), 1–13.

Avila, R. (2015). Uncertainty analysis and risk assessment. In C. Walther & D. K. Gupta (Eds.), *Radionuclides in the environment* (pp. 255–273). Cham: Springer International Publishing.

Aviso, K. B., Tan, R. R., & Culaba, A. B. (2010). Designing eco-industrial water exchange networks using fuzzy mathematical programming. *Clean Technologies and Environmental Policy*, *12*, 353–363.

Ayyub, B. M. (2014). *Risk analysis in engineering and economics* (2nd ed.). Boca Raton: CRC Press Taylor & Francis Group.

Badalpur, M., & Nurbakhsh, E. (2019). An application of WASPAS method in risk qualitative analysis: A case study of a road construction project in Iran. *International Journal of Construction Management*, 1–9.

Bakhtavar, E., Aghayarloo, R., Yousefi, S., Hewage, K., & Sadiq, R. (2019). Renewable energy based mine reclamation strategy: A hybrid fuzzy-based network analysis. *Journal of Cleaner Production*, *230*, 253–263.

Bakhtavar, E., Lotfian, R., & Mikaeil, R. (2019). Using grey GIS filtration to identify the potential area for cement plants in South Khorasan province. *International Journal of Mining and Geo-Engineering*, *53*, 151–155.

Bakhtavar, E., & Shirvand, Y. (2019). Designing a fuzzy cognitive map to evaluate drilling and blasting problems of the tunneling projects in Iran. *Engineering With Computers*, *35*, 35–45.

Bakhtavar, E., & Yousefi, S. (2018). Assessment of workplace accident risks in underground collieries by integrating a multi-goal cause-and-effect analysis method with MCDM sensitivity analysis. *Stochastic Environmental Research and Risk Assessment*, *32*, 3317–3332.

Bakhtavar, E., & Yousefi, S. (2019). Analysis of ground vibration risk on mine infrastructures: Integrating fuzzy slack-based measure model and failure effects analysis. *International Journal of Environmental Science and Technology*, *16*, 6065–6076. https://doi.org/10.1007/s13762-018-2008-0.

Bakhtavar, E., Yousefi, S., & Jafarpour, A. (2019). Evaluation of shaft locations in underground mines: Fuzzy multi-objective optimization by ratio analysis with fuzzy cognitive map weights. *Journal of the Southern African Institute of Mining and Metallurgy*, *119*(10), 855–864.

Becker, G. M. (2004). A practical risk management approach. In *PMI® global congress 2004—North America*.

Behraftar, S., Hossaini, S. M. F., & Bakhtavar, E. (2017). MRPN technique for assessment of working risks in underground coal mines. *Journal of the Geological Society of India*, *90*, 196–200.

Behzadian, M., Kazemzadeh, R. B., Albadvi, A., & Aghdasi, M. (2010). PROMETHEE: A comprehensive literature review on methodologies and applications. *European Journal of Operational Research*, *200*(1), 198–215.

Bian, T., Zheng, H., Yin, L., & Deng, Y. (2018). Failure mode and effects analysis based on D numbers and TOPSIS. *Quality and Reliability Engineering International*, *34*(4), 501–515.

Borges, V. (2017). Decision making under uncertainty and expressing uncertainty. In L. Walls, M. Revie, & T. Bedford (Eds.), *Risk, reliability and safety: Innovating theory and practice* (pp. 629–636). Taylor and Francis Group.

Braglia, M. (2000). MAFMA: Multi-attribute failure mode analysis. *International Journal of Quality & Reliability Management*, *17*, 1017–1033.

Brauers, W. K. M., & Zavadskas, E. K. (2006). The MOORA method and its application to privatization in a transition economy. *Control and Cybernetics*, *35*, 445–469.

Brauers, W. K. M., & Zavadskas, E. K. (2011). Multimoora optimization used to decide on a bank loan to buy property | Optimizavimas multimoora metodu imant banko paskola {ogonek} nekilnojamam turtuipirkti. *Technological and Economic Development of Economy*, *17*, 174–188.

Brown, R. M., McClelland, N. I., Deininger, R. A., & Tozer, R. G. (1970). A water quality index: Do we dare? *Water & Sewage Works*, *117*(10), 339–343.

Byrnes, J. P. (2002). The development of decision-making. *Journal of Adolescent Health*, *31*, 208–215. https://doi.org/10.1016/S1054-139X(02)00503–7.

Cantino, V., De Vincentiis, P., & Racca, G. (2011). Risk management. In O. Cretu, R. B. Stewart, T. Berends, & R. B. J. Stewart (Eds.), *Risk management for design and construction* (pp. 149–168). R.S. Means Company.

Carpitella, S., Certa, A., Izquierdo, J., & La Fata, C. M. (2018). A combined multi-criteria approach to support FMECA analyses: A real-world case. *Reliability Engineering and System Safety*, *169*, 394–402.
Cegan, J. C., Filion, A. M., Keisler, J. M., & Linkov, I. (2017). Trends and applications of multi-criteria decision analysis in environmental sciences: Literature review. *Environment Systems and Decisions*, *37*(2), 123–133.
Chang, K. H., Chang, Y. C., & Lee, Y. T. (2014). Integrating TOPSIS and DEMATEL methods to rank the risk of failure of FMEA. *International Journal of Information Technology and Decision Making*, *13*, 1229–1257.
Chang, K. H., Chang, Y. C., & Tsai, I. T. (2013). Enhancing FMEA assessment by integrating grey relational analysis and the decision making trial and evaluation laboratory approach. *Engineering Failure Analysis*, *31*, 211–224.
Chang, C. L., Wei, C. C., & Lee, Y. H. (1999). Failure mode and effects analysis using fuzzy method and grey theory. *Kybernetes*, *28*, 1072–1080.
Charnes, A., & Cooper, W. W. (1977). Goal programming and multiple objective optimizations: Part 1. *European Journal of Operational Research*, *1*, 39–54.
Chase, R. E. (Ed.), (2015). What is your risk appetite? In *Security leader insights for risk management* (pp. 37–39): Elsevier.
Chew, I. M. L., Aya, S., Tan, R. R., Chwan, D., & Foo, Y. (2011). Analysis of inter-plant water integration with indirect integration schemes through game theory approach: Pareto optimal solution with interventions. *Clean Technologies and Environmental Policy*, *13*, 49–62.
Chew, I. L. M., Tan, R. R., Chwan, D., Foo, Y., Shun, A., & Chiu, F. (2009). Game theory approach to the analysis of inter-plant water integration in an eco-industrial park. *Journal of Cleaner Production*, *17*(18), 1611–1619.
Chhipi-Shrestha, G., Hewage, K., & Sadiq, R. (2017a). Microbial quality of reclaimed water for urban reuses: Probabilistic risk-based investigation and recommendations. *Science of the Total Environment*, *576*, 738–751.
Chhipi-Shrestha, G., Hewage, K., & Sadiq, R. (2017b). Water-energy-carbon nexus modelling for an urban water system: A system dynamics approach. *Journal of Water Resources Planning and Management*, *143*, 1–11
Chhipi-Shrestha, G., Hewage, K., & Sadiq, R. (2018). Economic and energy efficiency of net-zero water communities: System dynamics analysis. *Journal of the Sustainable Water in the Built Environment*, *4*(3), 04018006.
Chhipi-Shrestha, G., Rodriguez, M., & Sadiq, R. (2019). Selection of sustainable municipal water reuse applications by multi-stakeholders using game theory. *Science of the Total Environment*, *650*, 2512–2526.
Daim, T., Kayakutlu, G., Suharto, Y., & Bayram, Y. (2014). Clean energy investment scenarios using the Bayesian network. *International Journal of Sustainable Energy*, *33*(2), 400–415.
de Almeida, A. T., Alencar, M. H., Garcez, T. V., & Ferreira, R. J. P. (2017). A systematic literature review of multicriteria and multi-objective models applied in risk management. *IMA Journal of Management Mathematics*, *28*(2), 153–184.
Dinius, S. H. (1987). Design of an index of water quality. *Water Resources Bulletin*, *23*(5), 833–843.
Dojlido, J., Raniszewski, J., & Woyciechowska, J. (1994). Water quality index—Application for rivers in Vistula river basin in Poland. *Water Science and Technology*, *30*(10), 57–64.
Eaton, M. T., & Barker, S. M. (2006). Artificial intelligence helps JEA optimize water resources. *Journal of American Water Works Association*, *98*(6), 40.
Elahi, B. (2018). Risk management process. In *Safety risk management for medical devices* (pp. 49–66). Elsevier.

Engemann, K. J., Miller, H. E., & Yager, R. R. (1993). A general methodology for decision making under uncertainty with a risk management applicaiton. *IEEE*, *8*, 194–199.

Fatemi, A., & Luft, C. (2002). Corporate risk management costs and benefits. *Global Finance Journal*, *13*(1), 29–38.

Forrester, J. W. (1961). *Industrial dynamics.* Cambridge, MA: MIT Press.

Forrester, J. W. (1968). *Principles of system dynamics.* Cambridge, MA: Productivity Press.

Govindan, K., & Jepsen, M. B. (2016). ELECTRE: A comprehensive literature review on methodologies and applications. *European Journal of Operational Research*, *250*(1), 1–29.

Gregoriades, A., & Mouskos, K. C. (2013). Black spots identification through a Bayesian Networks quantification of accident risk index. *Transportation Research Part C: Emerging Technologies*, *28*, 28–43.

Hwang, C., & Yoon, K. (1981). *Multiple attribute decision making: Methods and applications, a state of the art survey.* Berlin, Heidelberg: Springer-Verlag.

Ilangkumaran, M., Shanmugam, P., Sakthivel, G., & Visagavel, K. (2014). Failure mode and effect analysis using fuzzy analytic hierarchy process. *International Journal of Productivity and Quality Management*, *22*, 466.

International Organization for Standardization. (2020). *ISO 31000 risk management.* [Online]. Available: https://www.iso.org/iso-31000-risk-management.html [Accessed: 10-Jan-2020].

Jaafari, A. (Feb. 2001). Management of risks, uncertainties and opportunities on projects: time for a fundamental shift. *International Journal of Project Management*, *19*(2), 89–101.

Johansen, A., Eik-Andresen, P., & Ekambaram, A. (2014). Stakeholder benefit assessment—Project success through management of stakeholders. *Procedia—Social and Behavioral Sciences*, *119*(1877), 581–590.

Karamouz, M., Goharian, E., & Nazif, S. (2012). *Development of a reliability based dynamic model of urban water supply system: A case study.* Albuquerque, NM: American Society of Civil Engineers.

Karunathilake, H., Hewage, K., Mérida, W., & Sadiq, R. (Jan. 2019). Renewable energy selection for net-zero energy communities: Life cycle based decision making under uncertainty. *Renewable Energy*, *130*, 558–573.

Kasap, D., & Kaymak, M. (2007). Risk identification step of the project risk management. In *PICMET 2007 proceedings* (pp. 2116–2120).

Khan, F., Hashemi, S. J., Paltrinieri, N., Amyotte, P., Cozzani, V., & Reniers, G. (2016). Dynamic risk management: A contemporary approach to process safety management. *Current Opinion in Chemical Engineering*, *14*, 9–17.

Kisi, O., Shiri, J., & Nikoofar, B. (2012). Forecasting daily lake levels using artificial intelligence approaches. *Computational Geosciences*, *41*, 169–180.

Kosko, B. (1986). Fuzzy cognitive maps. *International Journal of Man-Machine Studies*, *24*, 65–75.

Lees, F. (2005). Inherently safer design. In *Lees' loss prevention in the process industries* (pp. 32/1–32/24). Elsevier.

Leontaris, G., Morales-Nápoles, O., & Wolfert, A. (2016). Planning cable installation activities for offshore wind farms including risk of supply delays. In L. Walls, M. Revie, & T. Bedford (Eds.), *Risk, Reliability and Safety: Innovating Theory and Practice* (pp. 660–666). Boca Raton, FL: CRC Press. https://doi.org/10.1201/9781315374987-99.

Leyton-Brown, K., & Shovan, Y. (2008). *Essentials of game theory: A concise, multidisciplinary introduction.* San Rafael: Morgan and Claypool.

Li, X., Li, H., Sun, B., & Wang, F. (2018). Assessing information security risk for an evolving smart city based on fuzzy and grey FMEA. *Journal of Intelligent Fuzzy Systems*, *34*, 2491–2501.

Li, C., Ren, J., & Wang, H. (2016). A system dynamics simulation model of chemical supply chain transportation risk management systems. *Computers and Chemical Engineering, 89*, 71–83.

Linares, P. (Aug. 2002). Multiple criteria decision making and risk analysis as risk management tools for power systems planning. *IEEE Transactions on Power Apparatus and Systems, 17*(3), 895–900.

Linkov, I., Satterstrom, F. K., Kiker, G., Batchelor, C., Bridges, T., & Ferguson, E. (2006). From comparative risk assessment to multi-criteria decision analysis and adaptive management: Recent developments and applications. *Environment International, 32*(8), 1072–1093.

Liu, H.-C., Chen, X.-Q., Duan, C.-Y., & Wang, Y.-M. (2019). Failure mode and effect analysis using multi-criteria decision making methods: A systematic literature review. *Computer & Industrial Engineering, 135*, 881–897.

Liu, H. C., Fan, X. J., Li, P., & Chen, Y. Z. (2014). Evaluating the risk of failure modes with extended MULTIMOORA method under fuzzy environment. *Engineering Applications of Artificial Intelligence, 34*, 168–177.

Liu, H. C., Liu, L., Liu, N., & Mao, L. X. (2012). Risk evaluation in failure mode and effects analysis with extended VIKOR method under fuzzy environment. *Expert Systems With Applications, 39*, 12926–12934.

Liu, K. F. R., & Lu, C.-F. (2009). BBN-based decision support for health risk analysis. In *2009 Fifth International Joint Conference on INC, IMS and IDC* (pp. 696–702).

Liu, H. C., Wang, L. E., You, X. Y., & Wu, S. M. (2019). Failure mode and effect analysis with extended grey relational analysis method in cloud setting. *Total Quality Management & Business Excellence, 30*, 745–767. https://doi.org/10.1080/14783363.2017.1337506.

Liu, H. C., You, J. X., Ding, X. F., & Su, Q. (2015). Improving risk evaluation in FMEA with a hybrid multiple criteria decision making method. *International Journal of Quality and Reliability Management, 32*, 763–782.

Liu, H. C., You, J. X., Li, P., & Su, Q. (2016). Failure mode and effect analysis under uncertainty: An integrated multiple criteria decision making approach. *IEEE Transactions on Reliability, 65*, 1–13.

Liu, H. C., You, J. X., You, X. Y., & Shan, M. M. (2015). A novel approach for failure mode and effects analysis using combination weighting and fuzzy VIKOR method. *Applied Soft Computing, 28*, 579–588. https://doi.org/10.1016/j.asoc.2014.11.036.

Lo, H. W., & Liou, J. J. H. (2018). A novel multiple-criteria decision-making-based FMEA model for risk assessment. *Applied Soft Computing, 73*, 684–696.

Lu, J., Zhang, G., & Ruan, D. (2008). Fuzzy multi-objective decision-making models and approaches. In C. Kahraman (Ed.), *Fuzzy Multi-Criteria Decision Making. Springer Optimization and Its Applications* (pp. 483–522). Boston, MA: Springer. https://doi.org/10.1007/978-0-387-76813-7_19.

Ma, J., Li, P., Lin, X., Zhu, W., & Yuan, X. (2015). Game theory method for multi-objective optimizing operation in microgrid. In *Proceedings of 2015 IEEE 12th international conference on networking, sensing and control* (pp. 421–425).

Madani, K. (2010). Game theory and water resources. *Journal of Hydrology, 381*(3–4), 225–238.

Mardani, A., Jusoh, A., Nor, K., Khalifah, Z., Zakwan, N., & Valipour, A. (2015). Multiple criteria decision-making techniques and their applications–a review of the literature from 2000 to 2014. *Economic Research-Ekonomska Istraživanja, 28*(1), 516–571.

Martani, C. (2015). *Risk management in architectural design: Control of uncertainty over building use and maintenance*. Cambridge, UK: University of Cambridge.

Martin, L. A. (1997). *The first step, D-4694*. Massachusetts Institute of Technology.

Metheny, M. (2013). Risk management. In *Federal cloud computing* (pp. 169–194). Elsevier.
Mohsen, O., & Fereshteh, N. (2017). An extended VIKOR method based on entropy measure for the failure modes risk assessment—A case study of the geothermal power plant (GPP). *Safety Science, 92*, 160–172.
Mounce, S. R., Boxall, J. B., & Machell, J. (2010). Development and verification of an online artificial intelligence system for detection of bursts and other abnormal flows. *Journal of Water Resources Planning and Management, 136*(June), 309–318.
Nash, J. (1950). Equilibrium points in n-person games. *Proceedings of the National Academy of Sciences of the United States of America, 36*, 1949–1950.
Nash, J. (1951). Non-cooperative games. *Annals of Mathematics, 54*(2), 286–295.
Nawarathna, B., George, B. A., & Malano, H. M. (2009). Future water supply and demand assessment in peri- urban catchments using system dynamics approach. In *18th WorldE IMACS/MODSIM congress*.
Olechowski, A., Oehmen, J., Seering, W., & Ben-Daya, M. (Nov. 2016). The professionalization of risk management: What role can the ISO 31000 risk management principles play? *International Journal of Project Management, 34*(8), 1568–1578.
Paltrinieri, N., Khan, F., Amyotte, P., & Cozzani, V. (2014). Dynamic approach to risk management: Application to the Hoeganaes metal dust accidents. *Process Safety and Environment Protection, 92*(6), 669–679.
Paltrinieri, N., Khan, F., & Cozzani, V. (2015). Coupling of advanced techniques for dynamic risk management. *Journal of Risk Research, 18*(7), 910–930.
Panahi, S., Khakzad, A., & Afzal, P. (2017). Application of stepwise weight assessment ratio analysis (SWARA) for copper prospectivity mapping in the Anarak region, central Iran. *Arabian Journal of Geosciences, 10*(22), 484.
Peleato, N. M., Legge, R. L., & Andrews, R. C. (2018). Neural networks for dimensionality reduction of fluorescence spectra and prediction of drinking water disinfection by-products. *Water Research, 136*, 84–94.
PMI. (2004). *A guide to the project management body of knowledge: PMBOK guide* (3rd ed.). Newtown Square, PA: Project Management Institute.
Polatidis, H., Haralambopoulos, D. A., Munda, G., & Vreeker, R. (2006). Selecting an appropriate multi-criteria decision analysis technique for renewable energy planning. *Energy Sources, Part B: Economics, Planning and Policy, 1*(2), 181–193.
Pourret, O., Naim, P., & Marcot, B. (Eds.), (2008). *Bayesian networks: A practical guide to applications*. John Wiley & Sons, Ltd: Chichester, UK.
Purpura, P. P. (2013). Resilience, risk management, business continuity, and emergency management. In *Security and loss prevention* (pp. 321–362). Elsevier.
Qi, C., & Chang, N.-B. (2011). System dynamics modeling for municipal water demand estimation in an urban region under uncertain economic impacts. *Journal of Environmental Management, 92*(6), 1628–1641.
Rah, J.-E., Manger, R. P., Yock, A. D., & Kim, G.-Y. (Nov. 2016). A comparison of two prospective risk analysis methods: Traditional FMEA and a modified healthcare FMEA. *Medical Physics, 43*(12), 6347–6353.
Ramík, J. (2000). Fuzzy goals and fuzzy alternatives in goal programming problems. *Fuzzy Sets and Systems, 111*, 81–86.
Romero-Gomez, P., Choi, C. Y., Woods, G., & Kang, D. (2011). A game theory approach for de-centralization of reclaimed water systems. In *Water distribution systems analysis 2010* (pp. 833–842). Reston, VA: American Society of Civil Engineers. https://doi.org/10.1061/41203(425)77.
Roy, J., Ranjan, A., Debnath, A., & Kar, S. (2016). An extended MABAC for multi-attribute decision making using trapezoidal interval type-2 fuzzy numbers. *ArXiv, 1607*, 1–14.

Ruparathna, R., et al. (Nov. 2017). Climate conscious regional planning for fast-growing communities. *Journal of Cleaner Production, 165*, 81–92.
Sabaei, D., Erkoyuncu, J., & Roy, R. (2015). A review of multi-criteria decision making methods for enhanced maintenance delivery. *Procedia CIRP, 37*, 30–35.
Sachdeva, A., Kumar, D., & Kumar, P. (2009). Multi-factor failure mode critically analysis using TOPSIS. *Journal of Industrial Engineering International, 5*(8), 1–9.
Sadiq, R., & Tesfamariam, S. (2007). Probability density functions based weights for ordered weighted averaging (OWA) operators: An example of water quality indices. *European Journal of Operational Research, 182*(3), 1350–1368.
Sakthivel, G., & Ikua, B. W. (2017). Failure mode and effect analysis using fuzzy analytic hierarchy process and GRA TOPSIS in manufacturing industry. *International Journal of Productivity and Quality Management, 22*, 466.
Sakthivel, G., Saravanakumar, D., & Muthuramalingam, T. (2018). Application of failure mode and effect analysis in manufacturing industry—An integrated approach with FAHP-fuzzy TOPSIS and FAHP-fuzzy VIKOR. *International Journal of Productivity and Quality Management, 24*(3), 398. https://doi.org/10.1504/IJPQM.2018.092984.
San Cristóbal, J. R. (Feb. 2011). Multi-criteria decision-making in the selection of a renewable energy project in spain: The Vikor method. *Renewable Energy, 36*(2), 498–502.
Sehlke, G., & Jacobson, J. (2005). System dynamics modeling of transboundary systems: The bear river basin model. *Ground Water, 43*(5), 722–730.
Shahriar, K., & Bakhtavar, E. (2009). Geotechnical risks in underground coal mines. *Journal of Applied Sciences, 9*, 2137–2143.
Sharif, M. N., Farahat, A., Haider, H., Al-Zahrani, M. A., Rodriguez, M. J., & Sadiq, R. (Jul. 2017). Risk-based framework for optimizing residual chlorine in large water distribution systems. *Environmental Monitoring and Assessment, 189*(7), 307.
Shi, J.-L., Wang, Y.-J., Jin, H.-H., Fan, S.-J., Ma, Q.-Y., & Zhou, M.-J. (2016). A modified method for risk evaluation in failure mode and effects analysis. *Journal of Applied Science and Engineering, 19*(2), 177–186.
Shin, J., Lee, S., & Yoon, B. (2018). Identification and prioritisation of risk factors in R & D projects based on an R & D process model. *Sustainability, 10*, 972.
Smith, D. G. (1990). A better water quality indexing system for rivers and streams. *Water Research, 24*(10), 1237–1244.
Soltani, A., Sadiq, R., & Hewage, K. (2016). Selecting sustainable waste-to-energy technologies for municipal solid waste treatment: A game theory approach for group decision-making. *Journal of Cleaner Production, 113*, 388–399.
Song, B., & Kang, S. (2016). A method of assigning weights using a ranking and nonhierarchy comparison. *Advances in Decision Sciences*, 1–9. https://doi.org/10.1155/2016/8963214.
Song, W., Ming, X., Wu, Z., & Zhu, B. (2013). Failure modes and effects analysis using integrated weight-based fuzzy TOPSIS. *International Journal of Computer Integrated Manufacturing, 26*, 1172–1186.
Szymański, P. (2017). Risk management in construction projects. *Procedia Engineering, 208*, 174–182.
Talabis, M., & Martin, J. (2012). Information security risk assessment: Reporting. In *Information security risk assessments* (pp. 195–231). Elsevier.
Tian, Z. P., Wang, J. Q., & Zhang, H. Y. (2018). An integrated approach for failure mode and effects analysis based on fuzzy best-worst, relative entropy, and VIKOR methods. *Applied Soft Computing, 72*, 636–646.
Triantaphyllou, E. (2000). Multi-criteria decision making methods. *Multi-criteria Decision Making Methods: A Comparative Study. Applied Optimization* (pp. 5–21). Boston, MA: Springer. https://doi.org/10.1007/978-1-4757-3157-6_2.

Tzotzos, G. T., Head, G. P., & Hull, R. (2009). Principles of risk assessment. In *Genetically modified plants* (pp. 33–63). Elsevier.

US EPA. (2001). Risk assessment guidance for superfund (RAGS) volume III—Part A: Process for conducting probabilistic risk assessment. In *Vol. III.* U.S. Environmental Protection Agency.

USCSB (United States Chemical Safety and Hazard Investigation Board). (2010). *Investigation report overview: Explosion and fire at the Macondo well.* Washington DC, Retrieved from https://www.hsdl.org/?view&did=754552.

Valderrama, J. G. S., & Useche, D. E. S. (2003). Calibration of water distribution networks under leaks environment, artificial intelligence application. In *World water & environmental resources congress 2003* (pp. 1–10). Reston, VA: American Society of Civil Engineers. https://doi.org/10.1061/40685(2003)89.

Wang, H., Cai, Y., Tan, Q., & Zeng, Y. (2017). Evaluation of groundwater remediation technologies based on fuzzy multi-criteria. *Water, 9*(443), 1–20.

Weber, P., Medina-Oliva, G., Simon, C., & Iung, B. (2012). Overview on Bayesian networks applications for dependability, risk analysis and maintenance areas. *Engineering Applications of Artificial Intelligence, 25*(4), 671–682.

Willuweit, L., & O'Sullivan, J. J. (2013). A decision support tool for sustainable planning of urban water systems: Presenting the dynamic urban water simulation model. *Water Research, 47*(20), 7206–7220.

Xu, X. (2001). The SIR method: A superiority and inferiority ranking method for multiple criteria decision making. *European Journal of Operational Research, 131*(3), 587–602.

Yager, R. R. (1988). On ordered weighted averaging aggregation operators in multicriteria decision making. *IEEE Transactions on Systems, Man, and Cybernetics, 18*(1), 183–190.

Yoon, K., & Hwang, C.-L. (1995). *Multiple attribute decision making.* Thousand Oaks, CA: SAGE Publications, Inc.

Zammori, F., & Gabbrielli, R. (2012). A multi criteria evaluation of the Risk Priority Number. *Quality and Reliability Engineering International, 28*(1), 85–104. https://doi.org/10.1002/qre.1217.

Zavadskas, E. K., Antuchevičienė, J., & Kar, S. (Eds.), (2019). *Multi-Objective and Multi-Attribute Optimisation for Sustainable Development Decision Aiding.* Basel: MDPI. https://doi.org/10.3390/books978-3-03921-143-2.

Zavadskas, E. K., & Turskis, Z. (2010). A new additive ratio assessment (ARAS) method in multicriteria decision-making. *Technological and Economic Development of Economy, 16*(2), 159–172.

Zhang, X., Feng, H., Mao, X., Zheng, M., & Wang, R. (2009). A dynamic water resources management approach in Beijing using system dynamics model. In *2009 international conference on management and service science* (pp. 1–4) Sep.

Zhang, S., Gao, H., Wei, G., Wei, Y., & Wei, C. (2019). Evaluation based on distance from average solution method for multiple criteria group decision making under picture 2-tuple linguistic environment. *Mathematics*, 7(3), 243.

Zhang, X. H., Zhang, H. W., Chen, B., Chen, G. Q., & Zhao, X. H. (Dec. 2008). Water resources planning based on complex system dynamics: A case study of Tianjin city. *Communications in Nonlinear Science and Numerical Simulation, 13*(10), 2328–2336.

Zhang, H., Zhang, X., & Zhang, B. (2009). System dynamics approach to urban water demand forecasting: A case study of Tianjin. *Tianjin University, 15*(1), 70–74.

Zhao, H., You, J. X., & Liu, H. C. (2017). Failure mode and effect analysis using MULTIMOORA method with continuous weighted entropy under interval-valued intuitionistic fuzzy environment. *Soft Computing, 21*, 5355–5367.

Further reading

Horton, R. K. (1965). An index number system for rating water quality. *Journal—Water Pollution Control Federation*, *37*(3), 300–306.

Ott, W. R. (1978). *Environmental indices: Theory and practice*. Ann Arbor, MI: Ann Arbor Science Publishers Inc.

Prati, L., Pavanello, R., & Pesarin, F. (1971). Assessment of surface water quality by a single index of pollution. *Water Research*, *5*, 741–751. Pergamon Press.

CHAPTER EIGHT

Human factors risk assessment

Norafneeza Binti Norazahar*
Centre of Hydrogen Energy, Institute of Future Energy, Universiti Teknologi Malaysia, Johor Bahru, Malaysia
School of Chemical and Energy Engineering, Faculty of Engineering, Universiti Teknologi Malaysia, Johor Bahru, Malaysia
*Corresponding author: e-mail address: norafneeza@utm.my

Contents

1. Perspective of human factors from other industries

Definitions of human factors can vary depending on organizations or the types of industry. According to the International Ergonomics Association (2020), the human factor is defined as a scientific study of human interaction with a machine or other elements of a system to optimize both human and system performance. The Health and Safety Executive (HSE, 2009) referred human factors to environmental, organizational, and job factors, and individual characteristics, which influence an individual's behavior at work. Consequently, human factors can affect safety and health.

In the aviation industry, organization or management understood human factors and its effects very well before integrating humans with

Methods in Chemical Process Safety, Volume 4
ISSN 2468-6514
https://doi.org/10.1016/bs.mcps.2020.02.005

the technology effectively. The study of human factors should consist of (i) collecting information about human abilities, limitations, and other characteristics and (ii) applying the knowledge to tools, equipment, systems, tasks, jobs, and environments. The understanding of human abilities and limitations is used to design equipment/machines and training, and develop policies or procedures. Human performance and safety are improved after introducing human factors in the aviation industry (Graeber, 1999). The implementation of human factors in the aviation industry is effective. The study of the human factor is primarily focused on technicians performing maintenance tasks. Later, the aviation industry extended the focus to engineering and technical organization (Federal Aviation Administration (FAA), 2019).

The human factor and its application are also well established in nuclear power plants. The nuclear power plant has a sophisticated industrial facility and requires the majority of the tasks to be performed by machines. The Nuclear Energy Agency defined human factors as the interface between a machine and an individual (or operator) (NEA, 1988). An individual's capabilities, limitations, and attitudes, as well as on the quality of instructions and training provided by the organization, determine the performance of individuals working with the machine in the nuclear power plant.

The human factor has been introduced in healthcare industry. According to the Medstar Health National Center for Human Factors in Healthcare, human factors can be described as a multi-disciplinary science that intersects psychology and engineering fields (Medstar Health, 2019). The healthcare industry emphasizes understanding the interaction among individuals and other elements of a system in the healthcare industry. Both psychological and physiological principles and theories are applied to the design of products, processes, and systems in the healthcare industry. The human factors in the healthcare industry have four areas of expertise, which are (i) human-technology interaction, (ii) cognitive engineering, (iii) physiological and psychological conditions, and (iv) physical ergonomics. The outcome of studying human factors and applying it effectively in healthcare could improve safety, efficiency, quality, and reliability, as well as reduce costs.

Professor Woods, together with Professor Dekker, who is well-known in human factors, defined human factors as a study on the intersection between people, technology, and work. The purpose of human factors is to examine areas where design and working conditions have the potential to introduce or produce human errors (Dekker, 2016a, 2016b, 2016c, 2016d, 2016e; Woods & Dekker, 2000).

2. Defining human factors for chemical process safety

The chemical process industry has many major accidents. The US Chemical Safety Board provides a link to accident cases and the investigation reports, which can be accessed through the website (https://www.csb.gov). Many investigation reports show that human factors cause major accidents.

Human factors in the chemical process industry are seriously discussed after the tragedy of the Piper Alpha. The Hon Lord Cullen (1990a, 1990b) has listed 106 recommendations. The recommendations were the subjects of safety cases, safety audit, independent assessment, legislations, safety committees and representatives, permit to work (PTW), incident reporting, control of the process, hydrocarbon inventory, risers and pipelines, fire and gas detection and emergency shutdown (ESD), fire and explosion protection, emergency response preparedness, evacuation, escape and rescue (EER), transportation modes for escape, personal survival, and finally, emergency training and drills. The recommendations cover all elements of human factors, which are the organizational factors, technical or equipment, regulations, procedures, and documentation.

Based on the Piper Alpha report, we can say that human factors are considered as an essential aspect in chemical process industries. Human factors in chemical process industries utilize scientific knowledge and fundamental principles from many disciplines to reduce the number of accidents (Attwood et al., 2007). The application of human factors in the process safety is expected to minimize the likelihood of human error, increase productivity and quality, and reduce the risk of work-related musculoskeletal disorders (Attwood et al., 2007).

Human factors can also be described as the interaction of individuals with each other, with facilities and equipment, and with management systems (IOGP, 2005). Interestingly, the IOGP includes elements of the working environment and the culture of the individual or society in the definition of human factors. These elements are directly or indirectly influenced the performance of individuals to interact with each other, equipment, or system.

Interaction is the keyword in defining human factors for the chemical process industry. The interaction is typically happened between human, individually or collectively, with machines. Individuals work with the machine or equipment for completing tasks or activities in any industry. Throughout the process, the interaction between human and machine or hardware could be influenced by other factors such as environmental conditions, standard operating procedures, resources, etc. Organizational factors, in particular, the work process, safety culture, leadership, vision, mission, goals, communication, coordination, and the number of workforces, could be included as an aspect of the human factors influencing human performance.

3. Performance shaping factors (PSFs) or performance influencing factors (PIFs)?

3.1 Performance shaping factors (PSFs)

Human factors consist of variable factors or aspects affecting human performance. The performance shaping factors have been introduced by Blackman, Gertman, and Boring (2008) to specify human factors. He and his colleagues defined PSF as an aspect of the individual's characteristics, environmental conditions, the organization, or tasks that enhance or affect human performance. The influence of PSF on an individual's performance can either increase or decrease the likelihood of human error (Blackman et al., 2008).

Fig. 1 illustrates placing PSFs in the human factors field. The performance shaping factor can be considered as part of human factors. There are eight PSFs which are:

(a) Time
(b) Stress
(c) Complexity
(d) Training and work experience

Fig. 1 PSFs inside the human factors.

(e) Procedures
(f) Human-machine interaction (HMI) or ergonomics
(g) Fitness for duty
(h) Work processes

3.1.1 Time

Time is the first performance shaping factor considered by Blackman et al. (2008). Each individual has 24 h a day. However, a shortage of time can affect the individual's ability to think clearly, consider alternatives, and perform the task. The shortage of time could be caused by other factors such as the number of tasks or activities at one time.

3.1.2 Stress and the stressors

Each of us experiences stress. Tasks at the workplace or daily tasks could cause stress to individuals. The level of stress is different from individual to another.

Blackman et al. (2008) describe stress as either negative or positive, motivating forces to an individual to perform. The level of undesirable conditions and circumstances can impede the individual from easily completing the task. Excessive workload, mental and/or physical stress, tasks requiring attention for a long period, and anxiety or nervousness associated with important events are examples of contributing factors to stress.

Stressors are also contributing to individuals experiencing stress. Stressors can induce stress, and later, it can affect the individual's mental or physical performance. The stressors are typically environmental factors. Weather conditions, excessive heat or cold, loud noise, poor ventilation and/or lighting, and radiation can be categorized as stressors.

Individuals without stress could experience boring and, consequently, have poor performance. A similar performance can also be observed to individuals experiencing high or extreme levels of stress. Individuals perform well when they receive a nominal level of stress. The Yerkes-Dodson law illustrates the effects of arousal or stress toward human performance, which is represented by a curvilinear graph (Cherry, 2019; Teigen, 1994).

3.1.3 Complexity

Ambiguous or complex task is one of the factors contributing to high chances of human errors. Individuals must take a long time and concerted effort to complete any complex tasks. According to Blackman et al. (2008), complexity can be described as difficulty of a specific task to be performed in the given context. The complexity can be categorized into three, which are tasks requiring mental effort, physical effort, and skill and apprehension (Blackman et al., 2008). Some tasks require individuals to perform mental calculations, memorize procedures or numbers, understand the system, as well as have in-depth knowledge. Individuals or operators working in the chemical processing industry are required to have exceptional skills, particularly critical thinking and troubleshooting for a nonroutine situation.

3.1.4 Training and work experience

Work experience is based on the years of the individual working in one organization or industry. Each individual must attend training related to his/her tasks at the workplace. The number of training and refresher training could influence an individual to react to both regular and abnormal events. The training should include the type of accident received by the individual, the amount of time passed since training, the systems involved in the task and scenario, and the novelty of scenario in which the individual involved in a training or operational setting (Blackman et al., 2008).

3.1.5 Procedures

According to Barge (2019), procedure refers to the existence and use of formal operating procedures for completing a task safely. Procedures are designed or provided by the management. Individuals using the procedures are sometimes experienced problems such as the procedures give wrong or inadequate information regarding a particular control sequence and the ambiguity of steps. The language used in the procedures is sometimes vague or confusing to be understood by individuals who probably have different education level or background.

3.1.6 Ergonomics or human-machine interaction (HMI)

Ergonomic can be described as a physical interaction of an individual with the equipment or machines to carry out the tasks. The design of equipment or machines can affect the physical interaction. Example of the design affecting the physical interaction are display and control panels, room layout, and quality and quantity of information provided by instrumentation are examples of physical interaction. There is also a need to emphasize ergonomics at the workplace due to the constant physical interaction of individuals with computers for completing tasks under consideration. Ergonomics should focus on the design of desks, chairs, the width of monitors, keyboards before creating a workspace. Poor ergonomics at the workplace could impact an individual's health and safety. Individuals working in awkward body postures and performing the same or similar tasks could experience musculoskeletal disorders (MSDs), which is a common ergonomic issue at the workplace.

3.1.7 Fitness for duty

Blackman et al. (2008) described fitness for duty as an individual's physical and mental fitness to perform the task at the required time. Factors contributing to the fitness for duty may include fatigue, sickness, drug use (either legal or illegal), overconfidence, distractions due to personal or family problems, boredom, and drowsiness. The factors are usually associated with individuals and not related to training, experience, or stress. Kim, Park, and Jung (2017) studied the quantitative measure for fitness for duty for individuals working in Korean nuclear power plants. He and his colleagues stated that fitness for duty should encompass the physical or mental capability of an individual to given tasks.

3.1.8 Work processes

The last PFS in this book is the work process, which refers to aspects of doing tasks. The aspects are team or crew dynamics, available staffing, inter-organizational impacts, safety culture, work planning, communication, and management support and policies (Blackman et al., 2008). Kim et al. (2017) suggested the work process should comprise coordination and communication, management support, supervision, procedure compliance, morale and motivation, strategy handling given situations, and corrective action programs. Individual can be affected by the work process; how the work is designed, planned, communicated, and implemented.

In summary, all PSFs can be connected. One PSF can influence another or more PSFs. An example of PSFs connected is illustrated in Fig. 2.

3.2 Performance influencing factors (PIFs)

The Health and Safety Executives (HSE) presented the performance influencing factors (PIFs). They defined PIFs as the characteristics of the job, the individual, and the organization which influencing human performance (HSE, 2019). The PIFs should be used for determining the likelihood of human failures. The PIFs are categorized into job, person, and organization. The list of PIFs for each category is as follows (HSE, 2019):

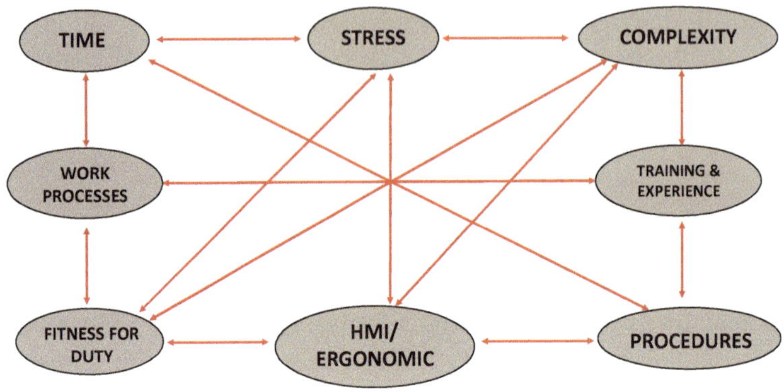

Fig. 2 The PSFs can be connected to each other.

(a) Job factors
- Clarity of signs, signals, instructions, and other information
- System/equipment interface (labeling, alarms, error avoidance/tolerance)
- Difficulty/complexity of the task
- Routine or unusual
- Divided attention
- Procedures inadequate or inappropriate
- Preparation for a task (for example, permits, risk assessments, checking)
- Time available/required
- Tools appropriate for the task
- Communication with colleagues, supervision, contractor, and others
- Working environment (noise, heat, space, lighting, ventilation)

(b) Person factors
- Physical capability and conditions
- Fatigue
- Stress
- Work overload/underload
- Competence to deal with circumstances
- Motivation against priority

(c) Organization factors
- Production against safety
- Level and nature of supervision or leadership

- Communication
- Manning levels
- Clarity of roles and responsibilities
- Consequences of failure to follow the rules or procedures
- Effectiveness of organizational learning
- Safety culture

4. Illustration of human factors contributing to accidents

HUMAN FACTORS ≠ HUMAN ERRORS

A Swiss cheese model is often used by many human factor analysts or researchers in discussing factors contributing to incidents or accidents (Reason, 1990). The Swiss cheese examines the failure of the system, rather than human failures. The slices of Swiss cheese represent the layers of defense that used to mitigate the hazards.

The Swiss cheese model can also illustrate the trajectory of hazards caused by human factors. Holes in the Swiss cheese model represent the weakness presented in the barriers, which can also be referred to as human factors. The hole from one layer affects the next layers causing the alignment of holes. The trajectory of hazards through these holes could cause individuals making wrong decisions or taking incorrect actions. Such condition later leads to an incident or accident. Thus, the Swiss cheese model can shift the focus of accident investigation from human blaming to no-blame investigation. The model can be used to identify the contributing factors that cause a person to make mistakes.

In the Swiss cheese model, there are two types of contributing factors, which are latent and active failures. Latent failures are the failures in the safety management systems. Latent failures could be caused by poor planning or designing equipment, ineffective implementation, and inadequate supervision or monitoring. The latent failures are stayed dormant for a long time and will only be identified after the incident or accident. On the other hand, the active failures could happen to front-line people such as operators. Active failures have an immediate impact or consequence on people. Fig. 3 shows the Swiss cheese model consisting of both latent and active failures.

There is one concern when using the Swiss cheese model for understanding and identifying factors leading to an incident or accident.

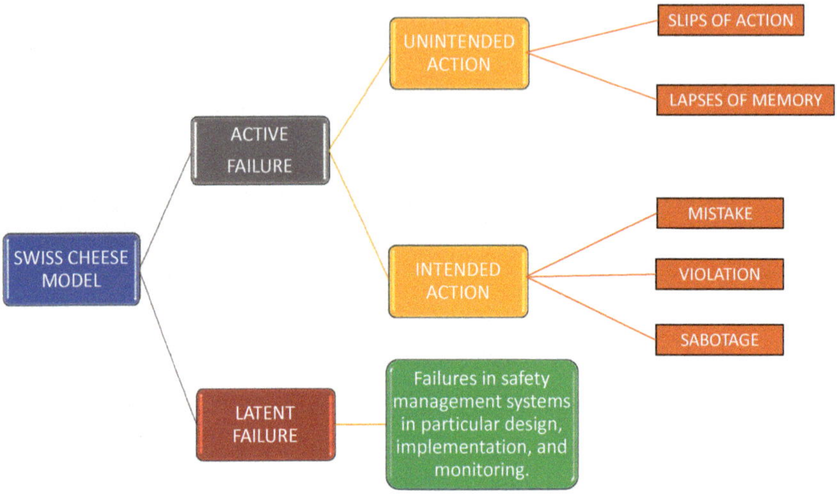

Fig. 3 The Swiss cheese model showing both the latent and active failures.

The illustration of linear trajectory contradicts with the real accidents, which happened in a non-linear model. Many human factors from different categories interact with machines/equipment as well as individuals. The Swiss cheese model cannot be used for simplifying the contributing factors to incident or accident.

5. Three generations of human reliability assessment

5.1 First generation of human reliability assessment

Human reliability assessment (HRA) techniques can be classified into the first, second, and third generation (Bell & Holroyd, 2009). The first generation of HRA is to evaluate the human error probability, which mainly includes both pre-processed tools and expert judgment approaches (Bell & Holroyd, 2009). The first generation must break down all tasks and analyze for any possible errors considering factors such as time, pressure, and equipment design. All human error probabilities (HEPs) are combined to get the nominal value of probability associated with tasks. The first-generation method is merely focused on the skill- and rule-based actions and is often criticized for failing to consider other elements such as the impact of context, organizational factors, and errors of commission (Bell & Holroyd, 2009). The technique of human error rate prediction (THERP), the human error assessment and reduction technique (HEART), and the simplified plant analysis risk human reliability assessment (SPAR-H) are the first generation of human reliability assessment.

5.2 Second generation of human reliability assessment

The second generation of human reliability assessment is developed to improve the drawbacks or limitations of the first-generation methods. The second-generation method incorporates cognitive elements and human behaviors in the model, which are assessed in a qualitative manner only (Pasquale, Iannone, Miranda, & Riemma, 2013). The cognitive reliability and error analysis method (CREAM) and a technique for human event analysis (ATHEANA) are the second-generation methods, which are widely used in nuclear power plant industries. Due to its practicality, these two methods are then applied in other industries as well. However, the second generation of human reliability assessment lacks empirical data for validating the human error probabilities.

5.3 Third generation of human reliability assessment

The third generation of human reliability assessment strongly depends on expert judgment. The absolute probability judgments (APJ) require one expert of APJ to make his/her judgments on the chances of a human error. The success likelihood index method using multi-attribute utility decomposition (SLIM-MAUD) is an approach to assessing human error probabilities using a structured expert judgment. The SLIM-MAUD method considers the performance shaping factors (PSFs) in identifying human errors and quantifying the human error probabilities (HEPs). The third-generation methods have a mathematical formula for validating the human error probabilities.

The human reliability assessment (HRA) methods are mainly estimated the human error probabilities. Minimizing the likelihood of human errors is one of suggestion to improve human performance of specific tasks in the industry. For effective human performance considering safety, the probability together with the potential consequences should be assessed by the human factors or HRA analysts.

6. Human factors risk assessment

Risks associated with human factors can be assessed based on the human error probability and its potential consequence. Fig. 4 shows the framework for conducting a risk assessment specifically for tasks, which often associated with human factors. The framework of human factors risk assessment is adapted from the hazard identification and risk assessment procedure (Crowl & Louvar, 2002). The risk assessment will begin with

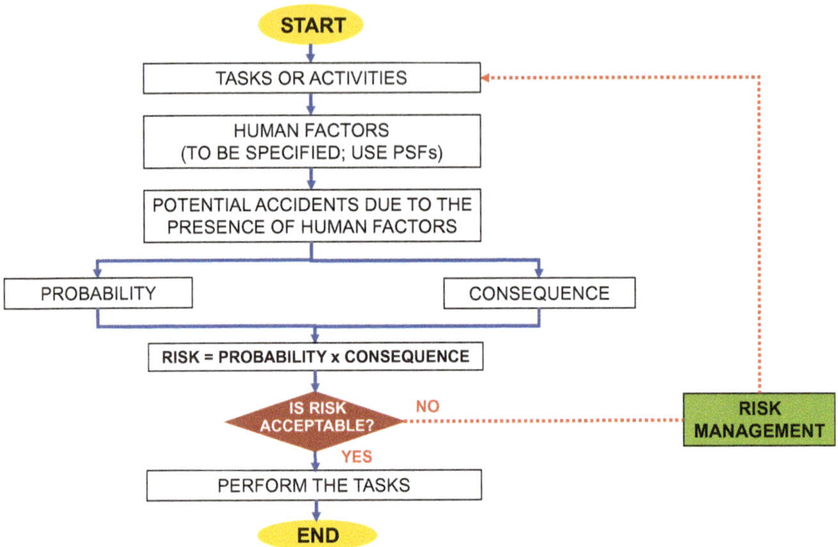

Fig. 4 The flowchart for assessing risk associated with human factors.

identifying tasks or activities. It must be followed by addressing the human factors and using the performance shaping factors to be specific. Any potential accidents for performing the tasks in the presence of human factors must be included in the risk assessment. Risk is calculated, taking the probability and consequence of accidents due to the presence of human factors in the task or activity. The risk is determined whether it is acceptable or not. For the risk assessment considering a semi-qualitative manner, the risk can be determined based on the risk matrix. The task with the presence of human factors can be performed when the risk is acceptable. However, the task with high-level risks should not be performed. Risk management will be applied for the task with a high level of risk. Risk management includes the minimization of the presence of human factors associated with tasks.

7. Conclusion

This chapter describes the importance of understanding human factors in chemical process safety before planning, designing, and operating or implementing equipment operated by individuals. Interaction is the keyword for understanding individuals working with equipment/machines at the workplace. Procedures, work process, the layout of the workspace, and other aspects related to the interaction should be prioritized in the phase of planning and designing. Human factors should be seen as contributing factors or causes leading to failures. Prioritizing human factors in the interaction between individuals and equipment/machines can improve performance and the productivity of the chemical processing industry.

References

Attwood, D., Baybutt, P., Devlin, C., Fluharty, W., Hughes, G., Isaacson, D., et al. (2007). Human factors methods for improving performance in the process industries. Hoboken, NJ: John Wiley & Sons, Inc.

Barge, R. (2019). What are performance shaping factors? Retrieved from https://www.ergonomicsblog.uk/performance-shaping-factors/. Retrieved date: November 30, 2019.

Bell, J., & Holroyd, J. (2009). Review of human reliability assessment methods. In *Health & Safety Laboratory, 78,* Retrieved from http://www.hse.gov.uk/research/rrpdf/rr679.pdf. Retrieved date: November 25, 2019.

Blackman, H. S., Gertman, D. I., & Boring, R. L. (2008). Human error quantification using performance shaping factors in the SPAR-H method. In *Paper presented at the proceedings of the human factors and ergonomics society annual meeting, 52,* (p. 1733). https://doi.org/10.1177/154193120805202109.

Cherry, K. (2019). The Yerkes-Dodson law and performance. Retrieved from https://www.verywellmind.com/what-is-the-yerkes-dodson-law-2796027. Retrieved date: November 30, 2019.

Crowl, D. A., & Louvar, J. F. (2002). Chemical process safety: Fundamental with applications. In *Upper Saddle River.* NJ: Prentice Hall, Inc.

Cullen, W. D. (1990a). The public inquiry into the Piper Alpha disaster. *Vol. 1*. London, UK: The Department of Energy.

Cullen, W. D. (1990b). The public inquiry into the Piper Alpha disaster. *Vol. 2*. London, UK: The Department of Energy.

Federal Aviation Administration (FAA). (2019). Chapter 14: Human factors. Retrieved from https://www.faasafety.gov/files/gslac/courses/content/258/1097/AMT_Handbook_Addendum_Human_Factors.pdf. Retrieved date: November 25, 2019.

Graeber, C. (1999). Human factors. Retrieved from http://www.boeing.com/commercial/aeromagazine/aero_08/human_textonly.html. Retrieved date: November 2, 2019.

Health and Safety Executive [HSE]. (2009). Reducing error and influencing behaviour. Health and Safety Executive Books.

Health and Safety Executive [HSE]. (2019). Performance influencing factors (PIFs). Retrieved from https://www.hse.gov.uk/humanfactors/topics/pifs.pdf. Retrieved date: November 29, 2019.

International Association of Oil & Gas Producers [IOGP]. (2005). Human factors: A means of improving HSE performance. Retrieved from https://www.iogp.org/bookstore/product/human-factors-a-means-of-improving-hse-performance/ Retrieved date: November 30, 2019.

International Ergonomics Association [IEA]. (2020). Definition and domains of ergonomics. Retrieved from https://www.iea.cc/whats/index.html. Retrieved date: December 22, 2018.

Kim, Y., Park, J., & Jung, W. (2017). A quantitative measure of fitness for duty and work processes for human reliability analysis. *Reliability Engineering and System Safety*, *167*, 595–601.
Medstar Health. (2019). What is human factors in healthcare? Retrieved from https://www.medicalhumanfactors.net/about-us/what-is-human-factors/. Retrieved date: November 29, 2019.
Pasquale, V. D., Iannone, R., Miranda, S., & Riemma, S. (2013). An overview of human reliability analysis techniques in manufacturing operations. In M. M. Schiraldi (Ed.), *Operations management*. IntechOpen, https://doi.org/10.5772/55065 Retrieved from: https://www.intechopen.com/books/operations-management/an-overview-of-human-reliability-analysis-techniques-in-manufacturing-operations. Retrieved date: December 1, 2019.
Reason, J. (1990). *Human error*. Cambridge, UK: Cambridge University Press.
Teigen, K. H. (1994). Yerkes-Dodson: A law for all seasons. *Theory & Psychology*, *4*(4), 525–547.
The Nuclear Energy Agency. (1988). NEA issue brief: An analysis of principal nuclear issues. Retrieved from https://www.oecd-nea.org/brief/brief-02.html. Retrieved date: November 9, 2019.
Woods, D., & Dekker, S. (2000). Anticipating the effects of technological change: A new era of dynamics for human factors. *Theoretical Issues in Ergonomics Science*, *3*(1), 272–282.

Video

Dekker, S. (2016a). Understanding human error part 1. Retrieved from https://www.youtube.com/watch?v=Fw3SwEXc3PU. Retrieved date: November 29, 2019.
Dekker, S. (2016b). Understanding human error part 2. Retrieved from https://www.youtube.com/watch?v=8R8nuAqpq-g. Retrieved date: November 29, 2019.
Dekker, S. (2016c). Understanding human error part 3. Retrieved from https://www.youtube.com/watch?v=rUmnI3Nq3V4. Retrieved date: November 29, 2019.
Dekker, S. (2016d). Understanding human error part 4. Retrieved from https://www.youtube.com/watch?v=crENdW2cGeU. Retrieved date: November 29, 2019.
Dekker, S. (2016e). Understanding human error part 5. Retrieved from https://www.youtube.com/watch?v=CwJePfJYcg8. Retrieved date: November 29, 2019.

Further reading

Edmonds, J. (2016). Human factors in the chemical and process industries—Making it work in practice. Elsevier.
Kim, J. W., & Jung, W. (2003). A taxonomy of performance influencing factors for human reliability analysis of emergency tasks. *Journal of Loss Prevention in the Process Industries*, *16*, 479–495.
Rekosuo, T. (2018). Human factors: How do stress and fatigue affect safety? *Power*, *162*(7), 48–49. Retrieved from https://www.powermag.com/human-factors-how-do-stress-and-fatigue-affect-safety/?pagenum=1. Retrieved date: November 30, 2019.
The US Chemical Safety Board (CSB). (2019). About the CSB. Retrieved from https://www.csb.gov/about-the-csb/. Retrieved date: November 30, 2019.

CHAPTER NINE

Acceptable risk: Regulatory aspects

Howard Pike*
Centre for Risk, Integrity and Safety Engineering (C-RISE), Faculty of Engineering, Memorial University, St. John's, NL, Canada
*Corresponding author: e-mail address: p79hlp@mun.ca

Contents

> *"The formulation of a problem is often more essential than its solution, which may be merely a matter of mathematical or experimental skills. To raise new questions, new possibilities, to regard old problems from a new angle requires creative imagination and marks real advances in science."*
>
> ***Einstein and Infeld (1938)***

This chapter reflects on the challenges and opportunities associated with establishing risk acceptance for risk assessment and management. In doing so, it borrows from and revisits themes and questions addressed in a paper authored under current publication.

Risk assessment and management have evolved from significant accidents associated with industrial activities—the same industrial activities that are created for our benefit to supply energy, food, and commodities. Risk assessment involves a formal and systematic analysis of the sources of harm or hazards, the consequences of exposure to such hazards, and barriers that prevent or mitigate exposure of such hazards. Such assessments attempt to forecast how the future will turn out if a particular course of action or inaction is undertaken. Risk assessment provides a basis for making informed decisions related to health and safety and the environment.

The view of risk acceptance as an activity added at the end of risk assessment is seriously deficient. What is needed for successful risk assessment and

Methods in Chemical Process Safety, Volume 4
ISSN 2468-6514
https://doi.org/10.1016/bs.mcps.2020.03.004

management is the consideration of risk acceptance at the very beginning of the process. Risk acceptance should lead to the determination of the scope and nature of risk assessment and subsequent risk management.

All human activity requires individuals and society to assume some level of risk. Risk assessment and management aim to identify and reduce the risks related to industrial activity. If risks cannot be eliminated, then how low is low enough in reducing the level of risk? The notation of acceptability immediately raises the question of acceptable to whom? At some point, someone needs to decide if risks are acceptable, but the general public has great difficulty with the concept of acceptable risk. The public has had notable conflicts over industrial development believed to be hazardous or environmentally unfriendly and with the implicit promise that industrial risks can be kept thoroughly under control. This promise has driven an increased expectation for a society free of involuntary risks. The primary objection to industrial activity is due to the potential involuntary nature of acceptable risk. This objection makes coping with risk situations potentially complex and controversial.

Society as a whole is continuously faced with the need to make decisions about hazardous or potentially hazardous activities. Risk acceptance cannot be divorced from broader issues in the community. It is intertwined in matters such as risk perception, fear of consequences, and their uncertainty as well as various other factors that influence and change society with time. For a variety of reasons, there has been an increasing requirement placed on governments and industry to inform the public about the potential exposure to hazards. The implication for risk analysis and evaluation is that the criteria and the degree to which risk might be accepted require both formal and informal processes for raising and collectively considering issues. The increasing importance placed on public opinion and expectations for public consultation underlies much of the conflict over industrial development. However, the practical difficulties of arriving at consensus decisions over the question of acceptable risk in society are considerable (NRC, 1996).

The appropriate level of effort to develop risk acceptance is situation-specific. The amount, content, and timing of effort that are appropriate to devote to risk acceptance need to be determined. A careful diagnosis of the nature of a hazard situation and the necessary information will be required to formulate the problem that the decision will address. Deliberation is essential at each step of the process that informs risk decisions, such as deciding which hazards to analyze and how to describe scientific uncertainty and disagreement. The principles that govern risk acceptance should leave enough flexibility to expand or simplify to suit the needs of the decision as the process evolves.

1. Criteria of risk acceptance

When identifying the criteria for risk acceptance, the description of the potentially hazardous situation needs to be accurate, thorough, and decision-relevant as possible, to address the concerns of the interested and affected parties. The information must be made understandable and accessible to the decision-makers and all parties. Diagnosis begins with surveying the kind of hazard and the state of knowledge. The affected parties and potential public reactions need to be identified. Risk acceptance requires different types of effort for different categories of decisions. For instance, unique and wide-impact situations tend to create strong needs for breadth, inclusion, and attention to process; in contrast, for many routine situations, narrow-impact choices, a simple, generic risk acceptance procedure may suffice. However, decisions to simplify the process should be taken with care, because an inappropriate or inflexible decision to use a narrow, routinized, or non-participatory method for risk characterization can undermine the decision-making process (NRC, 1996).

Although risk assessments are often completed for the benefit only of the decision-maker, it should be recognized that various other parties also use them. Such parties may be involved in the different phases of an activity such as those who are designing the installation, or those who ultimately have to operate the facility, in addition to those engaged in a public review of the proposed activity. A risk assessment should address what the interested and affected parties believe to be at risk in the particular situation, and incorporate their perspectives and knowledge. It may need to consider alternative sets of assumptions that may lead to divergent estimates of risk; to address social, economic, and ethical outcomes as well as consequences for health, safety, and the environment; and to consider issues for particular communities in addition to risks to whole populations. Under certain conditions, such as when trust in the responsible industry or regulator is low, they may need to make special efforts to ensure that the interested and affected parties accept as reasonable key underlying assumptions about risk-generating processes and risk estimation methods (NRC, 1996).

Important industrial activity decisions will always have proponents and opponents. Risk assessment results are afflicted with uncertainty, and parties with opposing interests will fight in many forums to get a project accepted or not. Often when that fight gets into the public arena, such a project becomes a political bone of contention. Although the work procedure for a risk assessment sounds straightforward, the execution is usually somewhat problematic. We depend on risk assessments to support optimum decision-making under uncertainty and must weigh the effort against the cost. We want to look into the future, but predictive tools are fallible, and the prediction results are helpful but uncertain. All assumptions, data, models, model parameters, calculations, and statements in a risk assessment include uncertainties.

The two main points of view of risk measures are related to individual and societal risks. The individual point of view is that a person, a system, or a facility can perform an activity, weighing the risks against the direct and indirect benefits. This point of view leads to acceptable individual risk, usually defined as the frequency or probability by which an individual may be exposed to a given level of hazard or experience specific consequences from such exposure. The specified level of consequence may be expressed in terms of loss of life, loss of the system, or suffering from certain illnesses. The societal point of view attempts to measure risk as either exposure or consequence of a large population to the number of lives lost, loss of life expectancy, changes in the health outcomes for a given population, or degradation of the environment. Societal risk is commonly expressed in terms of a total number of casualties, such as the relation between frequency and the number of people affected from a specified level of consequence in a given population from exposure to specified hazards. The particular level of consequence may be modeled by the frequency of exceedance curves of consequences (Modarres, 2006).

It should be recognized that risk cannot be eliminated, but it can be managed in such a way as to mitigate or reduce harm to the greatest extent practical. Risk reduction strategies are employed to reduce the risk to an acceptable level. One approach requires a reduction in the probability of accident occurrence by identifying and eliminating the conditions that cause accidents. Another strategy involves reducing the harmful consequences caused by the accidents that still occur by establishing accident mitigation and emergency response procedures, by reducing those causal factors likely to increase the injury and damage and by moving potential victims and assets out of the exposure zones.

There are four generally accepted categories of risk reduction strategies in order of preference: inherent, passive, active, and procedural (Crowl & Louvar, 2011). Inherent strategies identify and implement ways to eliminate or significantly reduce hazards, rather than to develop add-on protective systems and procedures. It is widely accepted that inherent strategies achieve their best effect when considered early in the life cycle of industrial activity. However, the strategy can be applied to reduce the risk of existing operations (Amyotte, Irvine, & Khan, 2018). There are four actions recognized as describing inherent strategies; Minimize (intensification); Substitute (substitution); Moderate (attenuation and limitation of effects); and Simplify (simplification and error tolerance).

Passive strategies add barriers that do not require the activation of any device. They will only perform as intended if they are not isolated from the activity they are designed to protect. Active strategies add shutdown systems to prevent accidents. Active strategies require event detection and device actuation to accomplish their intended function. Because they have moving parts that can fail when called upon to act, they must, therefore, be inspected, maintained, and tested regularly. Procedural strategies are the least effective way to deal with hazards in an activity. Nevertheless, procedures for tasks such as operations and maintenance are essential and are usually a regulatory requirement. It is, therefore, critical that procedures are correct, easy-to-understand, follow and enforce.

The application of these strategies is often described as a series of layers of protection surrounding an activity (Crowl & Louvar, 2011). Layers of protection are necessary because it is unlikely that a single strategy alone will effectively reduce the risk of the hazards. Each layer adds further defenses from the danger, thus reducing the risk from the activity. Inherently safer designs make these layers of protection more reliable and robust. Active and procedural layers of protection require constant maintenance and management to ensure that they continue to function as designed. If they are not managed correctly, the protection systems will degrade and increase the risks to an unacceptable level. These strategies are judged "acceptable" in the sense that there is no "undue risk." What "undue risk" means remains unquantified, which brings us back to risk acceptance.

If we are to judge how closely the estimated risk relates or represents reality, we must identify, quantify, and characterize uncertainties in the risk characterization. Uncertainty is a measure of the "goodness" of that estimate of risk, which is more than statistical error or inexactness of numbers. It is increasingly understood as a multidimensional concept stemming from the limitations of the assessment methods used, scarcity of data, ignorance, the use of assumptions, and limited social robustness of findings and methods. Uncertainty can manifest itself in different parts of the risk assessments such as context, system boundaries, indicator choice, model structure, parameters, and data.

When discussing uncertainty, it is important to distinguish between aleatory (stochastic) and epistemic (knowledge-related) uncertainty. Aleatory uncertainty refers to variation in large populations, is typically measured using frequentist probabilities. Aleatory uncertainty is linked with data, which can be improved by collecting more data or evidence to support the data. Epistemic uncertainty relates to a lack of knowledge about phenomena, usually translating into uncertainty about the parameters of a model used to describe some phenomenon or phenomena, is typically measured using subjective probability, and is considered reducible. Although, in many cases uncertainties can easily be distinguished, the two can occur combined due to inherent randomness. The uncertainty associated with analyzing the frequency of scenarios in probabilistic risk assessment increases with the challenging problem of data scarcity.

Paté-Cornell (1996) proposed six levels of treatment of both aleatory and epistemic types of uncertainty for risk analyses ranging from just noting there is a hazard, worst case, plausible upper-bound, best-estimate, first-order probabilistic risk analysis based on mean probabilities to second-order probabilistic risk analysis with a display of uncertainties. Aven and Zio (2011) have reviewed various methods to represent uncertainty along five main categories: the probabilistic analysis, probability bound analysis (combining probability analysis and interval analysis), imprecise probability (the robust Bayes statistics area), random sets (Dempster-Shafer belief functions), and possibility theory (a special case of the imprecise probability and random sets theories). They argue that a full risk uncertainty description is more than subjective probabilities and extends beyond the Bayesian approach. Berner and Flage (2016) stated that assumptions are part of the foundation of a risk assessment. They argue that the decision-maker should perform judgment of the assumptions and conservatism embodied due to uncertainty.

In the 1990s, the choice of a worst-case scenario versus a maximum-credible one was discussed at length (Khan, 2001). However, what is physically possible should not be ruled out too soon. Analysis and lowering of the uncertainty will support decision-making. A more credible threat may also stimulate resilience boosting measures (Kelly & Smith, 2009; Pasman & Rogers, 2018).

Public perception of risk can be heavily swayed by current events and their assessment of the worst-case scenario. An example of this can be taken from the aftermath of the September 11, 2001 attacks, when there was a significant shift from air travel to road travel. If one person gives up the relative safety of air travel and drives instead, they will almost certainly survive. But if millions of people take the same risk, it is just as likely that some of them suffer a loss as either an injury or death. Gigerenzer (2006) published a paper comparing road travel 5 years before the September 11 attacks and 5 years after. That study showed that the shift from air travel to road travel in America lasted 1 year, then traffic patterns returned to normal. The study showed that fatalities on American roads soared after September 2001 and returned to normal levels in September 2002. The number of Americans killed in vehicle collisions as a direct result of switching from air travel to road travel was calculated to be 1595 (Gigerenzer, 2006). That is more than one-half of the total death toll of September 11 and six times higher than the total number of people on board the flights involved in the attack (Gigerenzer, 2006). This example also illustrates the challenge involved in identifying and separating individual and societal risks.

The scope of the assessment, which, among other things, defines the system boundaries, becomes very important. If we narrow the scope too far, we may ignore or introduce other hazards which could have the effect of decreasing overall safety and environmental protection. A wide viewscreen, rather than a narrow one, is indispensable for overall safety and protection of the environment (Kletz, 2005).

While the risk of suffering harm is an inescapable aspect of all human activity, loss of life, serious illness, and degradation of the environment are no longer considered acceptable consequences from industrial development. Increased levels of awareness of environmental and development issues and greater engagement on the part of the public have led to a much keener interest in industrial risk management practices, policies, and effectiveness. Regulatory interventions and risk management use two widely recognized principles for risk reduction, namely, the "As Low As Reasonably Practicable" (ALARP) principle and the "Precautionary Principle" (PP) (TRB, 2018).

2. ALARP (As Low As Reasonably Practicable)

The ALARP principle requires that those responsible for the industrial activity and, indeed, the public's safety, reduce risks to levels that are "As Low as Reasonably Practicable." As such, the principle involves effective recognition of the fact that, while in most circumstances, risk can be reduced, beyond some point further risk-reduction is increasingly costly to implement (UKHSE, n.d.). The ALARP principle originated in the United Kingdom (UK). It is not a prescriptive risk management approach, but one that leaves considerable latitude to those responsible for the industrial activity to make effective decisions in deciding on the strategy to mitigate risks. The critical element of the ALARP principle is the term "reasonably practicable." The first formal definition of "so far as is reasonably practicable" was provided in the Court of Appeal judgment in Edwards v. National Coal Board, [1949] 1 All ER 743 (UKHSE, n.d.):

"'Reasonably practicable' is a narrower term than 'physically possible' ... a computation must be made by the owner in which the quantum of risk is placed on one scale and the sacrifice involved in the measures necessary for averting the risk (whether in money, time or trouble) is placed in the other, and that, if it be shown that there is a gross disproportion between them—the risk being insignificant in relation to the sacrifice—the defendants discharge the onus on them."

The first introduction into UK law of the concept that industrial risks should be as low as reasonably practicable (ALARP) was in the 1961 Factories Act. The concept of "reasonably practicable" lies at the heart of the UK health and safety system. It is a crucial part of the general duties of the Health and Safety at Work Act 1974 and many sets of health and safety regulations and those of Local Authorities (UKHSE, n.d.).

Canadian Occupational Health and Safety legislation refers to the concept of a duty of care for stakeholders of industrial activity. All stakeholders are required to take every precaution reasonable in their circumstances to avoid harm. Their circumstances include their position, authority, knowledge, and understanding of the risk. The general concept is of having exercised reasonable care or Due Diligence in their actions, which is very similar in concept to that of the UK ALARP. The harm has to be reduced to as low as reasonably practicable.

The ALARP approach sets an upper limit above which the risk must be reduced, or the activity must stop, and a lower limit below which resources expended produce negligible risk reduction. The strategy is to reduce the risk as low as reasonably practicably to or below the lower threshold. The critical words in ALARP are "low," "reasonably" and "practicable." Unfortunately, these are all relative terms—standards are not defined. "Reasonably" is also an emotive word, implying goodness, care, consideration, etc. What may be reasonable in some situations can be seen as inappropriate in others (Melchers, 2001). However, if the risk is above the maximum tolerable level, then the risk shall be reduced, or the activity giving rise to it discontinued whatever the cost.

What is the maximum tolerable level beyond which it is morally intolerable to impose risk? Flüeler and Seiler (2003) have identified four different categories:

Category 1: voluntary risk exposure to satisfy one's desires, e.g., dangerous sports

Category 2: high degree of self-determination, direct individual benefit, e.g., driving a car

Category 3: low degree of self-determination, personal gain, e.g., working conditions

Category 4: involuntary, imposed risk exposition, no direct benefit, e.g., local residence near a hazardous installation

The categories differ in the extent to which the individual is responsible for their exposure: participants in dangerous sports such as rock climbing are entirely responsible for their exposure; on the other hand if approval is granted to locate a hazardous facility in our neighborhood, there is little we can do about the risk, short of moving away. Flüeler and Seiler (2003) identify limits based on rates that Swiss society already accepts, ranging from 10^{-3} to 10^{-5} as being appropriate. For Category 4 risks, which are risks imposed and providing no direct benefit for the person exposed, the limit of tolerability is one death in 100,000 (10^{-5}); for Category 2 risks, such as driving a car, the limit of tolerability is one death in 1000 (10^{-3}). For Category 1 risks, for people who voluntarily engage in dangerous sports, it may be inappropriate to specify any risk level as intolerable, it could be argued, that would be an unwarranted interference with freedom of choice.

There are apparent financial/economic implications "'reasonable practicability' is not defined in legislation but has been interpreted in legal cases to mean that the degree of risk can be balanced against time, trouble, cost and physical difficulty of its risk reduction measures. Risks have to be reduced to the level at which the benefits arising from further risk reduction are disproportionate to the time, trouble, cost and physical difficulty of implementing further risk reduction measures" (Melchers, 2001). Baybutt (2014) concludes that the ALARP principle provides a useful framework within which to establish risk tolerance criteria. Cost-Benefit Analysis helps decide what risk reduction measures are warranted for the costs involved. With the use of a disproportionate factor in the analysis, the concept of reasonably practicable is formalized and provides justification and transparency for decisions made. In another paper it is argued that the ALARP principle should be interpreted dynamically, meaning that the interpretation of the gross disproportionate criterion ranges from one extreme, where decisions are made concerning expected values, to another extreme, in which the cautionary principle is given special weight with no reference to cost-benefit (cost-effectiveness) analyses (Abrahamsen, Abrahamsen, Milazzo, & Selvik, 2018).

The implementation of ALARP in risk management also requires the continuous analysis of the benefits gained from risk reduction versus the resource to be consumed to ensure that risk is reduced to the lowest level. Financial implications are recognized in pursuing any safety or environmental improvement to demonstrate ALARP. It is possible, in principle, to apply formal cost-benefit techniques to assist in making a judgment(s) of this kind. The assumption is that all factors involved can be converted to monetary values. Unfortunately, there are considerable difficulties and implied value judgments in evaluating monetary values for both benefits and costs. This problem is particularly acute for the analysis of hazardous facilities where the value of human life, the cost of suffering and deterioration of the quality of life, and degradation of the environment may play a significant role in the analysis (Melchers, 2001).

Risk tolerability cannot be divorced from broader issues in the community. It is intertwined in matters such as risk perception, fear of consequences, and their uncertainty, as well as various other factors that influence and change society with time. Societal risk tolerability can change very quickly when there is a major industrial accident. The implication for the ALARP approach might well be what was considered sufficiently "low" for a particular type of industrial activity before an "accident" might not be regarded as sufficient for another similar industrial activity after an accident. There will be very considerable societal and political pressures for changing the acceptance criteria (Melchers, 2001).

In most situations, deciding whether the risks are ALARP involves a comparison between the barriers those responsible for the industrial activity have in place or are proposing, and the measures that would typically be expected in such circumstances, i.e., recognized standards or relevant good practice. "Good practice" is defined in the general UK ALARP guidance (UKHSE, n.d.) as "those standards for controlling risk that are judged and recognized as satisfying the law, when applied to a particularly relevant case appropriately." In the United States, the Occupational Safety and Health Administration, OSHA, requires Recognized and Generally Accepted Good Engineering Practice (RAGAGEP).

Where the situation is complicated, it may be difficult to decide based on good practice alone. More analysis of the situation and processes is required.

There are some instances, often in high hazard industries or where there is a new technology with potentially serious consequences, where the situation is not clear-cut. In such cases, a more detailed comparison needs to be undertaken. The trouble is that risk and sacrifice are not usually measured in the same units. In these instances, a more formal Cost-Benefit Analysis (CBA) may provide additional insight to help come to judgment (UKHSE, n.d.).

ALARP does not represent zero risks. We have to expect the risk arising from a hazard to be realized sometimes and for harm to occur, even though the risk is ALARP. Of course, we should strive to make sure that those responsible for the industrial activity reduce and maintain the risks ALARP, and we should never be complacent. Nevertheless, we have to accept that risk from an activity can never be eliminated unless the activity is stopped. This also goes some way to explaining why risk assessments feed into contingency planning.

Risk assessments have a problem when dealing with low probability—high consequence events. Morally and practically, these events do not allow the luxury of a trial and error learning process. Then it is appropriate to invoke the "precautionary principle."

3. The precautionary principle (PP)

We have for generations been using the folk wisdom that "it is better to be safe than sorry" and "an ounce of prevention is worth a pound of cure," which embodies the precautionary approach or principle. The environmental debates of the 1970s and 1980s began expressing a precautionary approach to industrial development. In the early 1980s, the precautionary principle started to emerge in international law, most notably in the World Charter for Nature (UN, 1982), but it was not until after it was incorporated as "Principle 15" in the RIO Declaration (UNCED, 1992) that it achieved international prominence and acceptance. It has become a feature of close to one hundred international agreements and has been incorporated into scores of domestic environmental and public health laws worldwide (Tollefson & Thornback, 2008). However, different views exist of what precaution is, and the precautionary principle has different interpretations. The criticism that is most commonly raised is that the PP is unclear. It should be remembered that the precautionary principle is not a scientific hypothesis, theory, or methodology rule; rather, it is a principle for making practical decisions under conditions of scientific uncertainty (Resnik, 2003).

Sandin (1999) has analyzed the various definitions of the precautionary principle along four key dimensions: threat, uncertainty, action, and command. Under Sandin's analysis, threat refers to the nature of the imminent harm, particularly its seriousness and irreversibility, while uncertainty indicates our lack of knowledge as to whether and how this threat might materialize. Under most definitions of the principle, where both the threat and uncertainty meet defined thresholds, an action obligation is triggered to consider cost-effective measures to prevent degradation of health or the environment. The command dimension designates the legal status of the action obligation to be taken, which may be framed in either mandatory or permissive language.

According to Sandin, a challenge to utilizing the precautionary principle lies in the imprecision with which the dimensions of threat, uncertainty, action, and command are typically framed (Sandin, 1999).

The UNESCO World Commission on the Ethics of Scientific Knowledge and Technology (COMEST), has developed a working definition of the Precautionary Principle (PP), or Precautionary Approach, which goes some way to clarifying the imprecision and defining boundaries for its applicability (COMEST, 2005):

"When human activities may lead to morally unacceptable harm that is scientifically plausible but uncertain, actions shall be taken to avoid or diminish that harm. Morally unacceptable harm refers to harm to humans or the environment that is:

- threatening to human life or health, or
- serious and effectively irreversible, or
- inequitable to present or future generations, or
- imposed without adequate consideration of the human rights of those affected.

The judgment of plausibility should be grounded in scientific analysis. Analysis should be ongoing so that chosen actions are subject to review. Uncertainty may apply to, but need not be limited to, causality or the bounds of the possible harm.

Actions are interventions that are undertaken before harm occurs that seek to avoid or diminish the harm. Actions should be chosen that are proportional to the seriousness of the potential harm, with consideration of their positive and negative consequences, and with an assessment of the moral implications of both action and inaction. The choice of action should be the result of a participatory process."

Where the principle is part of the regulatory process, a regulator is empowered, and in some instances obliged to take it into account. Their response must be proportional to the risk and must adapt as knowledge of the threat becomes more certain. The Supreme Court of Canada's decision (2SCR241, 2001) in Spraytech is an illustration of how the precautionary principle can be applied and established the case law for the use of the precautionary principle. In that decision, the Court decided that a by-law that severely restricted the use of pesticides was a valid exercise of the Town of Hudson's authority to adopt by-laws to secure the health and general welfare of the people. The ruling by the majority of the Court used the principle as a means to provide justification and context for the Town's concerns with respect to pesticides rather than providing the legal basis for the validity of the by-law.

The most common way the principle finds its way into the regulatory process is through its implicit or explicit adoption in statutes. In Canada, the principle is found, in the preambles to Species At Risk Act (SARA), the Oceans Act, and the Canadian Environmental Protection Act (CEPA), in the purposes section of the old Canadian Environmental Assessment Act (CEAA) (s. 4), in the mandate section of the new Impact Assessment Act Bill C-69 (s.6 (2)), and as a mandatory strategic management principle under the Oceans Act (s. 30). It is expressed as a relevant consideration in the exercise of administrative duties vested in the Government of Canada and its agencies under CEPA and CEAA. It is also slowly finding its way into provincial legislation.

If there is considerable uncertainty about the nature of the hazards and the likelihood of them causing harm, then successful management of risk in any industrial activity must satisfy the premise that anything present in an activity which "presents the possibility of danger" is regarded as a "hazard." When that hazard may have consequences that may be irreversible and harmful, there is an overriding need to introduce control measures to address the risks (UKHSE, 2001).

The Health and Safety Executive in the United Kingdom has noted in their guidance that the quantum of risk can be evaluated by creating and evaluating credible scenarios (UKHSE, 2001):

"Uncertainty is overcome by constructing credible scenarios on how the hazards could be realized and thereby making assumptions about consequences and likelihood. The credible scenarios can range from a 'most likely' worst case to a 'worst case possible' depending on the degree of uncertainty. For example, by assuming that exposure to a putative carcinogenic chemical will cause cancer, the chemical becomes subject to a very stringent control regime."

The credible scenarios developed would form the basis for a risk assessment. The response that is required by the precautionary principle will depend on the outcome of the risk assessment. The overarching goal of the response is proportionality—the more significant and likely the threat, the higher the degree of precaution required. Where uncertainty exists, a margin of error should be left so that severe or irreversible harm is less likely to occur (Tollefson & Thornback, 2008).

The importance of continuing to monitor and analyze the chosen action based on operating experience, new evidence, and lessons learned should not be forgotten. The Precautionary Approach has the objective of achieving lower and more acceptable levels of risks but "is not based on 'zero risks,'" a risk scenario that is not reflective of human activity (COMEST, 2005).

4. ALARP versus PP: Which one to use when

Risk assessment is a model of decision-making used in making choices about environmental hazards and public health and safety. Risk assessment can be expressed in terms of answering three questions; what can go wrong, how likely is it, and what are the losses or consequences? Answering these questions identifies the hazard, consequences, and risk of an activity (Kaplan & Garrick, 1981). Risk assessment classifies and describes those various risks and looks at the implementation of barriers to reduce the probability of occurrence. Any additional response to a threat can be identified.

If there is considerable uncertainty about the nature of the hazards and the likelihood of them causing harm, then successful management of risk in an industrial activity must satisfy the premise that anything present in an activity which "presents the possibility of harm" is regarded as a "hazard." The use of risk assessment to determine the necessary control measures ensures that for hazards surmised to have consequences that may be irreversible and harmful, there is an overriding need to introduce control measures to address the risks.

Once the risks have been characterized, we can assess if the precautionary principle applies, or we have enough information to use the ALARP principle. The precautionary principle can be a rational procedural vehicle for decision-making in the face of uncertainty, provided that it instructs us to take reasonable precautions to deal with plausible threats. A dynamic interpretation of ALARP principle provides the flexibility to establish the criteria for making practical decisions. The importance of continuing to monitor and analyze the chosen action based on operating experience, new evidence, and lessons learned should not be forgotten. If the risk is above the maximum tolerable level, then the risk should be reduced, or the activity giving rise to it discontinued whatever the cost. This approach is illustrated in the decision flow diagram presented in Fig. 1.

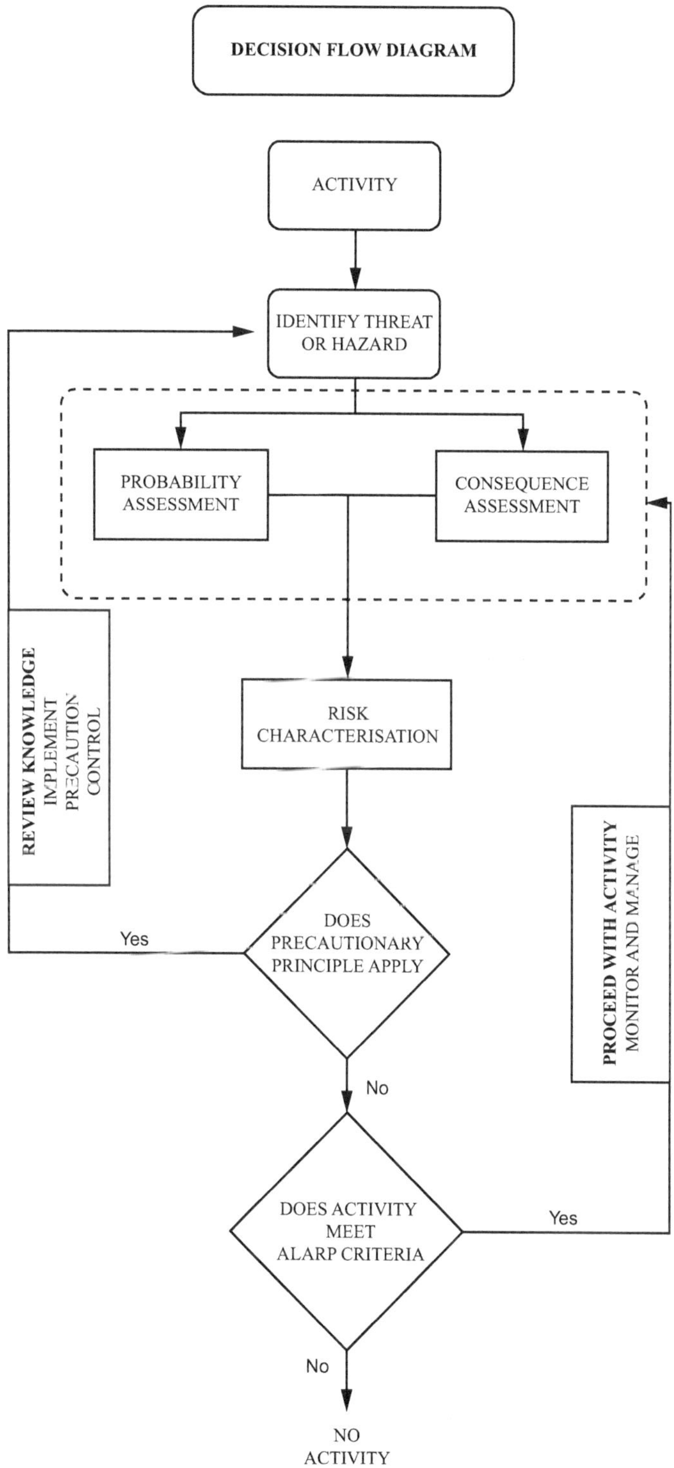

Fig. 1 The decision-making flow diagram.

ALARP and PP principles for risk acceptance provide the flexibility to expand or simplify as the risk assessment and risk management evolve. But, perhaps the essential aspect of risk acceptance is not the probability number, but the evidence and reasoning the process summarizes. The summary should tell the story of how the problem has been identified and addressed.

References

2SCR241. (2001). *114957 Canada Ltée (Spraytech, Société d'arrosage) v. Hudson (Town)—SCC Cases (Lexum).* Retrieved May 15, 2019, from https://scc-csc.lexum.com/scc-csc/scc-csc/en/item/1878/index.do.

Abrahamsen, E. B., Abrahamsen, H. B., Milazzo, M. F., & Selvik, J. T. (2018). Using the ALARP principle for safety management in the energy production sector of chemical industry. *Reliability Engineering and System Safety, 169*, 160–165. https://doi.org/10.1016/j.ress.2017.08.014.

Amyotte, P., Irvine, Y., & Khan, F. (2018). Chemical safety board investigation reports and the hierarchy of controls: Round 2. *Process Safety Progress, 37*(4), 459–466.

Aven, T., & Zio, E. (2011). Some considerations on the treatment of uncertainties in risk assessment for practical decision making. *Reliability Engineering and System Safety, 96*(1), 64–74. https://doi.org/10.1016/j.ress.2010.06.001.

Baybutt, P. (2014). The ALARP principle in process safety. *Process Safety Progress, 33*(1), 36–40. https://doi.org/10.1002/prs.11599.

Berner, C., & Flage, R. (2016). Strengthening quantitative risk assessments by systematic treatment of uncertain assumptions. *Reliability Engineering and System Safety, 151*, 46–59. https://doi.org/10.1016/j.ress.2015.10.009.

COMEST. (2005). *The precautionary principle.* Retrieved October 6, 2019, from https://unesdoc.unesco.org/ark:/48223/pf0000139578.

Crowl, D., & Louvar, J. (2011). *Chemical process safety—Fundamentals with applications* (3rd ed.). Boston, MA: Pearson Education.

Einstein, A., & Infeld, L. (1938). *The evolution of physics.* London, UK: Cambridge University Press.

Flüeler, T., & Seiler, H. (2003). Risk-based regulation of technical risks: Lessons learnt from case studies in Switzerland. *Journal of Risk Research, 6*(3), 213–231.

Gigerenzer, G. (2006). Out of the frying pan into the fire: Behavioral reactions to terrorist attacks. *Risk Analysis, 26*(2), 347–351.

Kaplan, S., & Garrick, B. J. (1981). On the quantitative definition of risk. *The Annals of Occupational Hygiene, 1*(1), 11–27.

Kelly, D. L., & Smith, C. L. (2009). Bayesian inference in probabilistic risk assessment—The current state of the art. *Reliability Engineering and System Safety, 94*(2), 628–643.

Khan, F. (2001). Use maximum-credible accident scenarios for realistic and reliable risk assessment. *Chemical Engineering Progress, 97*(11), 56–64.

Kletz, T. A. (2005). Looking beyond ALARP—overcoming its limitations. *Process Safety and Environmental Protection, 83*, 81–84.

Melchers, R. E. (2001). On the ALARP approach to risk management. *Reliability Engineering and System Safety, 71*(2), 201–208.

Modarres, M. (2006). *Risk analysis in engineering.* Boca Raton, FL: Taylor & Francis.

NRC. (1996). *Understanding risk informing decisions in a democratic society. In P. C. Stern & H. V. Fineberg (Eds.), National research council.* National Academies Press. https://doi.org/10.17226/5138.

Pasman, H., & Rogers, W. (2018). How trustworthy are risk assessment results, and what can be done about the uncertainties they are plagued with? *Journal of Loss Prevention in the Process Industries, 55*, 162–177.

Paté-Cornell, M. E. (1996). Uncertainties in risk analysis: Six levels of treatment. *Reliability Engineering and System Safety, 54*, 95–111. https://doi.org/10.1016/S0951-8320(96)00067-1.

Resnik, D. B. (2003). Is the precautionary principle unscientific? *Studies in History and Philosophy of Science Part C: Studies in History and Philosophy of Biological and Biomedical Sciences, 34*(2), 329–344.

Sandin, P. (1999). Dimensions of the precautionary principle. *Human and Ecological Risk Assessment (HERA), 5*(5), 889–907.

Tollefson, C., & Thornback, J. (2008). Litigating the precautionary principle in domestic courts. *Journal of Environmental Law and Practice, 19*, 33–58.

TRB. (2018). *Designing safety regulations for high-hazard industries. In Transportation Research Board—Special Report.* National Academies Press. https://doi.org/10.17226/24907.

UKHSE. (2001). *Reducing risks, protecting people.* Retrieved October 6, 2019, from https://www.hse.gov.uk/risk/theory/r2p2.pdf.

UKHSE. (n.d.). Risk management: ALARP at a glance. Retrieved May 15, 2019, from http://www.hse.gov.uk/risk/theory/alarpglance.htm

UN. (1982). *A/RES/37/7. World charter for nature.* Retrieved May 15, 2019, from https://www.un.org/documents/ga/res/37/a37r007.htm.

UNCED. (1992). *A/CONF.151/26 (Vol. I) Report of the United Nations Conference on Environment and Development.* Retrieved May 15, 2019, from https://www.un.org/documents/ga/conf151/aconf15126-1annex1.htm.

CHAPTER TEN

Advanced methods for environmental risk assessment in offshore operations

Ehsan Arzaghi[a], Zaman Sajid[b], Rouzbeh Abbassi[c,*]
[a]Faculty of Science and Engineering, School of Mechanical, Medical & Process Engineering, Queensland University of Technology, Brisbane, QLD, Australia
[b]Faculty of Engineering & Applied Science, Centre for Risk, Integrity and Safety Engineering, Memorial University of Newfoundland, St. John's, NL, Canada
[c]Faculty of Science and Engineering, School of Engineering, Macquarie University, Sydney, NSW, Australia
*Corresponding author: e-mail address: rouzbeh.abbassi@mq.edu.au

Contents

1. Introduction

Environmental Risk Assessment (ERA) is a systematic approach for assessing and managing the risk associated with human health and ecological entities caused by an event occurring in the environment. ERA is the characterization of adverse health effects that results from human and

Methods in Chemical Process Safety, Volume 4
ISSN 2468-6514
https://doi.org/10.1016/bs.mcps.2020.04.002

ecological exposures to environmental hazards (Abbassi, Dadashzadeh, Khan, & Hawboldt, 2012; Abbassi, Khan, & Hawboldt, 2010; Fahd, Khan, Hawboldt, & Abbassi, 2014). By increasing the public concern about the impacts of chemicals and hazards to the wildlife and ecosystem, there is considerable attention to ERA in past 3 decades. Responding to questions such as "is it safe for a proposed offshore facility to operate in a particular location" or is it safe to release the contaminants to the natural environment (e.g. ocean) is one of the main aims of conducting ERA.

In the past decade, various oil spill accidents in offshore drilling operations have occurred in fragile environments and harsh conditions have imposed high risks to marine ecosystems (Moroni, Pieri, & Tampucci, 2019). Oil spills to the marine environment due to the accidents are common and going to continue. The oil spills due to the collision between Sanchi oil tanker with CF Crystal (2018) and DP Deepwater Horizon disaster are examples since 2010 (Dadashzadeh, Abbassi, Khan, & Hawboldt, 2013; Li et al., 2019). For instance, approximately 40,000 m cubic of oil was released from Exxon Valdez in Alaska's Prince William Sound in 1989. More recently, the Deepwater Horizon (DWH) spill in 2010 resulted in the ejection of approximately 4.9 million barrels of oil in the Gulf of Mexico. As the largest accidental oil spill in the world, DWH caused 11 deaths and 17 injuries in addition to the disastrous impacts on the marine environment. These incidents have created a strong ongoing interest across both academia and offshore oil and gas industries to better understand the potential impacts of catastrophic oil spills, regardless of the form they take (Nelson & Grubesic, 2018). Contaminants released from the oil spill to the ocean environment is a critical challenge to the marine ecosystem. Different factors such as local environmental conditions (e.g. weather and sea conditions), water depth, type of the products during spills, and spill volume affecting the dispersion of the oil in marine environment, and produce different uncertainties in ERA of oil spills (Enache & Zagan, 2008; Heidari & Raie, 2018).

To conduct a comprehensive ERA, the potential hazards associated with oil spills, and the severity of the impact of oil spills in the marine environment need to be identified. There is a need to develop models reflecting oil spill concentration distributions considering the influential environmental, physical, biological and chemical factors (Ferguson, Solo-Gabriele, & Mena, 2020). So, this chapter is devoted to the following sections which help to better assess the risk associated with oil spills in marine environments. In Section 2, environmental hazards and associated risk in offshore operations are discussed. Section 3 explained the oil spills models and required data in the marine environment. Section 4 discusses the adverse effects of the exposure to oil spills on human health, and Section 5 explains some of the current advanced techniques applicable to environmental risk assessment. Finally, conclusions and future directions in the ERA of oil spills are explained in Section 6.

2. Environmental hazards and associated risks in offshore operations

Offshore petroleum operations have introduced significant risks to the natural environment and human lives. Researchers have shown that multiple factors impose various risks to the environment and human lives due to offshore operations. Those are the introduction of new technologies, an increase in mining activities, operational failures, environmental changes, and non-compliance with regulatory frameworks for offshore work (Yang, 2019). There are various environmental risks associated with such offshore operations, including disturbance of habitats from oil and gas drilling, underwater noise due to seismic surveys, mud discharges due to drilling and cuttings, and the potential of oil spills due to blowout of wells (Norwegian Ministry of the Environment, 2011). Such activities are not only causing environmental risks but are also adversely affecting marine species.

A study shows that seismic noise poses risks to minke whales, killer whales, fin whales, Atlantic white-beaked dolphins, and other marine mammals since the noise frequency of an oil exploration operation can travel over 3000 km in the ocean (Nieukirk et al., 2012). Marine and coastal ecosystems are the main support for economic prosperity, social well-being, and continuous population growth in coastal regions (De Groot et al., 2012). However, due to ever-increasing human activities which cause climate change, drilling incidents in offshore oil and gas operations, and runoff from agricultural lands, such benefits are on stake. Among these, oil and gas operations are major contributors to the environmental pollution of the offshore environment and the ecosystem (Islam & Tanaka, 2004). In the north-western Gulf of Mexico, there are more than 51,000 oil and gas wells and among them, nearly 27,000 abandoned wells which have never been checked for any environmental hazards (Neumann, 2010). An oil spill in offshore operations affects marine ecosystems in different ways. Some offshore areas are sensitive to toxicity levels of spilled oil and gas, while some (such as coral reefs) are vulnerable to physical impact. Below are four main environmental effects of an oil spill (Meng, 2016):

(1) Spilled oil can coat the feathers and fur of seabirds and marine mammals, which hinders their free movement. Oil can also be ingested by turtles and marine mammals, which could result in toxic effects.

(2) Eggs and larvae of marine species such as fishes have a greater possibility to be affected by toxins than adult marine species. The presence of polycyclic aromatic hydrocarbons can be carcinogenic for them.

(3) Limpets can sustain some degree of oiling; however, corals are quite sensitive. Shellfish may experience secondary health effects due to accumulated oil residues.

(4) Unlike plankton, individual planktonic organisms can exhibit toxic effects due to spilled oil. Due to mucilaginous coating, oil does not stick to seaweeds; however, it will affect the seagrasses.

Offshore operations such as oil and gas spills are the major anthropogenic source of hydrocarbon release in the marine environment, which results in hazardous environmental consequences for marine species, water flora and fauna. In a study conducted by Meng (2016), the author addressed the temporal and spatial attributes of the environmental hazards due to oil and gas operations in an offshore environment. The study was performed using cross variogram and spatial statistical models. Using the Gulf of Mexico as a reference, the work identified, assessed and mapped temporal and spatial aspects of the environmental hazards in the chosen offshore region. For offshore oil and gas operations, this study developed a spatial correlation matrix and quantified correlations among 12 environmental hazards. Those were environmental hazards due to factors such as explosions, fires, blowouts, incidents, and collision, causing fatalities, lost time accidents, pollution, emergency evacuations of the staff, injuries, and leak situations. The results of the study characterized the spatial behavior of 12 environmental hazards and identified the zones in the Gulf of Mexico with high environmental risks due to an oil spill.

Offshore operations have not only increased in the Gulf of Mexico but also in the North American Arctic. In the last decade, there has been seen substantial growth in marine traffic and offshore oil and gas operations in the Arctic (Brigham, 2010). Such rapid development has led to environmental hazards to its coastal communities, ecosystems and coastal infrastructures. A study on environmental security of the Arctic has been presented by Eicken et al. (2011). They studied environmental security of the Arctic in two aspects:

(1) Environmental safety of the Arctic: this included the prevention or impact minimization as a result of climate change and other human activities in the Arctic. It dealt with avoiding incidents such as oil spills in the Arctic which could cause long-term and short-term environmental hazards to the Arctic environment.

(2) Hazards due to environmental change in the Arctic: this included environmental change such as changes in weather which could pose risks to the well-being of Arctic communities, their infrastructure or individuals. This dealt with risks associated with climate change.

In their study, they developed a prototype coastal observation system to monitor environmental hazards and the effectiveness of emergency response in the sea ice environment in Barrow, Alaska. They used Hidden Markov modeling to perform early identification of hazardous events. This provided support for the operational decision-making process in the Arctic environment.

A fuzzy-based methodological framework was presented which assessed the environmental risks due to drilling waste disposed into the marine environment (Sadiq & Husain, 2005). Their methodology included the use of a fuzzy concept and hierarchical analysis to assess the environmental risk to marine species in offshore operations. The study developed a hierarchical structure model for environmental risks at three stages. The model was applied to a case study of drilling waste discharge. The study concluded there is a non-linear trend existed between risk and the remedy provided for drilling waste. A sensitivity analysis performed by the study revealed that environmental risks were highly dependent upon the chosen grade level of the risk and importance level of the environmental risk.

Johnston (2012) discusses the potential environmental risks posed to the Arctic due to shipping development and exploration of energy resources in the Arctic. This study also examines the impact of energy resource potential on a regional environment. Johnston (2012) argues that, due to its harsh environment, development in the Arctic region will be gradual rather than rapid, contradicting other studies. Moreover, the study shows that the equipment, contingency and emergency response plans of oil and gas companies operating in the Arctic have never been tested in case another oil spill occurs such as BP disaster in the Gulf of Mexico.

This introduces a level of uncertainty in managing potential accidents occurring in the Arctic, which could cause more intense environmental damage to the region than predicted. This study states that after the offshore BP oil spill and subsequent severe environmental damages, other companies must operate with caution to avoid another major environmental disaster.

In a study, Saud, Israni, and Goddard (2014) discussed the use of a bow-tie diagram for an environmental hazards study in offshore and onshore operations. Their study identified that a bow-tie diagram is a resourceful, useful and a real-time tool for the study of environmental hazards in the offshore industry. A bow-tie methodological framework provides a graphical summary of all incident scenarios due to a hazard. It shows comprehensive relationships among hazards, the top event, degradation factors and their controls, consequences, mitigating and preventive barriers (Lewis & Smith, 2010). Generally, bow-tie diagrams are used to represent hazards' controls for health, safety, and the environment (HSE), as discussed by Ferdous, Khan, Sadiq, Amyotte, and Veitch (2013). Saud et al. (2014) highlighted the use of environmental hazard identification (ENVID) in parallel to hazard identification (HAZID). The research findings of this study indicated that inclusion of ENVID to HAZID results helped to understand the potential causes of environmental impacts and correlating the results of HAZID to ENVID helped to understand the role of the various causes of environmental hazards and the barriers in place in environmental risk assessment.

Caia, Di Lullo, De Ghetto, and Guadagnini (2018) developed a methodological framework to assess the failure risk of a deepwater well barrier and the associated environmental hazards. They combined a decision tree and dynamic event tree to assess occurrence probability and post-blowout impact. The study used a Monte Carlo Simulation approach to deal with uncertainties in risk and its mitigation. They used Stochastic oil drift simulations (OS3D model based on OSCAR model) to assess the environmental impacts of a spill.

An advanced probabilistic model for the analysis of the oil pipeline network considering the emergency response and environmental hazards was presented by Zhang, Qin, Jiang, and Huang (2018). They used a Bayesian network (BN) to model accidents in oil pipelines and their subsequent environmental hazards. A BN is a graphical model that uses a probabilistic approach to represent risk variables and is based on Bayesian theory (Sajid, Khan, & Zhang, 2017, 2018). A study by Zhang et al. (2018) identified a probabilistic analysis model for consequences such as economic loss, causalities, social order influence and environmental hazards in the oil pipeline network.

Recently, a simultaneous risk and resilience-based approach to health, safety, and environmental (HSE) regulations for offshore operations were discussed by Yang (2019). The study aimed to identify hazards to HSE in offshore operations and to explore the availability of a regulatory framework for such operations. This research was based on legal analysis and a qualitative description of risk and resilience theories in Chinese laws and HSE regulations for offshore operations. This work identified a relationship between resilience and risk approaches to manage offshore operations. The study concluded that inclusion of resilience in the Chinese regulatory framework could help to minimize hazards to health, safety and the environment.

In order to minimize such hazards resulting from a spill scenario, it is essential to develop risk management and contingency plans prior to further exploration activities and any technological developments. This requires a deep understanding of the chemical and mechanical processes involved in oil fate and transport as well as the available techniques for modeling oil spills in the marine environment. It is only by accurately knowing the variation of oil with respect to time and location in the release scenario that the stakeholders may have adequate preparedness for tackling such complex incidents and control the extent of damage to the environment. The next section of this chapter intends to provide the fundamental knowledge required for oil spill modeling.

3. Oil spill models and data requirements

Nelson and Grubesic (2018) explained the three main sources of offshore oil spills include natural seepage, wellhead blowouts and tanker spills. Among these, the consequences of release due to blowouts and tanker spill may be catastrophic. These types of accidents may cause enormous quantities of hydrocarbon to be released into the environment creating outstanding challenges for the stakeholders including asset owners and operators. These difficulties mainly arise from the distance between accident locations and where civil protection teams and cleanup equipment are located and the prevalent seasonal and weather condition effects in offshore locations. Although accidents may result in the release of different hydrocarbons into the ocean, oil spills are the most disastrous and of primary concern for environmental risk management due to oil specific properties and the potential amount of release (NOAA, 2017).

In order to provide the technical requirement for response operations and development of contingency plans, modeling of oil spills is the key aspect of any risk assessment and mitigation strategies. The purpose of oil spill modeling is to predict the amount of distributing oil and its destination. Such models will be able to determine contaminant distribution through the transport of oil at the surface as well as transport due to current shear (Nordam, Nepstad, Litzler, & Röhrs, 2019).

The purpose here is to provide a brief description of the fundamental of oils spills in the marine environment that must be considered in modeling spill accidents and their associated risks surrounding communities that refer both to human settlements and ecological systems. The nuances of the available spill models are not the focus here, rather this brief review will focus on the more commonly utilized concepts of oil spill models (as detailed in the next sections). The outputs of these models are often used for risk and impact analysis and we attempt to detail the ways in which this is achieved. The focus on marine oil spills stems from the growth in the offshore oil and gas activities which account for approximately 30% of total oil production in 2015 as well as recent explorations of deeper, more complex and unexploited offshore locations including the arctic area (Bird et al., 2008; Gautier et al., 2009; Skogdalen & Vinnem, 2012).

3.1 Current approaches to oil spill modeling

Existing oil spill models commonly adopt a Lagrangian approach, where released oil is simulated by numerical particles which are transported due to wind, wave and current loads. The particles are also subject to buoyancy forces causing sinkage to rising movements. Hydrodynamic, wind and wave models are used to provide the necessary information for oil spill modeling. While wind and wave models are usually two dimensional to represent the surface transport processes, hydrodynamic models are largely three dimensional. Random walk methods have been commonly adopted for predicting transport. According to Nordam et al. (2019), random walks are considered to account for turbulent diffusion that compensates for unresolved eddies in the current data resulting in additional mixing processes. Spaulding (2017) has performed an extensive review on current generations including Oil Spill Contingency and Response Model (OSCAR) by Reed et al. (2000), Spill Impact Model Application Package/Oil Modeling Application Package (SIMAP/OILMAP) by National Oceanic and Atmospheric Association (NOAA) (NOAA, 2017) and GNOME/ADIOS

(Zelenke, O'Connor, Barker, Beegle-Krause, & Eclipse, 2012). He (Spaulding, 2017) asserts that while the advection and dispersion (transport) processes of all reviewed models are Lagrangian-based, each model uses individual algorithms to treat the fate processes. Fate processes involve modeling transfer of oil mass between different compartments in the environment (e.g. air, water column, sediment) and the change of the oil characteristics (e.g. density, viscosity and interfacial tensions). These algorithms describing the fate of oil are usually based on fundamental principles tested and validated by laboratory experiments and field observations; therefore often subject to significant limitations due to unavailability of data and spill scenario specifications. Oil spill models enable tracking the location of oil mass in space and time. This is a necessity for an assessment of spill impact in terms of identifying the species that are prone to contamination as well as the duration of exposure. In the next sections, the fundamentals of oil fate and transport models are discussed in more details.

3.2 Oil fate and transport

This section provides an overview of spill models based on fate and trans port processes of oil. While transport is considered both at surface and subsurface, the main fate processes accounted for in oil spill modeling include spreading, evaporation, entrainment, dissolution, emulsification, biodegradation and sediment-oil interaction (see Fig. 1). A commonly used Lagrangian approach to spill modeling enables capturing both

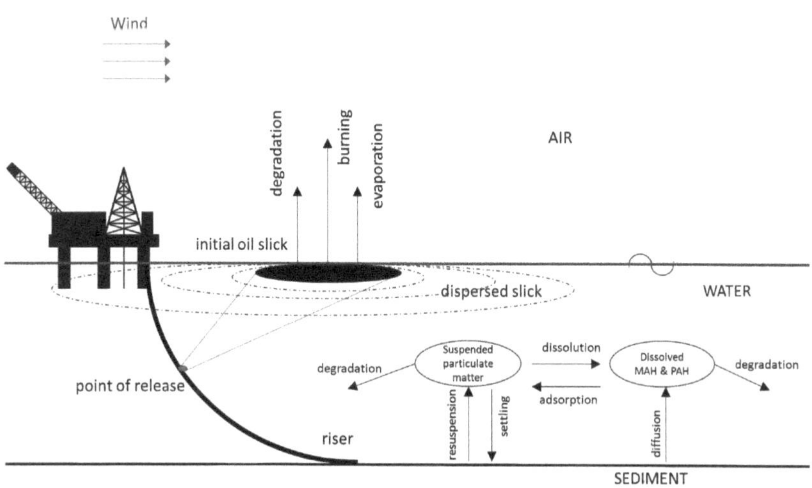

Fig. 1 Overview of oil fate processes in ocean environment.

temporal and spatial behavior of oil. Models must also incorporate oil chemical characteristics so that accurate estimations of mass balance among various compartments in achieved.

3.2.1 Surface and subsurface transport

A Lagrangian approach is usually taken for modeling transport of oil at surface and subsurface. A total current load can be used in the formulation of models as a summation of all environmental loadings from wind, wave, currents and turbulent diffusivity. The information about surface and subsurface currents can be adopted from hydrodynamic models where particular algorithms are developed for obtaining the percentage of oil drift on the surface. Dispersion and diffusion phenomena are modeled with random walk processes where dispersion coefficients are usually derived from hydrodynamic models (Nordam et al., 2019; Spaulding, 2017). A majority of available spill models adopt random walks to simulate oil motion at the surface or subsurface of the ocean. Spaulding (2017) has reviewed these methods that include GNOME model (Zelenke et al., 2012), OSCAR by SINTEF (Reed et al., 2000), OILMAP (Spaulding, Howlett, Anderson, & Jayko, 1992), SIMAP (McCay & Rowe, 2004) and LTRANS (North et al., 2008). Despite their effectiveness in estimating the large-scale behavior of spills, transport algorithms are challenged by the accuracy of predicting fine-scale structures which are an important aspect in the determination of spill distributions and transport. To address this issue, a great deal of effort has been dedicated to the development of techniques to understand the dynamics of highly non-linear and chaotic systems incorporating ocean and wind flows. These methods, mainly known as Lagrangian Coherent Structures (LCS), have been applied to several problems in the field of ocean and coastal engineering (Haller, 2002; Mathur, Haller, Peacock, Ruppert-Felsot, & Swinney, 2007; Peacock & Dabiri, 2010; Tang, Mathur, Haller, Hahn, & Ruggerio, 2010). Important progress for LCS has been the improvements in estimating surface oil transport in the DWH spill (Mezić, Loire, Fonoberov, & Hogan, 2010). Spaulding (2017) asserts that despite this achievement there has been no successful attempt to develop a method that enables the incorporation of LCS into spill models.

Another important task in surface and subsurface modeling of spills is to accurately determine the barriers of the spill. This is performed by Spaulding, Isaji, and Kim (2016), using a two-step methodology; step 1: at a spill initiation point the flow field is seeded with released oil (using outputs of hydrodynamic models) and then the Lagrangian motion of the particles is progressed for a specific period of time (depending on the scenario). The two methods that provide a good level of agreement in obtaining spill barriers have been relative dispersion (RD) and LCS Finite-Time Lyapunov Exponent (FTLE) methods. Step 2: In order to simulate the fine-scale turbulent processes, a dispersive component is added to the process. The RD of Lagrangian trajectories is computed by increasing dispersion coefficients in random walks/flights until the mixing barrier is no longer present. The dispersion coefficient that is associated with mixing barrier disappearance is an estimate of the upper bound. Xue, Zambon, Yao, Liu, and He (2015) have applied this method for developing RD maps of the Northern Gulf of Mexico.

3.2.2 Spreading

Spreading phenomena in oil spills have been usually modeled as changes to the area of a thick slick of oil with respect to time. This empirical method, first proposed by researchers (Mackay, Buist, Mascarenhas, & Petersen, 1980; Mackay, Paterson, & Trudel, 1980; Mackay, Shiu, Hossain, Stiver, & McCurdy, 1982) adopts a spreading coefficient to update the results when validating them against field observations. Concurrent to spreading, Lagrangian elements of spillage can undergo evaporation and entrainment into the water column until oil slick reaches a minimum thickness (i.e. predefined in the model, Spaulding, 2017) or the escaping of volatile elements from the slick stops. Some formulations have been criticized due to their assumption about the circular shape and constant thickness of slicks. This is in contrast with the actual observations which suggest spills tend to become thicker at the downwind and form an elongated shape, mainly due to wind and wave loads (Galt & Overstreet, 2009; Reed et al., 1999). The cause of this behavior has been investigated by Galt and Overstreet (2009) suggesting that the surface oil is often dispersed into the water column due to breaking waves and, once rising back to the surface because of buoyancy, a thicker slick will be shaped at the downwind. This is largely resulting from stronger currents near the free surface that push the rising heavier droplets (larger particles) toward the end tail of the slick. There is a great deal of complexity in modeling spreading of oil hence large efforts made in the research of this field. A more in-depth review of recent developments in oil spill spreading models is provided by Spaulding (2017).

3.2.3 Evaporation

Evaporation of spilled oil has been commonly modeled using pseudo-components representing chemical component groups, where the evaporation flux to the atmosphere is dependent on the properties of the component including molar volume, molecular weight and vapor pressure (Payne, 1984; Payne et al., 1984). Jones (1997) has established a relationship between these properties (molar weight, molecular volume and vapor pressure) to the boiling point of the component of interest. Then in his proposal, only the boiling point and initial volume fraction of components are required to model evaporation. Conventionally, crude oil is represented by six evaporating components that include volatile aliphatic, two semi-volatile aliphatic fractions as well as two petroleum aromatic hydrocarbons (PAH) fractions; however, more developments have been achieved in identifying more chemical component groups so that the oil is represented more accurately (McCay et al., 2015). There have been several attempts in the past to prove that oil evaporation is independent of wind speed, the turbulence of the airflow on the free surface containing the spill, slick thickness or spill area (Fingas, 2011; Fingas & Fieldhouse, 2014). Despite the empirical models presented to support this theory and provision of an evaporation model (for several hundreds of oil types) only as a function of time and temperature, the concept has been found to have few flaws the most important one being its high level of dependency on laboratory observations limiting the application to actual spill scenarios at sea (Spaulding, 2017).

3.2.4 Dissolution

The process in which soluble hydrocarbons enter the water column is known as dissolution. The hydrocarbons might be leaving an oil slick or oil droplets at the subsurface. While most soluble components are volatile, evaporation can decrease their concentration in the water column. Although, evaporation and dissolutions are competing phenomena, in most cases evaporation accounts for the larger proportion of mass loss from an oil slick. This is not the case when oil slick is generated under a surface ice over or when it is caused by subsurface droplets (i.e. having a larger surface area that facilitates dissolution). Hence, dissolution is of great importance when modeling oil spills in the ice-infested ocean environment. Mackay's (1977) model has been repeatedly used for dissolution from a surface oil slick. This model that is based on Raoult's Law considers the slick as a flat surface with the mass flux being dependent on component solubility and temperature.

3.2.5 Entrainment

Over the past few decades, a great deal of research has been dedicated to the development of oil entrainment models. These models have been to some extent more reliant or complementing the empirical methods proposed by researchers (Delvigne, 1993; Delvigne & Hulsen, 1994; Delvigne & Sweeney, 1988). The formulation laboratory and flume tank experiments have been used to estimate the entrainment rate of oil from sea surface as a function of breaking wave energy dissipation rate and fraction of the sea surface covered by oil and breaking waves. An experimental coefficient has also been considered to account for oil type and weathering processes. In their model, oil droplet rise velocity is governed by Stokes law, while oil release depth and particle size distribution are incorporated based on experimental observations. Further work was performed by Reed, Johansen, Leirvik, and Brørs (2009) to investigate the impact of oil viscosity on entrainment improving the formulation by increasing viscosity to non-Newtonian regions and adoption of Reynolds number to account for viscous effects. Upon validation using laboratory measurements, they suggested that lognormal distributions are the best representative of droplet size variability. In addition to the review by Spaulding (2017) on entrainment modeling, Johansen, Reed, and Bodsberg (2015) have investigated the major problems of the models presented by Delvigne (1993), Delvigne and Hulsen (1994), and Delvigne and Sweeney (1988).

3.2.6 Emulsification

A commonly adopted model for estimating the rate of emulsification have been provided by Mackay and Zagorski (1982) which is based on mousse viscosity and emulsification coefficients. Mackay et al. (1982), Mackay, Buist, et al. (1980), and Mackay, Paterson, et al. (1980) present the exponential increase of water content where maximum water fraction is determined by the user or from oil databases. This work was improved by Fingas (2011) and Fingas and Fieldhouse (2014) through the adoption of a stability index (SI) that determines whether oil emulsifies or not. SI that is dependent on density, viscosity, asphaltene and resin contents of oil can also specify the sort of emulsion process that is likely to occur. These types include unstable, entrained water-in-oil state, meso-stable and stable emulsions. Their model enables the resulting water content and water-in-oil viscosity to be estimated. While this scheme has been tested for more than 400 oil types providing accurate results with 90% confidence, its most accurate application is for unstable oils. In comparison with Mackay's model,

Fingas (2011) and Fingas and Fieldhouse (2014) show that his proposed approach outperforms in the prediction of oil properties for most types of oils (Spaulding, 2017).

3.2.7 Biodegradation

Microbial degradation occurs on the dissolved phase insoluble and semi-soluble aromatic compounds of oil (AR1-AR9), while for insoluble aliphatics and PAHs with high molecular weights, this mainly acts on the particulate phases. Li, McCay, and RPS ASA Group (2016) provide a detailed review of various biodegradation mechanisms for both soluble and insoluble compounds. It is known that soluble aliphatic compounds (AL1–AL8) and aromatics with high molecular weights undergo biodegradation process in the dispersed phase (i.e. floating oil droplets in the water column), where the attached microorganisms to the droplet surface cause degradation process. Hence, the estimation of biodegradation rates of these compounds in the entrained and dispersed oil will be an acceptable representative of occurring processes with microbial degradation. Spaulding (2017) suggests that biodegradation rates are usually predicted for droplets smaller than 70 μm which represent well-dispersed oil.

3.2.8 Sediment-oil interaction

It is inevitable for the dissolved aromatic compounds in the water column to experience adsorption to suspended particulates and settle on the seabed. In most spill models, the extent of adsorption is estimated using standard partitioning theory. This theory includes a dimensionless partition coefficient and concentration of suspended particulate matters (SPM) expresses as particulate mass per volume of water. Monte Carlo simulations (MCSs) have been repeatedly utilized to implement collision theory between SPM and oil droplets and evaluate the formation of oil-SPM aggregates (Khelifa, Ajijolaiya, et al., 2005; Khelifa, Hill, & Lee, 2003, 2005; Khelifa, Lee, Hill, & Ajijolaiya, 2004). Advancements have been also achieved with respect to modeling oil aggregation in the presence of dispersants resulting in oil settling rates for various aggregate types (Khelifa, Fingas, & Brown, 2008). This method is interesting to the oil spill modeling community as it has been validated using laboratory experiments, however, the computation effort required for conducting MCSs are not fully known and there is not sufficient information available on how to implement the method on other cases.

3.3 Data requirements for oil spill modeling

As discussed earlier, a great deal of information is required for modeling the processes involved in oil spill incidents. Nelson and Grubesic (2018) have performed a synthesis of the oil spill risk assessment literature particularly from an environmental point of view. In their work, they categorize oil spill models into ex-post and ex-ante approaches depending on whether they rely on real-world observations (ex-post) or utilize simulations to model the spill scenario (ex-ante). This classification is important as it determines the type of data available or required for modeling spills, while there may always be overlaps in data requirements and accessing quality data for spill modeling is a necessity to ensure the subsequent risk estimates are as realistic as possible.

An integral component of oil spill models is the type of hydrocarbon involved in the release scenario and its chemical compositions. Although this information is subject to change depending on the existing oil reserves at the extraction area under study, there are several databases available for extraction of essential information. For instance, NOAA has made an extensive library of various oil types and their properties available online. It should be noted that the type of hydrocarbon is not limited to crude oil and it may be of great interest to model spillage of processed fuels in the ocean environment.

The most critical data that is needed for oil spill modeling is on currents and tides at the oil release location under investigation. Three common sources exist for this type of data are oceanographic current and tide forecast (De Dominicis, Zodiatis, Lardner, & Geoscientific Model Development, 2013), climatological averages of current velocity and direction and real-time current and tide data (the latter often used with synthesized variability) (Beegle-Krause, 2001). Examples of available oceanographic models are the Intra-America Seas Nowcast Forecast System (IASNSF), the American Seas Navy Coastal Ocean Model (AmSEAS NCOM) (Nelson & Grubesic, 2018). Wind data which includes its direction and velocity is also a crucial input into spill models since it directly affects surface/subsurface transports, spreading and evaporation processes (Geng et al., 2016). Usually, oceanographic datasets are utilized to access this information, however, if lacked meteorological models and ocean buoy data can be of great use. In addition to this data, oil spill models need information on the amount and flow rate of release into the environment as well as its location, duration and depth. A number of coefficients, such as diffusivity, are also needed, depending on the process of interest.

3.4 New directions in oil spill modeling

Over the past years, there has been increasing attention to oil extraction from the Arctic ocean mainly due to some parts of this region becoming ice-free as a result of climate change. The extent of Arctic ice cover has been declining, with a 40% reduction observed from 1979 to 2007 and consequently the Northwest Passage which connects Europe and Asia became ice-free for the first time in the summer of 2007 (Giles, Laxon, & Ridout, 2008; Khon, Mokhov, Mojib, Semenov, & Park, 2010). These changes, along with the rising global energy demand, may facilitate exploring the substantial hydrocarbon reserves in the Arctic. While this can create outstanding opportunities to the energy industry and in turn the world economy, it can significantly add to the threat of oil spill during various activities including exploration/extraction processes and transport. The major risk factors such as extremely low temperatures will influence the properties of the structure material as well as the loads caused by the impact of drifting icebergs. In addition to the higher likelihood of accidents, the consequences of an oil spill in ice-infested waters will be exacerbated (DNV, 2014). These consequences may include distortion of the reproduction cycles of species and ecological changes. The criticality mainly lies in the slower decomposition of hydrocarbons in low temperatures making them more available to affect the marine life, greater sensitivity of the Arctic ecosystems with slower reproduction rates and more simple trophic structures and the fact the spills in such remote areas are more likely to remain longer, as the logistics challenges diminish the effectiveness of strategies for containing and cleaning up oil spills (Afenyo, Khan, Veitch, & Yang, 2016; Afenyo, Veitch, & Khan, 2016; Jonsson, Sundt, Aas, & Sanni, 2010; Nevalainen, Helle, & Vanhatalo, 2017).

A comprehensive environmental risk assessment, in particular for the Arctic region, is required to evaluate whether the proposed activities in the region are acceptable with respect to the industry's criteria as well as for developing effective contingency plans (Hasle, Kjellén, & Haugerud, 2009). This requires an analysis of the exposure in which the extent of contamination in the environment will be estimated, the exposed organisms and their possible responses to the stressor is identified (Afenyo, Khan, et al., 2016; Afenyo, Veitch, et al., 2016; Nazir, Khan, Amyotte, & Sadiq, 2008). To determine the level of contamination, the level of contaminant concentration must be quantified. Many researchers have adopted the fugacity concept (mostly level IV fugacity) for partition modeling (Nazir et al., 2008;

Sadiq, 2001; Yang, Khan, Garaniya, & Chai, 2015). A fugacity approach assists in efficient generation of time-series of contaminant concentration in every medium in the environment (i.e. ice, water column, sediment and air column), by simplifying the analysis due to continuity between the interfaces of the phases. Afenyo, Khan, et al. (2016) and Afenyo, Veitch, et al. (2016) developed a dynamic fugacity model to evaluate the exposure of four media to the oil released during Arctic shipping. The proposed model provides a profile for oil concentration in these media; however, it fails to consider the uncertainty of many parameters involved in the model which is essential for accurately estimating the environmental risk associated with the spill accident. Nazir et al. (2008) developed a methodology based on level IV fugacity for estimating the oil concentration in water and sediment medium and utilize Monte Carlo simulations (MCS) to incorporate the uncertainty of multimedia input parameters. Similarly, Afenyo, Khan, Veitch, and Yang (2017) proposed a probabilistic ERA model for Arctic marine oil spills. Their model provides probability distributions of the exposure concentration in different media as well as 95% percentile risk. However, due to the adoption of conventional methods that only propagate the uncertainty in input variables to model outputs, the probabilistic dependency of concentration on those inputs is neglected.

Arzaghi, Abaei, et al. (2018) and Arzaghi, Abbassi, Garaniya, Binns, and Khan (2018) proposed a Bayesian methodology for predicting the stochastic fate and transport of spilled oil in ice-infested regions based on level IV fugacity models. They have adopted a hierarchical Bayesian approach (HBA) to estimate the probability of time to reach a concentration (TRTC) based on the observations made numerical simulations. Their application of Bayesian method Markov Chain Monte Carlo (MCMC) simulations is able to deal with a wide range of information types and provide a useful estimation of model parameters when the data is sparse or the correlation between them is hard to perceive (Siu & Kelly, 1998). With the advent of Bayesian techniques in the field of probabilistic risk assessment (PRA), it is anticipated that these methods will find more applications in oil spill modeling and ERA of ocean oil releases.

4. Potential exposure related human health effects

Offshore operations not only have the risk of degrading the environment but can also be hazardous to human health. Oil spills offshore can be due to ill-maintained, aging, or sabotaged pipelines, or blowouts. Moreover, oil exploration and extraction in deeper water and a harsh environment (frozen and very stormy waters) are increasing the potential risks to the environment and humans (Jernelöv, 2010). The potential effects of an oil spill at an offshore facility on human health include the toxicological properties of spilled oil, which can pose risks to the coastal inhabitants and the individuals involved in cleanup and restoration operations (Laffon, Pásaro, & Valdiglesias, 2016).

A study reviewed the effects of oil exposure on human health (Aguilera, Méndez, Pásaro, & Laffon, 2010). This research work reviewed the results of major oil spills accidents on human health. Accidents analyzed were Erika (which occurred in South Penmarch, France), Exxon Valdez (Alaska, USA), Sea Empress (Milford Haven, UK), MV Braer (Southwest Shetland Islands, UK), Prestige (Galicia, Spain), Nakhodka (Northeast Oki Island, Japan), and Tasman Spirit (Karachi, Pakistan) oil spills. The study evaluated acute effects by considering autonomic (vegetative) nervous system symptoms, psychological effects, mucous and skin irritation. It suggested the biomonitoring of human populations exposed to oil spill accidents and especially for people involved in the oil cleanup activities. The authors argued that biomonitoring the human population would help to evaluate the immediate consequences of oil exposure to human health and to assess medium and long-term effects.

The British Columbia Ministry of Health (BC MOH) performed a human health risk assessment (HHRA), which was focused on assessing the potential impacts of offshore operations on human health in northeastern British Columbia (Krzyzanowski, 2012). The study was conducted in three phases; phase 1 was based on the identification of human health risks of oil and gas development in northeastern British Columbia. Phase 2 was based on performing a human health risk assessment of oil and gas activities, while phase 3 was based on communicating the results. Based on the findings of this study, the following health effects were of concern:

1. Cancers and respiratory diseases
2. Neurological, reproductive and acute effects
3. Cardiovascular effects and autoimmune disease

Based on the type of health outcome in the population exposed to an oil spill, one study identified three groups, namely (1) physical or physiological effects, (2) effects on mental health, (3) immunotoxicity (adverse effect on functioning of immune systems), genotoxic (properties of chemicals that damage the DNA and cause cancer), and endocrine toxicity (chemicals causing loss of body tissue function and development) (Laffon et al., 2016).

The study stressed performing a health risk assessment for prolonged periods to identify any harmful long-term exposure to human health. Researchers also identified that the risk evaluation in the epidemiologic studies should include the consideration of economic and social consequences of an oil spill for local residents, in addition to their exposure to oil chemicals.

Kwok et al. (2017) performed the GuLF STUDY (Gulf Long-term Follow-up Study) to assess human exposure to stressors and toxicants that lead to adverse health effects due to the Deepwater Horizon oil spill which occurred in 2010. The study helped to understand the relationship between exposure to an oil spill and its potential mental and physical effects on human health. The GuLF STUDY interviewed the people involved in oil spill response activities and cleanup activities between 2011 and 2013 and characterized their responses based on individual exposures. This questionnaire-based study revealed that the effect of toxic exposure on human health was dependent upon the job task of an individual, location, and the time when the exposure occurred. The study also revealed that, due to chemicals used in the oil spill response and cleanup (OSRC) operations after an oil spill, the workers were exposed to physical stressors such as high humidity and heat, musculoskeletal strain, psychological effects such as anxiety and depression, and financial burdens such as job loss.

In a more recent research, D'Andrea and Reddy (2018) studied the long-term adverse effects of the Deepwater Horizon Gulf oil spill. They examined the changes in platelets, white blood cells, hematocrit, hemoglobin, blood urea nitrogen, alkaline phosphatase (ALP), creatinine, alanine amino transferase (ALT) levels, aspartate amino transferase (AST) and cardiac as well as pulmonary functions for 88 subjects. The study was conducted for a 7-year follow-up for 44 oil spills cleanup workers who were exposed directly to the oil spill and 44 unexposed people. It was found that there was a significant reduction of platelet counts ($\times 10^3/\mu L$) in the workers exposed to the oil spill as compared to unexposed subjects. Moreover, a significant increase in hematocrit and hemoglobin was observed in exposed subjects than for those unexposed. Similarly, the levels of ALT, AST, and ALP were higher in people exposed to the oil spill than in those unexposed. The authors also reported the development of reactive airway dysfunction syndrome and chronic rhinosinusitis, as new symptoms, 7 years after the oil spill. Abnormalities in cardiac and pulmonary functions were seen in people exposed to the oil spill. The long-term follow-up for this work showed that there was a persistent alternation in hepatic, hematological, cardiac and pulmonary functions of the subjects. Table 1 summarizes various effects of oil spill accidents on human health.

Table 1 Effects of oil spill accidents on human health.

Methodological framework used in the study	Effects on human health	Reference
Prestige accident		
Impacts of oil spill on mental health, physical health, and functional capacity of the exposed population were analyzed. Health-related quality of life (SF-36) and clinical symptoms (SCL-36) were included in the data for more than 1 year after the oil spill accident	Individuals with higher exposure levels or living in close vicinity to oil spill had lower levels of mental health as compared to those with lower exposure levels and living further from the oil spill. The Health of women and fishers suffered more	Sabucedo, Arce, Senra, Seoane, and Vázquez (2010)
Effect of the oil spill was studied on the academic performance and class behavior of children and adolescents who lived on the Galician coast (Spain). 430 participants from three different age groups (5, 10, and 15 years) were interviewed	Oil spill effects were found to be scarce. Adolescents were seen to have more significant effects compared to other groups. After the oil spill accident, there was a decrease in their academic scores, while their scores of the intelligent behavior (verbal intelligence, creativity, and curiosity) and independence from the teacher were higher	Pérez-Pereira, Tinajero, Rodríguez, Peralbo, and Sabucedo (2012)
Hebei spirit accident		
The study evaluated the psychological health of the residents residing in Taean, involved in the cleanup of the Hebei Spirit (HS) oil spill. The method was designed to use questionnaires based on two scales: Psychological Well-being Index (PWI) scale which assessed psychosocial distress and the symptoms of depressive conditions, and suicidal impulses were assessed using CES-D (Center for Epidemiologic Studies-Depression) scale	The study concluded there were significant effects of HS oil spill on the psychological health of Taean residents. A comparison of results with unexposed general population revealed that residents of Taean were more likely (6.5 times) to have higher stress and depression (9.4–9.7 times). The study did not find any significant difference in the suicidal impulse rate. Factors associated with suicidal impulses, high stress, and depression were educational level, change in income, age, number of days involved in the cleanup, effects on daily activity, and hospital visits due to working on oil cleanup	Song et al. (2009)

Acute health effects of HS oil spill were investigated using a cross-sectional study. The study chose 10 seashore villages. Interviews were conducted for 10 males and females from each village. Questionnaires were based on cleanup activities, anxiety (STAI-X-1) and depression (CES-D), characteristics of the residents, their perception of hazardous oil, and physical symptoms. The logistic regression analysis was used for gender, smoking, age, education, perception of hazardous oil and anxiety	Residents were more likely to be engaged in cleanup if area was more contaminated and when it resulted in higher chances of exposure to oil. In the areas of heavy and moderately oil-soaked areas, there was an increased risk of nausea, fatigue, headache, dizziness, sore throat, runny nose, itchy skins, fatigue, hot flushing, limb tingling, cough, shortness of breath, rashes, and sore eyes. The study proposed monitoring the health of the residents by the long-term investigation	Lee et al. (2010)
Children's mental health in HS oil area was examined by conducting a cross-sectional questionnaire survey. The survey was conducted using Korean versions of the State Anxiety Inventory for children and children's Depression Inventory on 1362 children who were attending elementary schools in the vicinity of HS oil spill. Multiple logistic regression analysis was used to estimate symptom risks of state anxiety and depression for the age and gender of the children, including other covariates	Children living closer to the contaminated coastline showed significantly higher risks of depression as compared to those who were living at the farthest distance. Study did not find any association between distance from contaminated area to child's residence and anxiety symptoms	Ha et al. (2013)
The study analyzed the burden of disease (BOD) of the local residents living in the contaminated area, including their psychological and physical diseases due to HS oil spill. Disability adjusted life year (DALY) method was used	Years lived with Disability (YLD) attributable to spill oil were higher (7425 Disability Adjusted Life Year (DALYs)) in males than females (7299 DALYS) for the year 2008. The YLD of psychological diseases (post-traumatic stress disorder) and depression was found to be higher in males than in females. The YLD for asthma and allergies was higher in females than that for males. People in their 40s had greater effects of allergies and asthma while the burden of psychiatric illness was greater for people in their 20s. The study identified substantial adverse health effects in people living near the coastline where HS oil spill occurred	Kim et al. (2013)

Continued

Table 1 Effects of oil spill accidents on human health.—cont'd

Methodological framework used in the study	Effects on human health	Reference
The research analyzed the psychological health of 993 residents of Taean in South Korea after HS oil spill and determined vulnerability in the psychological symptoms of the residents. Questionnaires were developed to assess symptoms of depression, anxiety, post-traumatic stress (PTS), and suicidal ideation	Symptoms of depression, PTS, anxiety, and suicidal ideation were 22.0%, 19.5%, 4.2%, and 2.3%, respectively. Increased symptoms were observed in females who were older in age, had low family income, and were less educated. Risk of PTS was more in people living in the vicinity of the oil spill location and people working in fishery industries	Choi et al. (2016)
Deepwater Horizon accident		
Work assessed effects of Deepwater Horizon (DH) oil spill on the mental health of residents of southeastern Louisiana who were affected by the oil spill. Face-to-face and telephone interviews were conducted with 452 residents	Higher symptoms of depression, anxiety, and post-traumatic stress were observed in residents. There was a greater effect on the mental health of the residents due to disruption in their work, lives, social engagement, and family	Osofsky, Osofsky, and Hansel (2011)
The study assessed acute level of psychological distress (mood, anxiety and depression), perceived risk and mechanisms of adjustment (neurocognition, resilience and coping with the situation) in fishing community of Baldwin County, Alabama (Number of participants (N) = 23) directly affected and in Franklin County, Florida (N = 71) indirectly affected by DH oil spill. Participants with income stability (N = 47) were compared with those with spill-related income loss (N = 47)	No differences were observed in terms of adjustment, psychological distress, neurocognition, or environmental worry in both communities. Clinically significant depression and anxiety were observed in both groups. Participants who suffered loss due to oil spill had a worse score on anxiety or tension, fatigue, depression, total mood disturbance scales, and confusion, were less resilient, had higher depression rates as compared to those who had a stable income	Grattan et al. (2011)
The research assessed the mental health needs of total of 469 residents in the US states of Mississippi, after 5.5 months, and Alabama, 4 months after the DH oil spill accident. A cluster sampling methodology was used	15.4%–24.5% reported depressive symptoms. 21.4%–31.5% reported anxiety disorder 16.3%–22.8% reported ≥ 14 mentally unhealthy days in the last 30 days. Negative social engagements and negative quality of life were observed compared to those in the state's Behavioral Risk Factor Surveillance System (BRFSS) survey. 35.5%–38.2% reported exposure to spilled oil, while 32.1%–35.7% of all residents reported a decrease in their income after an oil spill	Buttke et al. (2012)

Mental health impacts of DH disaster were examined in the context of the influence of community attachment and sociological literature on the negative effects of fear and anxiety due to technological disaster. Survey data in three coastal Louisiana counties, which were proximate to the DH oil spill, were collected and analyzed	There were higher adverse effects on community attachment. These results were valid for people working in the seafood and fishing industry, people involved with the oil industry, and people having no immediate connection to either industry	Lee and Blanchard (2012)
Work assessed the role of the stigma of help-seeking behavior of the communities in rural area. The study was conducted in semi-structured focus groups 1 year after the DH oil spill, involving 21 mental health professionals. Interviews regarding the disaster were based on people in need, challenges, and recovery efforts	A number of stigma-related barriers were investigated by the clinician delivering services, including public stigma, self-stigma, and cultural implications of receiving aid. It was found that children were the main significant source to recognize and provide services to families in need. This nullified the previous stigma of help-seeking behavior	Locke and Werner (2013)
Questionnaires were given to the participants ($N=93$) in two coastal regions (Franklin County, FL, and Baldwin County, AL) 1 year after an oil spill. Profiles of Mood States (depression and anxiety) were performed. In addition, the impact of Event Scale (PTS symptoms such as hyperarousal, avoidance, and intrusion), and the Connor-Davidson Resilience Scale were also examined	Findings showed that there were mental health problems even 1 year after the disaster, and this was dominant in people who suffered income loss due to the accident. 89.66% of the people had significant anxiety scores, while 83.7% had substantial scores for clinical depression. Findings were not associated with education, age, gender, or psychiatric history	Morris Jr, Grattan, Mayer, and Blackburn (2013)
Predictors of psychological outcomes in residents ($N=219$) of South Louisiana were examined. Indicators such as religiosity, storm exposure, and stressors, mental health and perceived social support were measured and compared for the natural disasters (2005 Hurricanes Katrina and Rita) and man-made disaster (2010 British Petroleum DH oil spill)	A significant predictor of post-traumatic stress disorder (PTSD) was non-organizational religiosity. For depression symptoms, people with low income and those associated with coastal fisheries were significant predictors. Perceived social support was found to have a protective effect on the outcomes of mental health	Cherry et al. (2015)
National Institute of Environmental Health Sciences (NIEHS) DH Research Consortia program developed a program to address the impact of oil spill exposure on human health. Effects of stressors on a vulnerable population, mental health impacts, and exposure of spill oil contaminants were studied	The study provided an overview of the NIEHS project and outlined various articles and interviews on the impact of oil spills on human health. It reported the population's vulnerabilities to spilled oil, their mental health issues, and exposure of humans to the crude oil contaminants offshore	Croisant and Sullivan (2018)

5. Techniques for environmental risk assessment of oil release

Over the past few decades, the concerns around oil spill accidents in the ocean environment, which now includes the Arctic region (Giles et al., 2008), has prompted the stakeholders including the public, asset owners, governments and the International Maritime Organization (IMO), to review and amend existing regulations with respect to marine pollution (Chang, Stone, Demes, & Piscitelli, 2014; Orszulik, 2008). There can be substantial environmental damage caused by the release of oil in the sea, where the toxicity of chemicals may affect both humans and marine organisms with possible irreversible long-term consequences (Pula, Khan, Veitch, & Amyotte, 2006). Concerns of such have prompted a pressing need for methods and techniques that are able to estimate environmental risk. To perform a meaningful risk assessment, both accident likelihood and its consequences must be evaluated. The results of analyses can then be utilized for the amendment of in place policies as well as the development of risk management and contingency plans (i.e. prior and during operations).

The Petroleum Safety Authority (PSA) in Norway has provided a set of requirements for ERA of marine oil spills in its management regulations and associated guidelines in addition to those stated by the Norwegian Offshore Standardization Organization (NORSOK) (NORSOK Standard Norway, 1998). It should be noted that Norway is the only country that has had ERA requirements for more than a decade, hence it has been used by Vinnem (2014) in summarizing the main requirements that include:

- The assessment must include all scenarios identified by the HAZID that will contribute to the system risk level.
- The assessment must incorporate distributions for discharge rates and durations that are identified to vary across release scenarios as well as the concentration of all contaminants (as a function of the processes explained in previous sections) in various media (e.g. air, water column and sediment).
- The assessment must include the influence of action barriers such as detection likelihood and drain system into the scenario development.
- The assessment must perform exposure modeling for the sensitive entities in the surrounding environment during the critical period (i.e. based on release duration and amount).

- The assessment must include estimation of environmental consequences as a function of contaminant amounts and sensitivity of exposed entities (or more realistically for representatives of populations and habitats).
- The assessment must be performed in a consistent way for comparison with other scenarios at the same or other facilities.

It is only by following such detailed requirements that risk management plans and contingency plans can be developed with useful application to the offshore oil and gas industry. A range of methods have been adopted to conduct ERA some of which has seen major enhancements in terms of accuracy or scenario inclusion. In the next section a brief summary of these methods and several applications are reviewed.

5.1 Event tree method

Event tree (ET) represents visual models for a series of events (i.e. consequences) that may develop from a failure scenario. Top events in these structures are those which initiate the hazardous scenario and their occurrence likelihood is very important for modeling risk. ERA conducted by ETs is often a simple model of the effects that oil release may have to the environment and it is directly related to the duration/amount of release flow. Vinnem (2014) suggests that the factors that determine the duration are the presence of:

- The immediate killing of well before it develops into a full blowout.
- Capping; mechanical isolation of the flow.
- Bridging; self-stopping of flow in the reservoir.
- Relief well(s) drilling.

It is also very important to consider blowout ignition as it will result in the placement of movement restrictions on the personnel of installations making mechanical isolation of the blowout challenging to achieve. On the other hand, a lower amount of oil ends up into the environment due to a considerable proportion of it burning upon release. These possibilities of events must be considered while establishing the likelihood values and quantification of consequences. Various approaches to ET modeling have been adopted by previous researchers to conduct assessments of impacts from offshore oil and gas accidents (Caia et al., 2018; Lewis & Smith, 2010).

5.2 Statistical simulations

Statistical approaches such as Monte Carlo simulation (MCS) has been previously utilized probabilistic modeling and risk assessment of marine operations. Nazir et al. (2008) adopted this technique to develop a methodology for ERA of the oil spill from a riser. The proposed model is based on the US EPA framework and takes a fugacity-based approach to estimate the exposure to contaminants in the marine environment. MCSs were carried out to incorporate the uncertainty of multimedia input parameters, and the effect of stressor on organisms was obtained by using toxicity data found in the literature. Their method characterizes the ecological risk by transforming risk quotient (RQ) into probability distributions. French-McCay (2011) presents a biological effects model coupled with an oil fate model for supplying the required spatial and temporal estimation of oil concentrations. In their method, the long-term environmental effects are quantified using food web modeling and the risk of a spill is evaluated by MCSs.

In the context of Arctic oil spills where environmental damages may be substantially aggravated, there have been several attempts toward ERA based on both qualitative and quantitative approaches. Afenyo et al. (2017) proposed a probabilistic ERA model by implementing a fugacity-based fate and transport models and MCSs to develop probability distributions for the exposure concentration and to predict the 95th percentile risk. The presented probabilistic approach propagates the uncertainty of input variables through the model considering the presence of ice media and the complexity it creates in oil slick transport and dispersion. However, the probabilistic dependency of risk on the input parameters is neglected. Their proposed method does not provide a platform for ERA with respect to other important influencing parameters such as variance seasonal conditions.

5.3 Bayesian methods

As discussed earlier, Bayesian methods have been known to provide a promising approach to probabilistic risk assessment (PRA) of offshore structures and operations (Abaei et al., 2018; Arzaghi et al., 2017; Bhandari, Abbassi, Garaniya, & Khan, 2015; Khakzad, Khan, & Amyotte, 2013; Pui, Bhandari, Abbassi, & Garaniya, 2017). Nevalainen et al. (2017) used BN for analyzing the impacts of oil spills on the Arctic environment and provided a holistic view of such incident consequences. The authors assert that a food web approach must be adopted for improving the results of ERA. In their model, the influence of input parameters such as oil spill size and season on the acute and long-term ecological impacts are incorporated into the BN. However, they suggest that the model must be enhanced with the quantification of more problem variables. This is essential from a quantitative risk assessment viewpoint and to benefit from maximum capacity within BN. Arzaghi, Abaei, et al. (2018) and Arzaghi, Abbassi, et al. (2018) developed a framework based on Bayesian statistics for ERA of accidental release of oil in ice-infested areas. They adopted a fugacity approach to model the fate and transport of oil and predicted the 95th percentile Predicted Exposure Concentration ($PEC_{95\%}$) of pollutants in different media, accounting for the uncertainty of input variables. The 5th percentile Predicted No Effect Concentration ($PNEC_{5\%}$) is obtained from toxicity data for 19 species in the arctic region. Their model is extended to a Dynamic Bayesian Network (DBN) to evaluate the time-variable risk profile caused by seasonal changes.

6. Conclusions

- Fuzzy-based models help to assess environmental risks due to drilling wastes in offshore oil and gas operations. Such models also help to develop a hierarchical structure for environmental risk assessment.
- An inclusion of hazard identification (HAZID) into environmental hazard identification (ENVID) helps to understand various causes of environmental hazards and the potential barriers needed to mitigate environmental risks in offshore oil and gas operations.
- Significant advances are achieved in modeling oil entrainment and droplet size distributions. These models have reduced uncertainty in related coefficients based on the physical and chemical processes involved as well as laboratory experiments and historical spill events. The available models are now emerging for both surface and subsurface releases.
- Fugacity models have been finding more applications in fate and transport modeling of oil spills due to their capacity in providing a temporal variation of chemicals in the environment.
- Extreme environmental conditions indicate a significant increase not only in oil spill likelihood but also to the impacts on the environment, which highlights the need for more attention into planning appropriate risk mitigation strategies for operation in such regions.
- The risk of offshore oil and gas operations on human health should incorporate the social and economic consequences of an oil spill, in addition to the adverse toxicological impact of spilled oil on inhabitants, oil cleanup and habitat restoration crew.
- Application of BN to ERA of oil spills results in improving risk estimates due to incorporation of multiple influencing parameters and their probabilistic dependencies, in addition to increasing analysis efficiency.

References

Abaei, M. M., Arzaghi, E., Abbassi, R., Garaniya, V., Chai, S., & Khan, F. (2018). A robust risk assessment methodology for safety analysis of marine structures under storm conditions. *Journal of Ocean Engineering*, *156*, 167–178.

Abbassi, R., Dadashzadeh, M., Khan, F., & Hawboldt, K. (2012). Risk-based prioritisation of indoor air pollution monitoring using computational fluid dynamics. *Indoor and Built Environment*, *21*(5), 663–673.

Abbassi, R., Khan, F., & Hawboldt, K. (2010). Ecological risk-based performance evaluation of a waste stabilization pond. *Environmental Engineering and Management Journal*, *9*(6), 757–764.

Afenyo, M., Khan, F., Veitch, B., & Yang, M. (2016). Dynamic fugacity model for accidental oil release during Arctic shipping. *Marine Pollution Bulletin*, *111*(1), 347–353.

Afenyo, M., Khan, F., Veitch, B., & Yang, M. (2017). A probabilistic ecological risk model for Arctic marine oil spills. *Journal of Environmental Chemical Engineering*, *5*(2), 1494–1503.

Afenyo, M., Veitch, B., & Khan, F. (2016). A state-of-the-art review of fate and transport of oil spills in open and ice-covered water. *Ocean Engineering*, *119*, 233–248.

Aguilera, F., Méndez, J., Pásaro, E., & Laffon, B. (2010). Review on the effects of exposure to spilled oils on human health. *Journal of Applied Toxicology: An International Journal*, *30*, 291–301.

Arzaghi, E., Abaei, M. M., Abbassi, R., Garaniya, V., Binns, J., Chin, C., et al. (2018). A hierarchical Bayesian approach to modelling fate and transport of oil released from subsea pipelines. *Journal of Process Safety and Environmental Protection*, *118*, 307–315.

Arzaghi, E., Abaei, M. M., Abbassi, R., Garaniya, V., Chin, C., & Khan, F. (2017). Risk-based maintenance planning of subsea pipelines through fatigue crack growth monitoring. *Engineering Failure Analysis*, *79*, 928–939.

Arzaghi, E., Abbassi, R., Garaniya, V., Binns, J., & Khan, F. (2018). An ecological risk assessment model for Arctic oil spills from a subsea pipeline. *Marine Pollution Bulletin*, *135*, 1117–1127.

Beegle-Krause, J. (2001). General NOAA oil modeling environment (GNOME): A new spill trajectory model. In *International oil spill conference, edited* (pp. 865–871). American Petroleum Institute.

Bhandari, J., Abbassi, R., Garaniya, V., & Khan, F. (2015). Risk analysis of deepwater drilling operations using Bayesian network. *Journal of Loss Prevention in the Process Industries*, *38*, 11–23.

Bird, K. J., Charpentier, R. R., Gautier, D. L., Houseknecht, D. W., Klett, T. R., Pitman, J. K., et al. (2008). *Circum-Arctic resource appraisal: Estimates of undiscovered oil and gas north of the Arctic Circle*. US Geological Survey.

Brigham, L. W. (2010). *The fast-changing maritime Arctic*. Proceedings of the US Naval Institute, (pp. 54–59).

Buttke, D., Vagi, S., Bayleyegn, T., Sircar, K., Strine, T., Morrison, M., et al. (2012). Mental health needs assessment after the Gulf coast oil spill—Alabama and Mississippi, 2010. *Prehospital and Disaster Medicine*, *27*, 401–408.

Caia, A., Di Lullo, A. G., De Ghetto, G., & Guadagnini, A. (2018). Probabilistic analysis of risk and mitigation of deepwater well blowouts and oil spills. *Stochastic Environmental Research and Risk Assessment*, *32*, 2647–2666.

Chang, S. E., Stone, J., Demes, K., & Piscitelli, M. (2014). Consequences of oil spills: A review and framework for informing planning. *Ecology and Society*, *19*(2), 26.

Cherry, K. E., Sampson, L., Nezat, P. F., Cacamo, A., Marks, L. D., & Galea, S. (2015). Long-term psychological outcomes in older adults after disaster: Relationships to religiosity and social support. *Aging & Mental Health*, *19*, 430–443.

Choi, K.-H., Lim, M.-H., Ha, M., Sohn, J. N., Kang, J.-W., Choi, Y.-H., et al. (2016). Psychological vulnerability of residents of communities affected by the Hebei Spirit oil spill. *Disaster Medicine and Public Health Preparedness, 10*, 51–58.

Croisant, S., & Sullivan, J. (2018). Studying the human health and ecological impacts of the deep water horizon oil spill disaster: Introduction to this special issue of new Solutions. *New Solutions: A Journal of Environmental and Occupational Health Policy, 28*, 410–415.

Dadashzadeh, M., Abbassi, R., Khan, F., & Hawboldt, K. (2013). Explosion modeling and analysis of BP Deepwater Horizon accident. *Safety Science, 57*, 150–160.

D'Andrea, M. A., & Reddy, G. K. (2018). The development of long-term adverse health effects in oil spill cleanup workers of the Deepwater Horizon offshore drilling rig disaster. *Frontiers in Public Health, 6*, 117.

De Groot, R., Brander, L., Van Der Ploeg, S., Costanza, R., Bernard, F., Braat, L., et al. (2012). Global estimates of the value of ecosystems and their services in monetary units. *Ecosystem Services, 1*, 50–61.

Delvigne, G. A. L. (1993). Natural dispersion of oil by different sources of turbulence. In *International oil spill conference, edited* (pp. 415–419). American Petroleum Institute.

Delvigne, G. A. L., & Hulsen, L. J. M. (1994). Simplified laboratory measurement of oil dispersion coefficient—Application in computations of natural oil dispersion. In 1. *Proceedings of the 17th arctic and marine oil spill program, Environment Canada.* (pp. 173–187).

Delvigne, G. A. L., & Sweeney, C. E. (1988). Natural dispersion of oil. *Oil and Chemical Pollution, 4*(4), 281–310.

DNV, G. L. (2014). *The Arctic—The next risk frontier.* Høvik, Norway: DNV GL.

De Dominicis, M., Pinardi, N., Zodiatis, G., Lardner, R., & Geoscientific Model Development (2013). MEDSLIK-II, a Lagrangian marine surface oil spill model for short-term forecasting–Part 1: Theory. *Geoscientific Model Development, 6*(6), 1851–1869.

Eicken, H., Jones, J., Meyer, F., Mahoney, A., Druckenmiller, M. L., Rohith, M., et al. (2011). Environmental security in Arctic ice-covered seas: From strategy to tactics of hazard identification and emergency response. *Marine Technology Society Journal, 45*, 37–48.

Enache, I., & Zăgan, S. (2008). Risk assessment of oil marine pollution. In *Exposure risk assessment of chemical pollution—Contemporary methodology* . The Netherlands: Springer.

Fahd, F., Khan, F., Hawboldt, K., & Abbassi, R. (2014). Developing a novel methodology for ecological risk assessment of thiosalts. *Stochastic Environmental Research and Risk Assessment, 28*, 383–391.

Ferdous, R., Khan, F., Sadiq, R., Amyotte, P., & Veitch, B. (2013). Analyzing system safety and risks under uncertainty using a bow-tie diagram: An innovative approach. *Process Safety and Environmental Protection, 91*, 1–18.

Ferguson, A., Solo-Gabriele, M., & Mena, K. (2020). Assessment for oil spill chemicals: Current knowledge, data gaps, and uncertainties addressing human physical health risk. *Marine Pollution Bulletin, 150*, 110746.

Fingas, M. (2011). Models for water-in-oil emulsion formation. In *Oil spill science and technology* (pp. 243–273). Elsevier.

Fingas, M., & Fieldhouse, B. (2014). Water-in-oil emulsions: Formation and prediction 225. In *Handbook of Oil Spill Science and Technology.* Wiley.

French-McCay, D. (2011). Oil spill modeling for ecological risk and natural resource damage assessment. In *International oil spill conference proceedings (IOSC), edited*: American Petroleum Institute, abs415.

Galt, J. A., & Overstreet, R. (2009). *Development of spreading algorithms for the ROC, response options calculator.* Genwest Systems.

Gautier, D. L., Bird, K. J., Charpentier, R. R., Grantz, A., Houseknecht, D. W., Klett, T. R., et al. (2009). Assessment of undiscovered oil and gas in the Arctic. *Science, 324*(5931), 1175–1179.

Geng, X., Boufadel, M. C., Ozgokmen, T., King, T., Lee, K., Lu, Y., et al. (2016). Oil droplets transport due to irregular waves: Development of large-scale spreading coefficients. *Marine pollution Bulletin, 104*(1–2), 279–289.

Giles, K. A., Laxon, S. W., & Ridout, A. L. (2008). Circumpolar thinning of Arctic sea ice following the 2007 record ice extent minimum. *Geophysical Research Letters, 35*(22).

Grattan, L. M., Roberts, S., Mahan, W. T., Jr., McLaughlin, P. K., Otwell, W. S., & Morris, J. G., Jr. (2011). The early psychological impacts of the Deepwater Horizon oil spill on Florida and Alabama communities. *Environmental Health Perspectives, 119*, 838–843.

Ha, M., Jeong, W.-C., Lim, M., Kwon, H., Choi, Y., Yoo, S.-J., et al. (2013). Children's mental health in the area affected by the Hebei spirit oil spill accident. *Environmental Health and Toxicology, 28*, e2013010.

Haller, G. (2002). Lagrangian coherent structures from approximate velocity data. *Physics of Fluids, 14*(6), 1851–1861.

Hasle, J. R., Kjellén, U., & Haugerud, O. (2009). Decision on oil and gas exploration in an Arctic area: Case study from the Norwegian Barents Sea. *Safety Science, 47*(6), 832–842.

Heidari, P. A., & Raie, M. (2018). Probabilistic assessment of oil spill from offshore oil wells in Persian Gulf. *Marine Pollution Bulletin, 136*, 291–299.

Islam, M. S., & Tanaka, M. (2004). Impacts of pollution on coastal and marine ecosystems including coastal and marine fisheries and approach for management: A review and synthesis. *Marine Pollution Bulletin, 48*, 624–649.

Jernelöv, A. (2010). The threats from oil spills: Now, then, and in the future. *Ambio, 39*, 353–366.

Johansen, Ø., Reed, M., & Bodsberg, N. R. (2015). Natural dispersion revisited. *Marine Pollution Bulletin, 93*(1–2), 20–26.

Johnston, P. F. (2012). Arctic energy resources: Security and environmental implications. *Journal of Strategic Security, 5*, 13–32.

Jones, R. K. (1997). A simplified pseudo-component oil evaporation model. In *Arctic and marine oil spill program technical seminar, edited* (pp. 43–62). Canada: Ministry of Supply and Services.

Jonsson, H., Sundt, R. C., Aas, E., & Sanni, S. (2010). The Arctic is no longer put on ice: Evaluation of Polar cod (*Boreogadus saida*) as a monitoring species of oil pollution in cold waters. *Marine Pollution Bulletin, 60*(3), 390–395.

Khakzad, N., Khan, F., & Amyotte, P. (2013). Quantitative risk analysis of offshore drilling operations: A Bayesian approach. *Safety Science, 57*, 108–117. https://doi.org/10.1016/j.ssci.2013.01.022.

Khelifa, A., Ajijolaiya, L. O., MacPherson, P., Lee, K., Hill, P. S., Gharbi, S., et al. (2005). Validation of OMA formation in cold brackish and sea waters. In *Proceedings of the 28th Arctic and Marine Oilspill Program (AMOP), technical seminar on environmental contamination and response, at June 7–9, Calgary, Alberta.*

Khelifa, A., Fingas, M., & Brown, C. (2008). *Effects of dispersants on oil-SPM aggregation and fate in US coastal waters.* Final Report Grant Number: NA04NOS4190063.

Khelifa, A., Hill, P. S., & Lee, K. (2003). A stochastic model to predict the formation of oil-mineral aggregates. In *Arctic and marine oilspill program technical seminar, edited, 893–910: Environment Canada; 1999.*

Khelifa, A., Hill, P. S., & Lee, K. (2005). A comprehensive numerical approach to predict oil-mineral aggregate (OMA) formation following oil spills in aquatic environments. In *International oil spill conference, edited* (pp. 873–877). American Petroleum Institute.

Khelifa, A., Lee, K., Hill, P. S., & Ajijolaiya, L. O. (2004). Modelling the effect of sediment size on OMA formation. In *Arctic and marine oilspill program technical seminar, edited, 383–396: Environment Canada; 1999.*

Khon, V. C., Mokhov, I. I., Mojib, L., Semenov, V. A., & Park, W. (2010). Perspectives of northern sea route and northwest passage in the twenty-first century. *Climatic Change, 100*(3–4), 757–768.

Kim, Y.-M., Park, J.-H., Choi, K., Noh, S. R., Choi, Y.-H., & Cheong, H.-K. (2013). Burden of disease attributable to the Hebei Spirit oil spill in Taean, Korea. *BMJ Open, 3*, e003334.

Krzyzanowski, J. (2012). Environmental pathways of potential impacts to human health from oil and gas development in Northeast British Columbia, Canada. *Environmental Reviews, 20*, 122–134.

Kwok, R. K., Engel, L. S., Miller, A. K., Blair, A., Curry, M. D., Jackson, W. B., et al. (2017). The GuLF STUDY: A prospective study of persons involved in the Deepwater Horizon oil spill response and clean-up. *Environmental Health Perspectives, 125*, 570–578.

Laffon, B., Pásaro, E., & Valdiglesias, V. (2016). Effects of exposure to oil spills on human health: Updated review. *Journal of Toxicology and Environmental Health, Part B, 19*, 105–128.

Lee, M. R., & Blanchard, T. C. (2012). Community attachment and negative affective states in the context of the BP Deepwater Horizon disaster. *American Behavioral Scientist, 56*, 24–47.

Lee, C. H., Kang, Y. A., Chang, K. J., Kim, C. H., Hur, J. I., Kim, J. Y., et al. (2010). Acute health effects of the Hebei oil spill on the residents of Taean, Korea. *Journal of Preventive Medicine and Public Health, 43*, 166–173.

Lewis, S., & Smith, K. (2010). Lessons learned from real world application of the bow-tie method. In *6th global congress on process safety* (pp. 22–24). TX: American Institute of Chemical Engineers San Antonio.

Li, Z., & McCay, D. F. (2016). *Review of hydrocarbon biodegradation rates for use in modeling oil fate in seawater.* Wakefield, RI: RPS ASA Group.

Li, Y., Yu, H., Wang, Z. Y., Li, Y., Pan, Q. Q., Meng, S. J., et al. (2019). The forecasting and analysis of oil spill drift trajectory during the Sanchi collision accident, each ChinaSea. *Ocean Engineering, 187*, 106231.

Locke, C. R., & Werner, D. (2013). Stigma of help-seeking behavior following the Deepwater Horizon oil spill. *Contemporary Rural Social Work Journal, 5*, 3.

Mackay, D. (1977). *Mathematical model of the behavior of oil spills on water with natural and chemical dispersion.* EPS-3-EC-77-19, Environment Canada.

Mackay, D., Buist, I., Mascarenhas, R., & Petersen, S. (1980). *Oil spill processes and models.* Report EE-8 Canada: Environmental Protection Service.

Mackay, D., Paterson, S., & Trudel, K. (1980). *A mathematical model of oil spill behaviour: Environment Canada.* Environmental Protection Service, Environmental Impact.

Mackay, D., Shiu, W. Y., Hossain, K., Stiver, W., & McCurdy, D. (1982). *Development and calibration of an oil spill behavior model.* Toronto Univ (ONTARIO) Dept of Chemical Engineering and Applied Chemistry.

Mackay, D., & Zagorski, W. (1982). *Water-in-oil emulsions.* Environment Canada manuscript report EE-34, Ottawa, Ontario Canada.

Mathur, M., Haller, G., Peacock, T., Ruppert-Felsot, J. E., & Swinney, H. L. (2007). Uncovering the Lagrangian skeleton of turbulence. *Physical Review Letters, 98*(14),144502.

French McCay, Jayko D.P., Li K., Z., Horn., Kim M., Isaji Y., Crowley T., Spaulding D., et al. 2015. Technical reports for Deepwater Horizon water column injury assessment—WC_TR14: Modeling oil fate and exposure concentrations in the deepwater plume and cone of rising oil resulting from the Deepwater Horizon oil spill. DWH NRDA WaterColumn Technical Working Group Report. https://www.doi.gov/deepwater horizon/adminrecord.

McCay, D. F., & Rowe, J. J. (2004). Evaluation of bird impacts in historical oil spill cases using the SIMAP oil spill model. In *Arctic and marine oilspill program technical seminar, edited* (pp. 421–452). Citeseer.

Meng, Q. (2016). The spatiotemporal characteristics of environmental hazards caused by offshore oil and gas operations in the Gulf of Mexico. *Science of the Total Environment, 565*, 663–671.

Mezić, I., Loire, S., Fonoberov, V. A., & Hogan, P. (2010). A new mixing diagnostic and Gulf oil spill movement. *Science, 330*(6003), 486–489.

Moroni, D., Pieri, G., & Tampucci, M. (2019). Environmental decision support systems for monitoring small scale oil spills: Existing solutions, best practices and current challenges. *Journal of Marine Science and Engineering*, 7, 19.

Morris, J. G., Jr., Grattan, L. M., Mayer, B. M., & Blackburn, J. K. (2013). Psychological responses and resilience of people and communities impacted by the Deepwater Horizon oil spill. *Transactions of the American Clinical and Climatological Association, 124*, 191.

National Oceanic and Atmospheric Administration's Office of Response and Restoration (NOAA). (2017). *Largest oil spills affecting US waters since 1969.* National Oceanic and Atmospheric Association.

Nazir, M., Khan, F., Amyotte, P., & Sadiq, R. (2008). Subsea release of oil from a riser: An ecological risk assessment. *Risk Analysis, 28*(5), 1173–1196.

Nelson, J. R., & Grubesic, T. H. (2018). Oil spill modeling: risk, spatial vulnerability, and impact assessment. *Progress in Physical Geography: Earth and Environment, 42*(1), 112–127.

Neumann, J. (2010). *The Gulf of Mexico is home to 27,000 abandoned oil and gas wells.* https://gawker.com/5581170/the-gulf-of-mexico-is-home-to-27000-abandoned-oil-and-gas-wells.

Nevalainen, M., Helle, I., & Vanhatalo, J. (2017). Preparing for the unprecedented—Towards quantitative oil risk assessment in the Arctic marine areas. *Marine Pollution Bulletin, 114*(1), 90–101.

Nieukirk, S. L., Mellinger, D. K., Moore, S. E., Klinck, K., Dziak, R. P., & Goslin, J. (2012). Sounds from airguns and fin whales recorded in the mid-Atlantic Ocean, 1999–2009. *The Journal of the Acoustical Society of America, 131*, 1102–1112.

Nordam, T., Nepstad, R., Litzler, E., & Röhrs, J. (2019). On the use of random walk schemes in oil spill modelling. *Marine Pollution Bulletin, 146*, 631–638.

NORSOK Standard Norway. (1998). *Risk and emergency preparedness analysis, Z-013.* Oslo: Standard Norway.

North, E. W., Schlag, Z., Hood, R. R., Li, M., Zhong, L., Gross, T., et al. (2008). Vertical swimming behavior influences the dispersal of simulated oyster larvae in a coupled particle-tracking and hydrodynamic model of Chesapeake Bay. *Marine Ecology Progress Series, 359*, 99–115.

Norwegian Ministry of the Environment. (2011). *First update of the integrated management plan for the marine environment of the Barents Sea–Lofoten area.* Oslo: Norwegian Ministry of the Environment.

Orszulik, S. T. (2008). *Environmental technology in the oil industry.* Springer.

Osofsky, H. J., Osofsky, J. D., & Hansel, T. C. (2011). Deepwater Horizon oil spill: Mental health effects on residents in heavily affected areas. *Disaster Medicine and Public Health Preparedness, 5*, 280–286.

Payne, J. R. (1984). Weathering of petroleum in the marine environment. *Marine Technology Society Journal, 18*, 24–42.

Payne, J. R., Kirstein, B. E., McNabb, G. D., Jr., Lambach, J. L., Redding, R., Jordan, R. E., et al. (1984). Multivariate analysis of petroleum weathering in the marine environment–sub Arctic 21. In *Environmental Assessment of the Alaskan Continental Shelf, Final Report of Principal Investigators.*

Peacock, T., & Dabiri, J. (2010). *Introduction to focus issue: Lagrangian coherent structures.* American Institute of Physics.

Pérez-Pereira, M., Tinajero, C., Rodríguez, M. S., Peralbo, M., & Sabucedo, J. M. (2012). Academic effects of the prestige oil spill disaster. *The Spanish Journal of Psychology, 15*, 1055–1068.

Pui, G., Bhandari, J., Abbassi, R., & Garaniya, V. (2017). Risk-based maintenance of offshore managed pressure drilling (MPD) operation. *Journal of Petroleum Science and Engineering, 159*, 513–521.

Pula, R., Khan, F. I., Veitch, B., & Amyotte, P. R. (2006). A grid based approach for fire and explosion consequence analysis. *Process Safety and Environmental Protection, 84*(2), 79–91. https://doi.org/10.1205/psep.05063.

Reed, M., Daling, P. S., Brakstad, O. G., Singsaas, I., Faksness, L.-G., Hetland, B., et al. (2000). OSCAR2000: a multi-component 3-dimensional oil spill contingency and response model. In *Arctic and marine oil spill program technical seminar, edited, 663–680: Environment Canada; 1999.*

Reed, M., Johansen, Ø., Brandvik, P. J., Daling, P., Lewis, A., Fiocco, R., et al. (1999). Oil spill modeling towards the close of the 20th century: overview of the state of the art. *Spill Science and Technology Bulletin, 5*(1), 3–16.

Reed, M., Johansen, O., Leirvik, F., & Brørs, B. F. (2009). Numerical algorithm to compute the effects of breakingwaves on surface oil spilled at sea. Final report submitted to the Coastal Response Research CenterIn *Vol. 10968*, (p. 131).

Sabucedo, J. M., Arce, C., Senra, C., Seoane, G., & Vázquez, I. (2010). Symptomatic profile and health-related quality of life of persons affected by the prestige catastrophe. *Disasters, 34*, 809–820.

Sadiq, R. (2001). *Drilling waste discharges in the marine environment: A risk based decision methodology.* Memorial University of Newfoundland.

Sadiq, R., & Husain, T. (2005). A fuzzy-based methodology for an aggregative environmental risk assessment: A case study of drilling waste. *Environmental Modelling & Software, 20*, 33–46.

Sajid, Z., Khan, F., & Zhang, Y. (2017). Integration of interpretive structural modelling with Bayesian network for biodiesel performance analysis. *Renewable Energy, 107*, 194–203.

Sajid, Z., Khan, F., & Zhang, Y. (2018). A novel process economics risk model applied to biodiesel production system. *Renewable Energy, 118*, 615–626.

Saud, Y. E., Israni, K. C., & Goddard, J. (2014). Bow-tie diagrams in downstream hazard identification and risk assessment. *Process Safety Progress, 33*, 26–35.

Siu, N. O., & Kelly, D. L. (1998). Bayesian parameter estimation in probabilistic risk assessment. *Reliability Engineering & System Safety, 62*(1), 89–116.

Skogdalen, J. E., & Vinnem, J. E. (2012). Quantitative risk analysis of oil and gas drilling, using Deepwater Horizon as case study. *Reliability Engineering and System Safety, 100*, 58–66.

Song, M., Hong, Y. C., Cheong, H. K., Ha, M., Kwon, H., Ha, E. H., et al. (2009). Psychological health in residents participating in clean-up works of Hebei spirit oil spill. *Journal of Preventive Medicine and Public Health, 42*, 82–88.

Spaulding, M. L. (2017). State of the art review and future directions in oil spill modeling. *Marine pollution Bulletin, 115*(1–2), 7–19.

Spaulding, M. L., Howlett, E., Anderson, E., & Jayko, K. (1992). *OILMAP: A global approach to spill modeling.* (No. EC/EPS-93-01710).

Spaulding, M. L., Isaji, T., & Kim, Y. H. (2016). Selection of dispersion coefficients for use in Lagrangian spill transport models to preserve underlying flow dynamics and transport barriers. *39th AMOP technical seminar on environmental contamination and response, edited, 22–34.*

Tang, W., Mathur, M., Haller, G., Hahn, D. C., & Ruggerio, F. H. (2010). Lagrangian coherent structures near a subtropical jet stream. *Journal of the Atmospheric Sciences, 67*(7), 2307–2319.

Vinnem, J.-E. (2014). Offshore risk assessment. In *Vol. 2. Springer Series in Reliability Engineering*.

Xue, Z., Zambon, J., Yao, Z., Liu, Y., & He, R. (2015). An integrated ocean circulation, wave, atmosphere, and marine ecosystem prediction system for the South Atlantic Bight and Gulf of Mexico. *Journal of Operational Oceanography*, *8*(1), 80–91.

Yang, Y. (2019). Reforming health, safety, and environmental regulation for offshore operations in China: Risk and resilience approaches? *Sustainability*, *11*, 2608.

Yang, M., Khan, F., Garaniya, V., & Chai, S. (2015). Multimedia fate modeling of oil spills in ice-infested waters: An exploration of the feasibility of fugacity-based approach. *Process Safety and Environmental Protection*, *93*, 206–217.

Zelenke, B., O'Connor, C., Barker, C. H., Beegle-Krause, C. J., & Eclipse, L. (2012). *General NOAA operational modeling environment (GNOME) technical documentation.*

Zhang, C., Qin, T. X., Jiang, B., & Huang, C. (2018). A comprehensive probabilistic analysis model of oil pipelines network based on Bayesian network. *IOP Conference Series: Earth and Environmental Science*, *113*,012083.

CHAPTER ELEVEN

Economic approaches for making prevention and safety investment decisions in the process industry

Chao Chen[a], Genserik Reniers[a,b,c,*]

[a]Safety and Security Science Group, Faculty of Technology, Policy and Management, TU Delft, Delft, The Netherlands
[b]Faculty of Applied Economics, Antwerp Research Group on Safety and Security (ARGoSS), University Antwerp, Antwerp, Belgium
[c]CEDON, KULeuven, Campus Brussels, Brussels, Belgium
*Corresponding author: e-mail address: g.l.l.m.e.reniers@tudelft.nl

Contents

Methods in Chemical Process Safety, Volume 4
ISSN 2468-6514
https://doi.org/10.1016/bs.mcps.2020.03.003

1. Introduction

The chemical and process industries are central to the global economy and play a prominent role in creating and maintaining modern-day life. There is a long tradition of forming clusters in the chemical industry due to various reasons, such as benefits of scale, exchange of material streams, and optimization of energy streams (Reniers, Sörensen, Khan, & Amyotte, 2014; Swuste, van Nunen, Reniers, & Khakzad, 2019). A chemical industrial cluster is a geographically limited concentration of chemicals-using companies and service providers operating in the chemical industrial sector (Reniers et al., 2014). Chemical industrial clusters can be found around the world, consisting of tens of different chemical plants and chemical (and other) logistic service providers situated in each other's vicinity. Hundreds and even thousands of installations are located in this conglomerate of chemical plants to transfer, store, and process large amounts of hazardous (e.g., flammable, explosive and toxic) substances (Chen, Reniers, & Khakzad, 2020b; Chen, Reniers, & Zhang, 2018). As a result, even a minor disruption (accidental events, Natech events or intentional events) can trigger a chain of major accidents (fires, explosions or toxic releases) and cause performance degradation of chemical plants, resulting in substantial consequences.

As mentioned in the abstract, in light of risk in chemical industrial areas, safety management is usually conducted by selecting among a vast array of different safety measures to reduce the frequency and mitigate the consequences of undesired events. Besides, different safety management principles may be used in different stages of the entire life of a chemical plant, such as inherent safety design, passive barriers, active barriers and procedural actions for emergency response. However, it is impossible to take all possible safety measures. Moreover, a severe accident may induce catastrophic impacts on a process plant and lead to huge financial consequences. In brief, the economic issues related to risk should not be ignored in the decision-making process of prevention measures in the process industry.

Economics reminds us that protection resources are limited and the resources allocated to one item are not available for others (Poole, 2008). This is called "opportunity costs." Economic analysis can facilitate the decision-making on safety investments in prevention and may thereby avoid losses, increasing profitability and be part of a business strategy leading to long-term profitability and to sustainable and intrinsically healthy organizations (Reniers & Van Erp, 2016). Economic models, therefore, are usually used to optimize the allocation of budgets so as to maximize the protection effectiveness, the layout optimization of process plants (Castell, Lakshmanan, Skilling, & Banares-Alcantara, 1998), design optimization of process facilities, safety barrier optimization (Eslami Baladeh, Cheraghi, & Khakzad, 2019), etc. Cost-benefit analysis and cost-effectiveness analysis are the most used economic tools in prevention investment decision-making and safety investments in the process industry (Chen, Reniers, & Khakzad, 2020a; Janssens, Talarico, Reniers, & Sörensen, 2015; Khakza & Reniers, 2019; Reniers & Audenaert, 2009; Reniers & Sorensen, 2013; Villa, Reniers, Paltrinieri, & Cozzani, 2017; Wang, Li, Qin, & Zhang, 2018).

This chapter aims to present economic models that may be used in decision-making on prevention and safety investments. First, we introduce the traditional decision-making process on safety investment and possible prevention principles and options in Section 2. Section 3 expounds on cost analysis including the costs of prevention strategies and the cost of accidents. A cost-benefit analysis on the basis of cost analysis and the principle of assumed benefit is elaborated in Section 4. Section 5 presents a cost-effectiveness analysis of prevention strategies. Based on these economic models, economic optimization is illustrated in Section 6 to find out the best way to prevent accidents. Finally, conclusions drawn from this work are presented in Section 7.

2. Decision-making on process safety

Making a decision in relation to process safety investment is often complex and difficult due to the uncertainties involved and the risks these uncertainties entail (Reniers & Van Erp, 2016). It depends upon not only technical aspects but also political, psychological and social concerns. During the complex and difficult decision-making process, clearly identifying the risks and the performance of risk reduction measures is vital. The uncertainties also affect the consequences of different decisions thus the performance of alternatives needs to be estimated and examined for decision-making. Moreover, decision-making on safety also depends on the risk attitude of decision-makers. A risk aversion factor (a) can be used to represent the risk attitude of decision-makers, as follows:

$$R_i = f_i \times C_i^a \tag{1}$$

where R is the risk of an undesired scenario i, f is the frequency (probability) of this scenario and C are the consequences of the scenario. If a decision-maker is a risk-averse, $a > 1$ while $a < 1$ if he/she is risk-seeking. $a = 1$ represents a risk-neutral decision-maker. In terms of an organization, the large amount of risks related to different scenarios can be summed to obtain the accumulated risk. Following risk analysis, safety managers can make decisions based on their emotions and intuition, safety legislation, as well as the available safety resources and budget. There are two main options: risk acceptance (consisting of risk retention and risk transfer) and risk reduction (consisting of risk avoidance and risk control) (Fuller & Vassie, 2004).

2.1 Risk criteria

The consequences (*C*) of an accident can be categorized into casualties, economic loss, and environmental pollution, etc. (Chen et al., 2020a; Chen, Reniers, & Khakzad, 2019). Decision-makers pay most attention to fatalities thus the risk of fatalities usually is used to determine risk levels for decision-making on prevention policy. For this purpose, individual risk and societal risk are calculated.

2.1.1 Individual risk

In the Netherlands, individual risk is defined as the probability that an average unprotected person is killed when he/she is permanently present at a certain location, due to an accident related to a hazardous activity (Bottelberghs, 2000). The UK's health and safety executive (HSE) provides a different definition of "individual risk": the risk that a typical user of development is exposed to a dangerous dose or worse of toxic substance, heat or blast overpressure (HSE, 1984). The two definitions describe the individual risk as a location-based risk which is not dependent on whether people or residences are present. But injury induced by a dangerous dose is considered in the latter definition besides death. In this chapter, we adopt the former definition since legislators often use levels of individual risk with fatality as a regulatory approach to setting risk criteria allowing an activity or not. Locations around a hazard with equal individual risk levels are shown as risk contours (iso-risk curves) for the application of land-use planning. Fig. 1 shows two examples of individual risk represented by risk contours for a hazardous installation (Fig. 1). An individual risk (fatality risk) of 10^{-6} from industrial activities is considered to be a very low level and thus used as a criterion in many countries (Jonkman, van Gelder, & Vrijling, 2003; Reniers & Van Erp, 2016). From the perspective of legislators and safety supervisors, an individual risk lower than the legislative criterion is considered to be acceptable but "ALARP" (As Low As Reasonably practicable) indicating that additional safety measures should be taken when practicable.

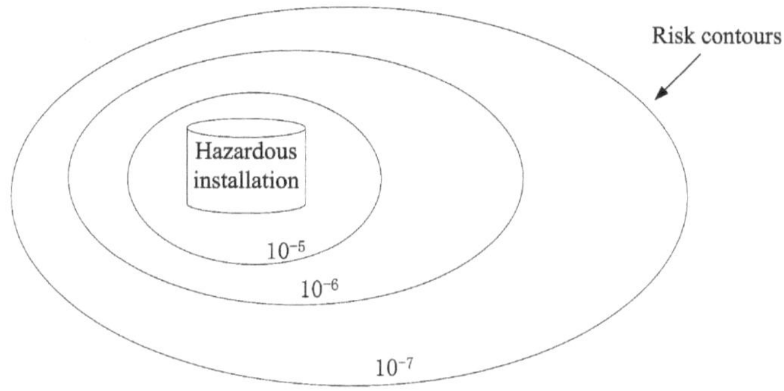

Fig. 1 Examples of iso-risk contours showing the distribution of individual risk based on a point source of a hazardous installation.

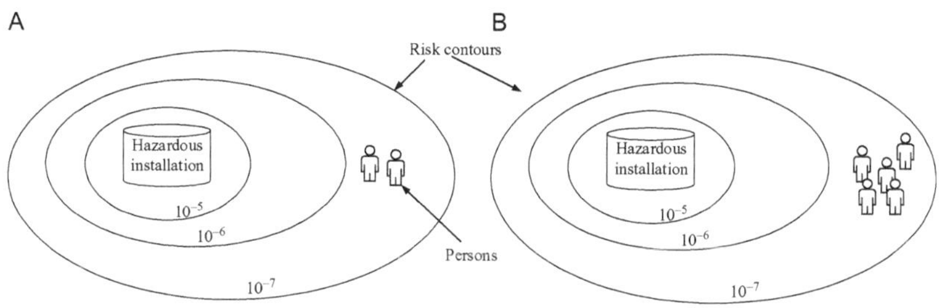

Fig. 2 The effects of the number of persons exposed to a hazardous installation on fatality risk.

2.1.2 Societal risk

The individual risk depends on the geographic position and is characterized by the so-called iso-risk contours on a geographical map of the establishment, regardless of the risk difference induced by the number of people present in the area. However, it is evident that the level of risk will not be identical if the numbers of persons are different in the same establishment. As shown in Fig. 2, the individual risk in Fig. 2A is equal to that in Fig. 2B but the expected number of fatalities in Fig. 2B is much greater than that in Fig. 2A if an accident occurs at the hazardous installation.

To address the exposed population, the so-called societal risk is defined as the probability that a group of N persons or more is simultaneously killed due to an accident at the hazardous activity (Bottelberghs, 2000). It is a location-independent risk characterization method that depends on both hazardous activity and the surrounding population. Societal risk is represented by the form of an "F-N curve" in which the exceeding probability is a function of the number of people killed due to a major accident at the establishment, usually plotted on double logarithmic scales. Thus the societal risk will be zero if no people are present around the hazardous activity while the individual risk may be rather high. A "boundary line" representing a risk level following legislation (societal risk criterion) is often used to be compared with the F—N curve to decide to allow an activity or not. The points of on the F—N curve below the boundary line are tolerable whereas it is unacceptable if the points are above the boundary line.

Besides the criteria based on fatality risk, uncommon-used criteria such as the risk criteria based on economic loss or environmental damage can also be found (Barlettani, Fruttuoso, & Lombardi, 1997; Diamantidis, 2017; Jonkman et al., 2003). However, there seems to be a lack of integrated risk criteria addressing individual risk, societal risk, economic loss, and environmental damage.

2.2 Risk acceptance

2.2.1 Risk retention

An organization may retain the responsibility for a specified risk intentionally or unintentionally. Intentional retention with risk information and knowledge can be regarded as a manner of self-insurance or auto-financing and can be seen as financial planning of potential losses by an organization's own resources. On the other hand, unintentional retention without enough risk information and knowledge is caused by insufficient hazard identification and inadequate risk assessment and management, which should be avoided. The advantage of self-insurance is saving safety investment costs due to no intermediaries expecting payments for costs, profits, or commissions. However, intentional retention will only be an option for large organizations able to set up an administrate and insurance fund and it may not suitable for dealing with the risk of events for which only extremely few historical data are available (Reniers & Van Erp, 2016).

2.2.2 Risk transfer

Risk transfer concerns shifting the responsibility for a specified risk to a third party. Insurance is an important option for an organization to transfer risks. If an accident occurs that is insured, the insurance company will provide an indemnity for the organization. As a result, the risk can be totally or partly transferred to an insurance company via insurance. In that case, an organization's perceived risks can be significantly reduced by providing a premium to an insurance company. However, not all risks are insurable, and the insured risks may have a large part of uninsured consequences that might not be possible to be compensated. Covered Risks by insurance typically include employer's liability, public liability, product liability, fire, and transport. Uninsured risks mainly consist of equipment repairs, employee sick pay, loss of production, staff replacement, damage to corporate image, and accident investigations (Reniers & Van Erp, 2016). Besides, insurance transfers the risks only by financial compensation but the life loss in an accident cannot be regained. Therefore, prevention measures are needed for chemical companies with a specific risk higher than the risk criteria.

2.3 Risk reduction

Risk reduction in the process industries aims to reduce the hazard of a process, the probability or consequences of an accident scenario, or both, using a wide variety of strategies, techniques, procedures, policies, and systems (Bollinger & Crowl, 1997). Risk reduction consists of risk avoidance and risk control. Bollinger and Crowl (1997) divided risk reduction strategies into four categories: inherent risk avoidance on the one hand, and passive, active and procedural (risk control) on the other hand. For more information, the reader can be referred to Bollinger and Crowl (1997) or to Reniers and Van Erp (2016). In light of these risk reduction strategies, selecting a combination of prevention and protection measures under the constraint of safety budget to maximize the reduced risk or find out the most profitable way for accident prevention is the main mission of the field of economics related to safety in the process industry. The residual contents of this chapter introducing some widely used microeconomic tools that may be implemented by risk management, including cost analysis, cost-benefit analysis, cost-effectiveness analysis, economic optimization and the law of diminishing returns.

3. Cost analysis

Cost analysis is a basic tool in economics to support decision-making. Not only safety investments represent a cost but also accidents bear a cost. The total annual cost of safety in the process industry thus consists of the annual cost of safety investment and the costs of expected monetary loss from accidents.

3.1 The cost of prevention strategies

To implement a prevention strategy (a combination of safety measures) or to update existing protection systems, an economic analysis is recommendable in industrial practice since companies are always confronted with budget limitations (Reniers & Sörensen, 2013). In this section, the various costs related to a prevention strategy that a company may decide to implement, are illustrated. The prevention and protection costs consist of investments that occur at presents such as initial costs, installation costs, and the costs that occur throughout the whole remaining lifetime of the facility (Reniers & Brijs, 2014). A cost analysis should include direct and indirect economic costs of applying safety measures. Eight categories of costs related to safety barriers (Reniers & Van Erp, 2016) are listed in Table 1. The initial costs and installation costs don't occur yearly and thus don't need to be discounted to present values while the operation-, maintenance-, inspection-, logistics-, as well as transport- and contractor costs which occur throughout the life cycle of the facility, should be discounted to present values.

The present value of costs caused by the implementation of one safety measure (C_m) is the sum of the initiation costs, installation costs, and the present value of six other cost types, as follows:

$$C_m = C_{m,\text{ini}} + C_{m,\text{ins}} + \frac{(1+r)^y - 1}{r(1+r)^y} \left(C_{m,\text{ope}} + C_{m,\text{mai}} + C_{m,\text{ins}} + C_{m,\log} + C_{m,\text{con}} + C_{m,\text{oth}}\right) \tag{2}$$

Table 1 Categories of prevention and protection costs.

Cost category	Subcategories
Initiation	Investigation, selection and design material, training, changing guidelines and informing
Installation	Production loss, start-up, equipment, installation team
Operation	Utilities consumption and labor Utilities
Maintenance	Material, maintenance team, production loss, start-up
Inspection	Inspection team
Logistics and transport	Transport and loading/unloading of hazardous materials, storage of hazardous materials, drafting control lists, relative documents
Contractor	Contractor selection, training
Other	Office furniture, insurance, and stationery items

Adopted from Reniers, G. L., & Van Erp, H. N. (2016). *Operational safety economics: A practical approach focused on the chemical and process industries.* John Wiley & Sons.

where $C_{m,\ ini}$ is the initiation cost of measure m, $C_{m,\ ins}$ is the installation cost of measure m, $C_{m,\ ope}$ is the annual operation cost of measure m, $C_{m,\ mai}$ is the annual maintenance cost of measure m, $C_{m,\ ins}$ is the annual inspection costs of measure m, $C_{m,\ log}$ is the annual logistics and transport cost of measure m, $C_{m,\ con}$ is the annual contractor cost of measure m, $C_{m,\ oth}$ are other annual costs of measure m, r is the discount rate, y is the number of years that the protection measure can operate. In case of more information for the cost calculation of subcategories listed in Table 1, readers are referred to Reniers and Van Erp (2016).

In terms of a prevention strategy n with M safety measures, the total annual present value of costs due to the use of the prevention strategy is estimated as:

$$C_I{}^n = \sum_{m=1}^{M} C_m / y \tag{3}$$

where $C_I{}^n$ is the annual present cost of the investment in prevention strategy n, M is the total number of safety measures taken in the strategy to reduce the risk of accidents.

3.2 The costs of accidents

Possible accident scenarios in a chemical industrial area should be considered in accident cost analysis. In terms of multiple hazardous installations in an industrial area, domino effects should also be considered to avoid underestimating the risk and thus the costs of accidents. Considering S accident scenarios in a chemical industrial area being identified, the costs related to these accidents under a prevention strategy can be simplified as the sum of the product of the frequency and the losses caused by accident scenarios, as follows:

$$C_A{}^n = \frac{(1+r)^y - 1}{ry(1+r)^y} \sum_{s=1}^{S} \left(f_s^n \cdot L_s^n \right) \tag{4}$$

where $C_A{}^n$is the annual cost of accident losses under the prevention strategy n. $f_s{}^n$ is the frequency of accident scenario s under the prevention strategy n. L_s^n is the loss caused by the accident scenario s under the prevention strategy n.

The loss assessment of major accidents in the process industry should take into account casualties, economic losses, environmental loss as well as any other possible damage and detriment. Both the direct losses that are immediately visible and tangible and the indirect losses that are intangible and invisible are of significance for cost analysis of major accidents (Jallon, Imbeau, & de Marcellis-Warin, 2011; Reniers & Van Erp, 2016). The direct losses consist of the losses caused by damage to installations, products and equipment, medical expenses, paying fines and an insurance premium's rise while the indirect losses include capacity losses, production scheme problem losses, recruitment and wage costs among others (Gavious, Mizrahi, Shani, & Minchuk, 2009). The quantification of indirect losses is more difficult since due to them being invisible, usually resulting in underestimation (Jallon et al., 2011). One simple method to estimate the indirect losses is using a ratio of indirect to direct losses based on the assessment results of direct losses. The ratio varies in the literature, inducing difficulties for users to choose a suitable value. For example, a widely used ratio of 4 is proposed according to an analysis of 7500 accidents while that of a range of 20 to 1 depending on different industrial *sectors* and methods used, was also found (Dorman, 2000).

In this chapter, we adopt the loss assessment method proposed by Reniers and Brijs (2014) to account for the losses of major accidents in chemical industrial areas (Reniers & Van Erp, 2016). Therefore, the total losses caused by the accident scenario *s* can be estimated as a sum of 11 contributions, as follows:

$$L_s = L_{s,sup} + L_{s,dam} + L_{s,leg} + L_{s,ins} + L_{s,hum} + L_{s,env} + L_{s,per} + L_{s,med} + L_{s,int} + L_{s,rep} + L_{s,inv} \tag{5}$$

where $L_{s,\,\mathrm{sup}}$ is the supply chain loss, $L_{s,\,dam}$ is the damage loss, $L_{s,\,leg}$ is the legal loss, $L_{s,\,ins}$ is the insurance loss, $L_{s,\,hum,}$ is the human loss, $L_{i,\,env}$ is the environmental loss, $L_{s,\,per}$ is the personnel loss, $L_{s,\,med}$ is the medical loss, $L_{s,\,int}$ is the intervention loss, $L_{s,\,rep}$ is the reputation loss, $L_{s,\,inv}$ is the accident investigation and clean up loss, $L_{s,\,sec}$ is the security-related loss which is different from accidental losses, such as the losses caused by psychological and political effects. The loss of each category can be calculated by adding up the subcategories presented in Table 2.

3.3 Economic value of human life

A specific issue in accident cost analysis is the monetary value of human life. Some people regard it as unethical to even assign a price to human life. However, not addressing the economic value of human life

Table 2 Categories of accident losses.

Loss category	**Subcategories of costs**
Supply chain	Production, start-up, schedule
Damage	Damage to own material/property, other companies' material/property, surrounding living areas, public material/property
Legal	Fines, interim lawyers, specialized lawyers, internal research team, experts at hearings, legislation, permit, and license
Insurance	Insurance premium
Human	Compensation victims, injured employees, recruitment
Environmental	Environmental damage
Personnel	Productivity of personnel, training of new or temporary employees, wages
Medical	Medical treatment at location, medical treatment in hospitals and revalidation, using medical equipment and devices, medical transport
Intervention	Intervention
Reputation	Share price
Investigation and clean up	Accident investigation, clean up

Adapted from Reniers, G. L., & Van Erp, H. N. (2016). *Operational safety economics: A practical approach focused on the chemical and process industries*. John Wiley & Sons.

may lead to lower economic costs and thus results in underestimating the risk of accidents (Dorman, 2009; Jonkman et al., 2003). Although the economic value of human life cannot be directly monetized, it may be assumed that their valuation into money can be derived from the values of those matters that can be monetized directly (Reniers & Van Erp, 2016). Four main methods of valuation of human life based on this principle are available in literature, as follows:

3.3.1 Human capital approach

The human capital approach is to estimate the present value of the income stream foregone due to premature death. According to this approach, the value of human life increases with his/her earning capability and expected longevity, whereas it decreases with his/her age and the interest rate selected for calculating the present value. The Expected future incomes are widely used to compensate wrongful deaths, but they are generally far lower than the values obtained from those approaches that assign utility values to a risk reduction. The method based on future earnings is objectionable because it only considers human's capacity to produce, neglecting the value that individuals place on their own lives (Dorman, 2009).

3.3.2 Quality-adjusted life years

The economic value of human life can be referred to as the costs of saving an extra life. The costs of saving an extra life can be calculated by dividing the investment with the reduction of the expected number of fatalities. Besides this purely quantitative approach of life-saving, also the quality related to life saved can be considered. To do this, the quality-adjusted life years (QALY) method can be considered. It is based on the number of years of life that would be increased by an intervention. The QALY is an indicator widely used in health care, related to the burden of disease, considering both the quality and the quantity of lives saved (Jonkman et al., 2003; Reniers & Van Erp, 2016).

3.3.3 Willingness to pay and willingness to accept

The willingness to pay (WTP) refers to an individual's willingness to pay (WTP) for a reduction of the risk of a loss. This approach is based on the trade-off people make between risk and money. It depends on how much people are willing to pay to reduce the risk of any loss, for instance premature death. In theory, the WTP should be identical to the willingness to accept (WTA) which refers to how much an individual is prepared to accept in compensation for the loss of a benefit, but WTP values are always lower than WTA values. Nonetheless, this approach usually obtains higher values than the HCA method (Dorman, 2009; Reniers & Van Erp, 2016).

3.3.4 Value of statistical life

The WTP for one individual can be transferred into the value of a statistical Life (VoSL), which expresses the statistic of the WTPs over a large group of people. The VoSL method is also a sensitive and debated subject since it raises emotional feelings among different people. In recent decades, a large number of studies have been carried out worldwide to determine the VoSL. Different countries propose different values of a statistical life, as illustrated in Table 3 for some countries.

Table 3 A list of VoSLs used in different countries (Reniers & Van Erp, 2016).

Country	VoSL (million euro)	Department
Australia	3.0	Department of Prime Minister and Cabinet
The Netherlands	5.8	Ministerie van Infrastructuur en Waterstaat
United Kingdom	2.4	Health and Safety Executive
USA	7.0	U.S. department of Transportation Analysis

3.4 The total annual cost of safety

The total annual cost of safety in the process industry in terms of prevention strategy n (C^n) can be represented as the sum of the annual cost of safety investment ($C_I{}^n$) and the costs of expected monetary loss from accidents ($C_A{}^n$), as follows:

$$C^n = C_I{}^n + C_A{}^n \tag{6}$$

Based on a cost analysis, safety decision-makers can find out the prevention strategy n with the minimal total annual cost C^n. However, profits of prevention strategies are not considered in such approach. The next section discusses cost-benefit analysis, taking total costs and benefits into consideration in the economic exercise.

4. Cost-benefit analysis

A cost-benefit analysis (CBA) is a systematic and analytical evaluation process of comparing benefits and costs in the same units, usually money. Although the CBA cannot demonstrate whether one safety or security investment is intrinsically better than another, a CBA allows decision-makers to improve their decisions by adding appropriate information on costs and benefits to certain prevention or mitigation investment decisions. An investment is recommended when the total net present value (NPV) of all cash flows is positive, and an investment is usually rejected when the NPV is negative. Therefore the NPV can be an indicator for assessing investment strategies (Quah & Haldane, 2007; Reniers & Van Erp, 2016).

4.1 Net benefit analysis

The benefit of a protection strategy is estimated as the difference in the costs of expected monetary loss from accidents without and with the implementation of safety measures. In order to calculate the benefits of a protection strategy n, a baseline ($n=0$) should be defined. The baseline can be the strategy without any safety measure, or the initial strategy before a protection upgrade. In that case, the benefits of a protection strategy n can be defined, as follows:

$$B^n = C_A{}^0 - C_A{}^n \tag{7}$$

where B^n is the hypothetical benefit of prevention strategy n. C^0 is the cost of expected monetary loss of baseline prevention strategy 0. In that case, the net present value given a protection strategy n (NPV^n) can be expressed as the difference of the total protection benefit and the total protection cost of strategy n in y years, as follows:

$$NPV^n = y(B^n - C_I{}^n) \tag{8}$$

A prevention strategy n is usually recommended if the net present value (NPV) equals or exceeds zero ($NPV \geq 0$), otherwise, it is considered to be not cost-effective or inefficient (Reniers & Van Erp, 2016; Stewart & Mueller, 2013). The NPV is able to help decision-making as regards prevention strategies, addressing the uncertainty of different accident scenarios.

4.2 Disproportion factor analysis

In terms of low-frequency and high-consequence events such as domino effects with catastrophic damage, the cost-benefit analysis based on net present value might not be satisfactory for decision-making on prevention strategies in which the hypothetical benefits of protection strategies might be less than the costs for prevention. To take this into account, a disproportion factor (DF) or so-called gross disproportion factor that can be used to deal with high impact low probability accidents is presented in this section.

The ratio of the costs to the benefits is defined as the proportion factor (PF). The PF of a protection strategy n can be represented as:

$$PF^n = \frac{C_I{}^n}{B^n} \tag{9}$$

A disproportion factor (DF) is a threshold to determine whether the protection measure is grossly disproportionate or not. A prevention investment is "reasonably practicable" when its costs are proportionate to the benefits ($PF < DF$). In the case of $PF \geq DF$, a further risk reduction would be too costly compared with the extra benefit gained from the protection measure. The DF reflects a bias in favor of safety over costs if it is higher than 1. Goose (2006) stated that DF is usually less than 10 and should never be greater than 30.

Theoretically, the *DF* increases with an increased risk level and it should be infinite when the risk level reaches the unacceptable region, meaning that the risk must be reduced no matter the costs (Talarico & Reniers, 2016).

A calculation method involving three factors to estimate the value of *DF* was proposed by Goose (2006). The factors are referred to "how bad," "how risky" and "how variable," and they are probably dependent on each other. Data or information needed for determining the three factors can be derived from an *FN* curve. For information on how to construct an *FN* curve, the reader is referred to Reniers and Van Erp (2016).

The "how bad" factor depends upon the average number of fatalities per event, and can be calculated, as follows:

$$How\ bad = \log_{10}(N_{av}) \tag{10}$$

$$N_{av} = \frac{EV}{\sum FR} \tag{11}$$

where EV is the average number of deaths expected per year, ΣFR is the sum of failure rates of all events per year, N_{av} is the average number of fatalities per event, represented by the ratio of N_{av} to ΣFR.

The "how risky" factor represents the effects of EV on the DF:

$$How\ risky = \log_{10}\left(10^5 \times EV\right) \tag{12}$$

The "how variable" factor depends on the ratio of the maximum potential fatalities for a single event (N_{max}) to the average number of fatalities per event (N_{av}):

$$How\ variable = \log_{10}\left(\frac{N_{\max}}{N_{av}}\right) \tag{13}$$

The *DF* can then be calculated by adding 3 (dimensionless) to the product of the three "how" factors:

$$DF = \log_{10}(N_{av}) \times \log_{10}\left(10^5 \times EV\right) \times \log_{10}\left(\frac{N_{\max}}{N_{av}}\right) + 3 \tag{14}$$

5. Cost-effectiveness analysis

Cost-effectiveness analysis (CEA) serves to determine the best allocation of available prevention strategies, to optimize the quantity and/or quality of safety. Different from a cost-benefit analysis, the approach gives an idea of whether an investment is "affordable" on top of being an optimal investment. A cost-effectiveness analysis does not strictly require the monetization of benefits, but always needs to compute cost-effectiveness ratios (CERs) and use these ratios to select strategies that are most effective (Reniers & Van Erp, 2016). As a result, cost-effectiveness can avoid three well-known problems associated with CBA: (i) the ethical problems associated with monetization of certain costs and benefits (e.g., the value of human life), (ii) not all impacts of a prevention investment can be monetized, and (iii) the relationship between intermediate impacts and certain end-goals or preferences is not fully clear (Boardman, Greenberg, Vining, & Weimer, 2017).

A CEA always includes the monetary cost of a protection strategy and the effectiveness corresponding to the benefits associated with the prevention strategy. The effectiveness can be any benefits induced by prevention strategies that decision-makers prefer to use, such as lives saved, amount of chemical spill avoided, employees' competencies improved, the reduction of social risk and the reduction of individual risk. The cost-effectiveness ratio (CER) with respect to process safety thus can be defined as the ratio of the cost of a prevention strategy n and the effectiveness of the strategy on prevention accidents, as follows:

$$CER_n = \frac{C_I{}^n}{Eff^n} \tag{15}$$

where CER^n represents the cost-effectiveness ratio of prevention strategy n. Eff^n represents the additional prevention effectiveness of strategy n. According to the formula in Eq. (15), the CER can be considered as the cost per unit of process safety. In order to decide whether an additional prevention strategy is efficient, the CER obtained by economic evaluations is always compared with a pre-determined threshold. The threshold represents the maximum value decision-makers are willing to pay for a unit of safety and CER beyond the threshold prevention measures are considered to be no longer cost-effective. Although cost-effectiveness ratios are undoubtedly informative and significant in assessing the effectiveness of prevention measures, it is difficult to determine a specific value for organizations. For example, the quality-adjusted life years (QALY) is often used in cost-effectiveness analysis for allocating health Care Resources. The maximum value of costs that decision-makers are willing to pay for a unit of QALY varies with different countries and different methods for estimating the value of the threshold are available in the literature. The World Health Organization's

Commission on Macroeconomics in Health suggested cost-effectiveness thresholds based on multiples of a country's per-capita gross domestic product (GDP). But in practice, the GDP-based thresholds in decision-making lack country specificity and lead to a wrong decision on how to spend health-care resources (Bertram et al., 2016; Eichler, Kong, Gerth, Mavros, & Jönsson, 2004; Weinstein & Stason, 1977).

Cost-effectiveness analysis can also be used to comparative analysis in the decision-making process by using the so-called incremental cost-effectiveness ratios (ICERs). Consider two prevention options, n_1 and n_2. The CER of investment n_1 relative to investment n_2, ICER_{12}, is given by the following formula:

$$ICER = \frac{C_I^{n_1} - C_I^{n_2}}{Eff^{n_1} - Eff^{n_2}} \tag{16}$$

if option n_2 is the current situation (e.g., doing nothing, no investment), Eq. (16) is identical to Eq. (15) and the option with the minimal value of ICER is the most cost-effective protection strategy and may be selected.

6. Economic optimization

The cost-benefit analysis in Section 4 demonstrates that a protection strategy is recommended if the so-called net present value (NPV) is equal to or higher than zero or in case of high impact low-frequency risks, the costs are proportionate to the benefits. However, companies usually face budget limitations and are expected to maximize their profits when it comes to decision-making on safety investments. This section aims to determine the most profitable or cost-effective protection strategy with budget limitations.

6.1 Optimization based on NPV

The optimized allocation of safety resources in chemical industrial areas can be regarded as the strategy with the highest NPV, as follows:

$$\begin{cases} \max\limits_{n} NPV^n \\ C^n \leq C_{\text{Budget}} \\ n \in \{1, 2, 3, \ldots, N\}, N \in Z \end{cases} \tag{17}$$

Eq. (17) indicates that the net present value of benefits (NPV) from the possible prevention strategies should be maximized within the constraint of prevention budget (C_{Budget}). The monetary cost of a prevention strategy n from N possible combinations of safety measures should not exceed the C_{Budget}.

6.2 Optimization based on proportion factor

As elaborated in Section 4.2, the proportion factor (*PF*) and disproportion factor (*DF*) can be determined and implemented to help understand a protection strategy's financial impact and thus determine whether the strategy can be recommended or not. Therefore, The optimal prevention strategy can be regarded as the protection strategy with the minimal value of the proportion factor (*PF*), as follows:

$$\begin{cases} \min\limits_{n} \ PF^{n} \\ C^{n} \leq C_{\text{Budget}} \\ n \in \{1, 2, 3, \ldots, N\}, N \in Z \end{cases} \tag{18}$$

6.3 Optimization based on cost-effectiveness ratio

Compared with cost-benefit analysis, cost-effectiveness analysis is more flexible since decision-makers can select the risk reduction indicators that they are willing to use, such as lives saved, amount of chemical spill avoided, the reduction of societal risk and the reduction of individual risk. For example, societal risk (SR) may be selected as the indicator if the decision-making concerns multiple fatalities. As a result, the optimal option would be the protection strategy with minimal costs per societal risk, as follows:

$$\begin{cases} \min\limits_{n} \ ICER^{n,0} \\ C^{n} \leq C_{\text{Budget}} \\ n \in \{1, 2, 3, \ldots, N\}, N \in Z \end{cases} \tag{19}$$

$$ICER^{n,0} = \frac{C_I{}^{n} - C_I{}^{0}}{SR^{0} - SR^{n}} \tag{20}$$

where C_I^0 represents the costs of the current prevention strategy and it will be equal to 0 if no safety investment has been done. SR^0 denotes the current societal risk and SR^{n} represents the societal risk associated with the prevention strategy n ($SR^0 > SR^{\text{n}}$).

6.4 Marginal hypothetical benefit analysis

In economics, marginal utility is the satisfaction a consumer gains from consuming one more unit of a good or service. Similar to any goods or services, the marginal safety or security utility can be regarded as the satisfaction obtained from one unit of additional special protection measures (Reniers & Van Erp, 2016). The marginal safety utility of a certain type of safety measures decreases with increasing more of this type of measure. From a viewpoint of producers, diminishing returns indicates the decrease in marginal output (impact) from increasing one unit of input (Anderson & Mittal, 2000). Safety resources also follow the law of diminishing returns: the marginal return in safety benefits of a safety measure decreases when furtherly increasing the amount of the taken measure (Fuller & Vassie, 2004). In other words, additional investment in a safety measure becomes less cost-effective than the preceding investment.

Hence, the investment in (for instance) detection measures follows the law of diminishing marginal returns: increasing additional measures or equipment results in smaller increases in protection benefits. In that case, when the investment in safety measures reaches a certain amount, further increasing the number of detection measures will become less and less cost-effective. The law of diminishing marginal returns is also suitable for the investment in safety barriers such as fireproof coatings and water delivery systems (Chen et al., 2019). The benefits obtained from an investment in fireproof coatings or water delivery systems decreases with increasing the number of tanks with these safety barriers. Thus, to be more efficient for prevention accidents, different types of safety measures should be taken, such as detection measures, active barriers, passive barriers, etc. In other words, the total protection budget available should be allocated to a wide range of protection measures.

According to the law of diminishing marginal returns, we can obtain a curve of diminishing marginal rate of return for each type of protection measures. A mathematical function for each curve of diminishing marginal rate can be determined via a numerical fitting. The type of function form of such curves is given, as follows:

$$y = \frac{ux}{x + v} \tag{21}$$

The parameters of u and v are constants and depend on the exact shape of a curve. The parameter u represents the maximal value that the curve is approximating. The parameter v indicates how fast the curve is approximating the maximum value, displaying the level of efficiency of the measures (Reniers & Van Erp, 2016). Therefore, to achieve the most cost-effective protection strategy, the optimization under budget restriction can be expressed:

$$\begin{cases} \max \sum_{w=1}^{w=W} \dfrac{u_w x_w}{x_w + v_w} \\ x_1 + x_2 + x_3 + \ldots + x_w = C_{Budget} \end{cases} \tag{22}$$

where w represents a type of protection measure and the total number of measures is W. To obtain the solution, the Lagrange method can be implemented (Hoy, Livernois, McKenna, Rees, & Stengos, 2011).

$$L = \sum_{w=1}^{w=W} \frac{u_w x_w}{x_w + v_w} - \lambda\left(C_{Budget} - \sum_{w=1}^{w=W} x_w\right) \tag{23}$$

Carrying out all the first-order partial derivatives for Eq. (23), the optimization problem can be transferred to solve an equation set with $W+1$ equations. In this way, we can obtain the optimal investment allocation for each type of protection measure under the restriction of budget limitations.

6.5 Optimization algorithms

There may be thousands or even millions of prevention strategies in a large chemical industrial area. At the same time, there is a limited budget. It is unreasonable to assess the benefits or effectiveness of all protection strategies by an exhaustive method. Therefore, the allocation of safety resources in chemical industrial areas with respect to linear optimization can be regarded as the so-called Knapsack problem, well-known in the field of Operations Research (Reniers & Sorensen, 2013; Villa et al., 2017). In terms of complex safety issues such as knock-on risks and domino effects, a chemical industrial area with large quantities of hazardous installations may be regarded as an interdependent system. Hence, the optimization may be a non-linear problem. In order to obtain the optimal prevention strategy in such situation, a sequential optimization algorithm called "PROTOPT" can be adapted for this problem, as shown in Fig. 3 (Chen et al., 2020a).

Besides, advanced evolutionary algorithms such as genetic algorithms may be used to solve both non-linear optimization models and multi-objective optimization models in safety investment, determining the optimal investment to maximize the protection benefits. Genetic algorithms (GAs) have been proven to be an effective tool to deal with large scale difficult problems (NP-hard problems). GA is a stochastic global search method that mimics the process of natural selection by simulating the evolution of a population of solutions. The intrinsic parallel design allows the algorithm to explore rapidly a large solution space and helps in avoiding to be trapped in local minima.

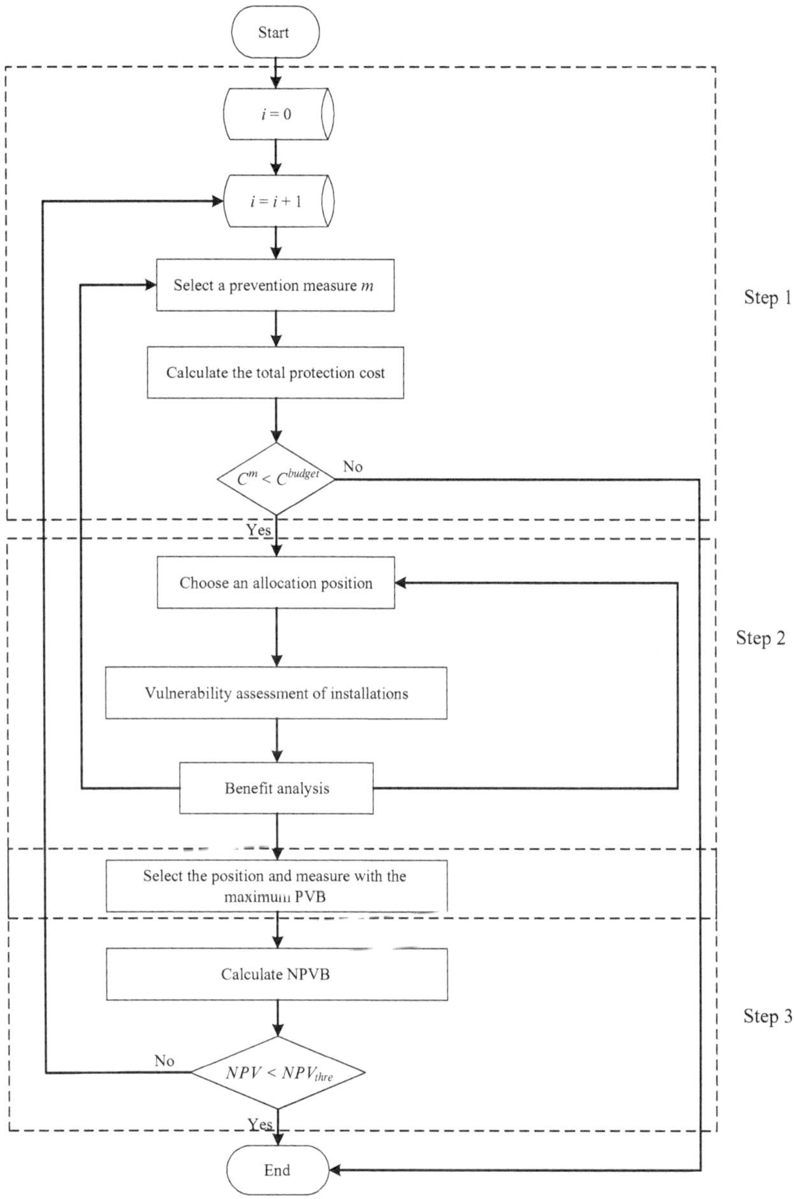

Fig. 3 The "PROTOPT" algorithm to achieve the optimal prevention strategy.

Therefore, it was used to optimize the layout of the process plant, the allocation of safety measures and multi-objectives in safety-based system design (Busacca, Marseguerra, & Zio, 2001; Caputo, Pelagagge, & Palumbo, 2011; Castell et al., 1998).

7. Conclusions

In the process industry, safety management is a significant work since many hazardous installations are located in these chemical plants to transfer, store, and process large amounts of hazardous substances. Making the decision on safety investments and the allocation of safety resources thus plays a dominant role in safety management. Although quantifying the costs and benefits of safety is rather challenging, it is a necessary work w.r.t. the effectiveness of a company's prevention policy as well as the company's long-term profitability. In this chapter, the well-known decision-making methods on process safety are presented in a nutshell, including risk criteria, risk acceptance and risk reduction. In order to improve the decision-making on process safety, several economic tools are introduced: cost analysis, cost-benefit analysis, cost-effectiveness analysis and economic optimization. By the application of these economic tools, the optimal prevention strategy, under the restriction of budget limitations, can be obtained, making decision-making on protection strategies more profitable and cost-effective.

References

Anderson, E. W., & Mittal, V. (2000). Strengthening the satisfaction-profit chain. *Journal of Service Research*, *3*, 107–120.

Barlettani, M., Fruttuoso, G., & Lombardi, P. (1997). An ecological approach for risk and impact ranking. Society of Risk Analysis, Stockholm: Proceedings New Risk Frontiers.

Bertram, M. Y., Lauer, J. A., De Joncheere, K., Edejer, T., Hutubessy, R., Kieny, M. P., et al. (2016). Cost-effectiveness thresholds: Pros and cons. *Bulletin of the World Health Organization*, *94*, 925–930. https://doi.org/10.2471/BLT.15.164418.

Boardman, A. E., Greenberg, D. H., Vining, A. R., & Weimer, D. L. (2017). *Cost-benefit analysis: Concepts and practice*. Cambridge University Press.

Bollinger, R. E., & Crowl, D. A. (1997). Inherently safer chemical processes: A life cycle approach. [Wiley-AIChE].

Bottelberghs, P. (2000). Risk analysis and safety policy developments in the Netherlands. *Journal of Hazardous Materials*, *71*, 59–84.

Busacca, P. G., Marseguerra, M., & Zio, E. (2001). Multiobjective optimization by genetic algorithms: Application to safety systems. *Reliability Engineering & System Safety*, *72*, 59–74.

Caputo, A. C., Pelagagge, P. M., & Palumbo, M. (2011). Economic optimization of industrial safety measures using genetic algorithms. *Journal of Loss Prevention in the Process Industries*, *24*, 541–551. https://doi.org/10.1016/j.jlp.2011.01.001.

Castell, C., Lakshmanan, R., Skilling, J., & Banares-Alcantara, R. (1998). Optimisation of process plant layout using genetic algorithms. *Computers & Chemical Engineering, 22*, S993–S996.

Chen, C., Reniers, G., & Khakzad, N. (2019). Integrating safety and security resources to protect chemical industrial parks from man-made domino effects: A dynamic graph approach. *Reliability Engineering and System Safety, 191*. https://doi.org/10.1016/j.ress.2019.04.023.

Chen, C., Reniers, G., & Khakzad, N. (2020a). Cost-benefit management of intentional Domino effects in chemical industrial areas. *Process Safety and Environmental Protection, 134*, 392–405. https://doi.org/10.1016/j.psep.2019.10.007.

Chen, C., Reniers, G., & Khakzad, N. (2020b). A thorough classification and discussion of approaches for modeling and managing domino effects in the process industries. *Safety Science, 125*, 104618. https://doi.org/10.1016/j.ssci.2020.104618.

Chen, C., Reniers, G., & Zhang, L. (2018). An innovative methodology for quickly modeling the spatial-temporal evolution of domino accidents triggered by fire. *Journal of Loss Prevention in the Process Industries, 54*, 312–324. https://doi.org/10.1016/j.jlp.2018.04.012.

Diamantidis, D. (2017). A critical view on environmental and human risk acceptance criteria. *International Journal of Environmental Science and Development, 8*, 62–68.

Dorman, P. (2000). The economics of safety, health, and well-being at work: An overview. [ILO Geneva].

Dorman, P. (2009). Markets and mortality: Economics, dangerous work, and the value of human life. Cambridge University Press.

Eichler, H. G., Kong, S. X., Gerth, W. C., Mavros, P., & Jönsson, B. (2004). Use of cost-effectiveness analysis in health-care resource allocation decision-making: How are cost-effectiveness thresholds expected to emerge? *Value in Health, 7*, 518–528.

Eslami Baladeh, A., Cheraghi, M., & Khakzad, N. (2019). A multi-objective model to optimal selection of safety measures in oil and gas facilities. *Process Safety and Environmental Protection, 125*, 71–82. https://doi.org/10.1016/j.psep.2019.02.024.

Fuller, C., & Vassie, L. H. (2004). Health and safety management: Principles and best practice. [Pearson Education].

Gavious, A., Mizrahi, S., Shani, Y., & Minchuk, Y. (2009). The costs of industrial accidents for the organization: Developing methods and tools for evaluation and cost–benefit analysis of investment in safety. *Journal of Loss Prevention in the Process Industries, 22*, 434–438.

Goose, M. H. (2006). Gross disproportion, step by step–a possible approach to evaluating additional measures at COMAH sites.

Hoy, M., Livernois, J., McKenna, C., Rees, R., & Stengos, T. (2011). Mathematics for economics. MIT press.

HSE. (1984). Risk criteria for land-use planning in the vicinity of major industrial hazards. [Health and Safety Executive].

Jallon, R., Imbeau, D., & de Marcellis-Warin, N. (2011). Development of an indirect-cost calculation model suitable for workplace use. *Journal of Safety Research, 42*, 149–164. https://doi.org/10.1016/j.jsr.2011.05.006.

Janssens, J., Talarico, L., Reniers, G., & Sörensen, K. (2015). A decision model to allocate protective safety barriers and mitigate domino effects. *Reliability Engineering & System Safety, 143*, 44–52. https://doi.org/10.1016/j.ress.2015.05.022.

Jonkman, S. N., van Gelder, P. H. A. J. M., & Vrijling, J. K. (2003). An overview of quantitative risk measures for loss of life and economic damage. *Journal of Hazardous Materials, 99*, 1–30. https://doi.org/10.1016/s0304-3894(02)00283-2.

Khakzad, N., & Reniers, G. (2019). Low-capacity utilization of process plants: A cost-robust approach to tackle man-made domino effects. *Reliability Engineering & System Safety, 191*, https://doi.org/10.1016/j.ress.2018.03.030.

Poole, R. W. (2008). Toward risk-based aviation security policy. [OECD/ITF Joint Transport Research Centre Discussion Paper].

Quah, E., & Haldane, J. (2007). *Cost-benefit analysis*. Routledge.
Reniers, G. L., & Audenaert, A. (2009). Chemical plant innovative safety investments decision-support methodology. *Journal of Safety Research*, *40*, 411–419. https://doi.org/10.1016/j.jsr.2009.09.007.
Reniers, G., & Brijs, T. (2014). Major accident management in the process industry: An expert tool called CESMA for intelligent allocation of prevention investments. *Process Safety and Environmental Protection*, *92*, 779–788. https://doi.org/10.1016/j.psep.2014.02.003.
Reniers, G., & Sörensen, K. (2013). Optimal allocation of safety and security resources. *Chemical Engineering Transactions*, *31*, 397–402.
Reniers, G. L., & Sorensen, K. (2013). An approach for optimal allocation of safety resources: Using the knapsack problem to take aggregated cost-efficient preventive measures. *Risk Analysis*, *33*, 2056–2067. https://doi.org/10.1111/risa.12036.
Reniers, G. L. L., Sörensen, K., Khan, F., & Amyotte, P. (2014). Resilience of chemical industrial areas through attenuation-based security. *Reliability Engineering & System Safety*, *131*, 94–101. https://doi.org/10.1016/j.ress.2014.05.005.
Reniers, G. L., & Van Erp, H. N. (2016). Operational safety economics: A practical approach focused on the chemical and process industries. John Wiley & Sons.
Stewart, M. G., & Mueller, J. (2013). Terrorism risks and cost-benefit analysis of aviation security. *Risk Analysis*, *33*, 893–908. https://doi.org/10.1111/j.1539-6924.2012.01905.x.
Swuste, P., van Nunen, K., Reniers, G., & Khakzad, N. (2019). Domino effects in chemical factories and clusters: An historical perspective and discussion. *Process Safety and Environmental Protection*.
Talarico, L., & Reniers, G. (2016). Risk-informed decision making of safety investments by using the disproportion factor. *Process Safety and Environmental Protection*, *100*, 117–130. https://doi.org/10.1016/j.psep.2016.01.003.
Villa, V., Reniers, G. L. L., Paltrinieri, N., & Cozzani, V. (2017). Development of an economic model for counter terrorism measures in the process-industry. *Journal of Loss Prevention in the Process Industries*, *49*, 437–460. https://doi.org/10.1016/j.jlp.2017.06.001.
Wang, Y. F., Li, B., Qin, T., & Zhang, B. (2018). Probability prediction and cost benefit analysis based on system dynamics. *Process Safety and Environmental Protection*, *114*, 271–278. https://doi.org/10.1016/j.psep.2018.01.002.
Weinstein, M. C., & Stason, W. B. (1977). Foundations of cost-effectiveness analysis for health and medical practices. *New England Journal of Medicine*, *296*, 716–721.

CHAPTER TWELVE

Inherently safer design principles in risk management

Kayleigh Rayner Brown[a], Albert Addo[a], Faisal I. Khan[b], Paul R. Amyotte[a,*]

[a]Department of Process Engineering & Applied Science, Dalhousie University, Halifax, NS, Canada
[b]Centre for Risk, Integrity and Safety Engineering (C-RISE), Faculty of Engineering & Applied Science, Memorial University of Newfoundland, St. John's, NL, Canada
[*]Corresponding author: e-mail address: paul.amyotte@dal.ca

Contents

1. Introduction

The purpose of this chapter is to encourage the use of inherently safer design (ISD) for eliminating hazards and reducing risk when outright elimination is not feasible. We therefore attempt to answer the question: why use

Methods in Chemical Process Safety, Volume 4
ISSN 2468-6514
https://doi.org/10.1016/bs.mcps.2020.02.006

inherently safer design for risk management? Beginning with the enduring quote by the late Trevor Kletz, "what you don't have, can't leak," evidence of the value of inherently safer design has increased over the past 40 years since the publication of the paper by the same name (Kletz, 1978). Many specific examples of alternate designs and approaches that use less hazardous chemicals, reduce hazardous inventories, reduce the complexity of processes, or reduce the severity of processing conditions, have been developed—resources such as CCPS (Center for Chemical Process Safety) (2009) (updated in 2019) and Kletz and Amyotte (2010) describe many of them.

However, much of the evidence and examples of how inherently safer design could improve process safety come from lessons learned after incidents and losses; Kletz and Amyotte (2019) describe the engineering lessons of case histories and the role of case histories in enhancing process safety knowledge. This brings us to the primary motivation for the current chapter—the fact that major accidents are still occurring. Within the past year, there have been several incidents that have received public attention. One such event occurred at the Philadelphia Energy Solutions Refinery on June 21, 2019, attracting attention due to a series of explosions in a hydrofluoric acid alkylation unit. This incident caused the release of hydrocarbons and hydrofluoric acid, and sent a 38,000-lb_m (17,300-kg) vessel fragment a distance of 2100 ft. (640 m) across the Schuylkill River, according to the most recent US Chemical Safety Board (CSB) report available at this time (CSB (Chemical Safety Board), 2019a). Fortunately, there were no fatalities, but losses in process safety should not depend on good fortune. We need to ask ourselves how can we stop major accidents like these from occurring in the first place. Amyotte et al. (2016) discuss seven concepts that all play a role in preventing major process accidents—inherently safer design is one such concept.

This chapter is comprised of several sections, each of which provides motivation and practical suggestions for using inherently safer design to manage risk. Section 2 provides an overview of inherently safer design and the hierarchy of controls. It is our belief that with an understanding of these fundamental concepts, anyone can incorporate inherently safer design into their everyday lives in some way—for example, by using less hazardous household cleaning products or by minimizing the storage of hazardous materials in personal dwellings. Internalizing the concepts of ISD can be a catalyst for change in both individuals and organizations. Hopkins (2009) has stated "safety culture…operate[s] at both the individual and organisational level…"

The above quote typifies Section 3, which discusses how inherently safer design can be incorporated into organizations, specifically through elements of a process safety management framework. It is noteworthy that the theme of the process safety session at the 2019 Canadian Chemical Engineering Conference held in Halifax, Nova Scotia was "Process Safety Management—Not Just for the Chemical Process Industries." It is our hope that readers in a broad range of industries will find something of value in the concepts discussed in Section 3 (and throughout the entire chapter), and will find applicability in their organizations—just as aspects of other fields such as learning from high-reliability organizations have relevance for the chemical process industries (Hopkins, 2009).

Section 4 describes a protocol, or method, that can be used to incorporate inherently safer design into process hazard analysis (PHA), specifically within the bow tie methodology. This tool promotes the incorporation of ISD in practical applications by allowing users to explicitly consider it in their collection of barriers and safeguards. The need for such tools is embodied in the following personal anecdote from one of the current authors (KRB). During the period of manuscript preparation, KRB was working in the role of graduate teaching assistant for the senior-year capstone design course in the Chemical Engineering Program at Dalhousie University. One of the projects being overseen was particularly well-suited for examination of inherently safer design features during the process hazard analysis phase. When discussion of process safety for this particular project arose with the team of undergraduate students, the suggestion was therefore made to explicitly consider ISD. Their response was quick and genuine: "Yes…but how?"

Section 5 outlines another tool for incorporating inherently safer design into hazard analysis, this time with respect to dust hazard analysis (DHA). This section includes a brief discussion of DHA fundamentals and how inherently safer design can be incorporated into DHAs to address hazards and risks associated with combustible dusts.

This chapter also incorporates *Opportunities for Reflection* at the end of each of Sections 2–5, echoing the additions to the latest edition of *What Went Wrong? Case Histories of Process Plant Disasters and How They Could Have Been Avoided* (Kletz & Amyotte, 2019), which has been updated with *Opportunities for Reflection* chapters at the end of each main book part. These chapters pose the following questions: "Was there one, single statement or concept that resonated with you—that made you pause and take note? How will you use your key message from this part of the book to effect process safety performance and fewer incidents in your work?" Here, we continue this momentum and encourage readers of the current chapter to complete a similar reflection exercise at the conclusion of each section. We thus propose another set of questions that are used in the "*Start-Stop-Continue*" exercise found in a wide range of settings from project management to team building to personal development: What would you like to *start* doing? What would you like to *stop* doing? What would you like to *continue* doing?

After reading a section, readers may recognize that they would like to *start* something new in their organization—perhaps implementing an inherently safer design training workshop or distributing specific resources such as the Center for Chemical Process Safety (CCPS) Process Safety Beacon. Readers may identify something as an opportunity for improvement or a gap within their organization—something they would like to *stop* doing—perhaps normalization of deviance, ignoring early warning signs, or other aspects related to process safety culture. Perhaps an organization already considers inherently safer design as part of hazard analysis or management of change, but it may seem more of a "check-the-box" activity that can be improved. Lastly, *continue* captures the positive things that are happening—what has been going well, what is working well, or "what went right"; this aligns with a recent publication titled the same (Amyotte, 2019). All readers—no matter their role, experience level or setting—are capable of being agents of change for the widespread, systematic adoption of the principles of ISD in risk management.

2. Fundamentals of inherently safer design

The discussion in this section is based primarily on Kletz and Amyotte (2010) with relevant excerpts. CCPS (Center for Chemical Process Safety) (2009, 2019) are also valuable references that include examples of ISD industrial applications and case studies.

Inherently safer design (ISD) is the component of process safety that focusses on avoiding hazards or reducing their likelihood or severity rather than relying on the use of add-on devices and procedures (Kletz & Amyotte, 2010). ISD is the foundation of risk management within the hierarchy of controls, which is the system used to eliminate or reduce the risk arising from identified hazards. As illustrated in Fig. 1, the hierarchy of controls—in order of increasing effectiveness—is comprised of the following categories:

- Administrative,
- Active engineered,

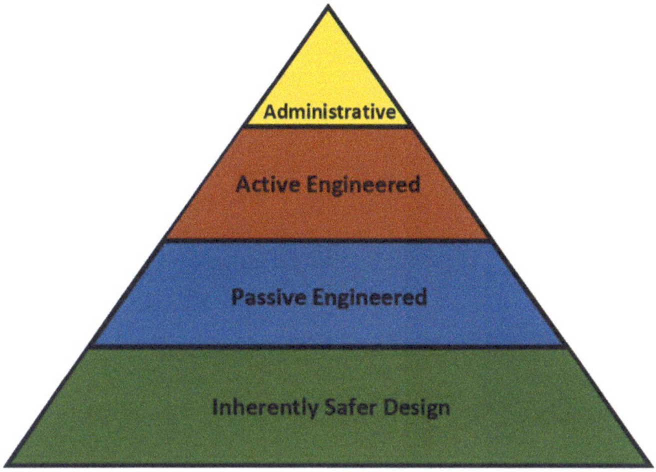

Fig. 1 Hierarchy of controls.

- Passive engineered, and
- Inherently safer design (ISD).

There are various graphical representations of the hierarchy of controls—for example, NIOSH (National Institute for Occupational Safety and Health) (2015) shows the hierarchy as an inverted pyramid, with elimination and substitution on the top as most preferred, and moving downward with administrative measures and personal protective equipment (PPE) at the pyramid base as the least preferred controls. In this chapter, the hierarchy of controls is represented as a pyramid with ISD on the bottom as it is the foundational base for risk management. ISD should be considered first, prior to adding on subsequent measures in the form of passive, active and administrative controls. This approach to risk management is aligned with the preferred order of controls outlined in Kletz and Amyotte (2010) and Amyotte (2013).

ISD is based on four main principles: minimization, substitution, moderation, and simplification. The use of these measures is referred to as inherently safer design because hazards are avoided due to the inseparable characteristics of the process, rather than a dependence on add-on safety equipment and human intervention. Minimization refers to the reduction or elimination of the hazard, which, for example, may refer to the quantity of a toxic chemical used or stored, equipment inventory, or the size of equipment. Substitution involves replacing a hazardous chemical or process with a less hazardous alternative. Moderation reduces a hazard by using less hazardous forms of materials or process conditions. Moderation also refers to facility design that reduces the effect of a loss of containment of material or energy (Khan & Amyotte, 2003). Examples include using less hazardous operating temperatures or pressures, and chemical concentration or form. Limitation of effects and avoiding knock-on (domino) effects are sub-principles of moderation. Simplification reduces hazards by minimizing the complexity of equipment or a process; it encompasses design related to addressing human factors and reducing hazards associated with maintenance and operations (Kletz & Amyotte, 2010). Sub-principles of simplification are making incorrect assembly impossible, making status clear, tolerance of misuse, and ease of control.

Broad-based examples of ISD with respect to these principles are given below, based on CCPS (Center for Chemical Process Safety), 2009 and Kletz and Amyotte (2010) with relevant excerpts.

Minimization:

- Minimize the inventory of hazardous material through equipment selection (CCPS (Center for Chemical Process Safety), 2009).
- Remove deadleg piping (CCPS (Center for Chemical Process Safety), 2009).
- Use process intensification to reduce inventories (Kletz & Amyotte, 2010).

Substitution:

- Use alternate chemicals that are less hazardous (CCPS (Center for Chemical Process Safety), 2009; Kletz & Amyotte, 2010).
- Use alternate processes that are less hazardous (CCPS (Center for Chemical Process Safety), 2009; Kletz & Amyotte, 2010).
- Use more corrosion resistant materials of construction (Kletz & Amyotte, 2010).

Moderation:

- Make operating conditions less severe (e.g., lower temperatures and pressures by using a catalyst) (Kletz & Amyotte, 2010).
- Use less concentrated hazardous raw materials to reduce the hazard potential (e.g., aqueous ammonia or methylamine instead of the anhydrous material) (CCPS (Center for Chemical Process Safety), 2009).
- Use limitation of effects—avoid hazardous equipment and operations (e.g., use closed loop sample stations to limit sampling procedure of hazardous materials; avoid glass sight glasses) (Kletz & Amyotte, 2010).
- Avoid domino/knock-on effects—use facility siting considerations/ exclusion zones around process plant (Kletz & Amyotte, 2010).

Simplification:

- Use double-walled pipe for extra corrosion/erosion allowance (CCPS (Center for Chemical Process Safety), 2009).

- Minimize the number of bends in piping (potential erosion points) (CCPS (Center for Chemical Process Safety), 2009).
- Reduce the potential of a hazard by eliminating liquid accumulation points or vibration stress (CCPS (Center for Chemical Process Safety), 2009).
- Design equipment with an MAWP (maximum allowable working pressure) to contain the maximum pressure generated without reliance on pressure relief systems, even if the "worst credible event" occurs (CCPS (Center for Chemical Process Safety), 2009).
- Design equipment that does not allow for incorrect assembly or at least allows an incorrect assembly to be apparent/determined (Kletz & Amyotte, 2010).

While inherently safer design is commonly referred to as being common sense, it is not yet as broadly and consistently adopted as one would expect given its proven effectiveness. This lack of adoption of ISD is due to a number of reasons, including but not limited to: (i) lack of awareness and knowledge of the concept, (ii) lack of understanding of the concept of ISD and its potential, (iii) absence of a set of specific methodologies to follow during application, (iv) lack of methods for measuring ISD characteristics, (v) difficulty of implementation in existing facilities, and (vi) absence of real case studies on successful ISD implementation (Hendershot, 2011; Jafari et al., 2018; Khan & Amyotte, 2003).

It is important to note that ISD is not a standalone concept. For it to be consistently and systematically used, ISD needs to be effectively integrated into an appropriate process safety management (PSM) system (Amyotte, Goraya, Hendershot, & Khan, 2007). The incorporation of ISD within a PSM framework is discussed in the next section of this chapter. Having concluded the current section, readers are invited to consider the first *Opportunity for Reflection* exercise in this chapter. With respect to the fundamentals of inherently safer design:

- What would you like to *start* doing?
- What would you like to *stop* doing?
- What would you like to *continue* doing?

3. Incorporation of ISD in process safety management systems

Process safety management (PSM) is defined as "the application of management principles and systems to identify, understand, avoid, and control process-related hazards in order to prevent, mitigate, prepare for, respond to, and recover from process-related incidents" (CSA, 2017; CSChE (Canadian Society for Chemical Engineering), 2012). The PSM framework published in 2012 by the Canadian Society for Chemical Engineering (CSChE) is composed of 12 elements, each containing several components. The elements are as follows: (i) *Accountability: objectives and goals*, (ii) *Process knowledge and documentation*, (iii) *Capital project review and design procedures*, (iv) *Process risk management*, (v) *Management of change*, (vi) *Process and equipment integrity*, (vii) *Human factors*, (viii) *Training and performance*, (ix), *Incident investigation*, (x) *Company standards, codes and regulations*, (xi) *Audits and corrective actions*, and (xii) *Enhancement of process safety knowledge* (CSChE (Canadian Society for Chemical Engineering), 2012). Previous work completed by Amyotte, Goraya, et al. (2007) demonstrated the link between these PSM elements and the explicit consideration of ISD within an organization. Research on incorporation of inherent safety principles specifically for hazards associated with combustible dusts has also been undertaken (Amyotte, Khan, & Dastidar, 2003; Amyotte, Pegg, & Khan, 2009; Amyotte, Pegg, Khan, Masaharu, & Yingxin, 2007). Another helpful reference is Amyotte and Lupien (2017) in Volume 1 of *Methods in Chemical Process Safety*, which presents an extensive discussion of safety management systems and concepts in PSM.

In 2017, the Canadian Standards Association (CSA) published a PSM standard (building on the CSChE 12-element system)—the Z767-17 *Process Safety Management* standard (CSA, 2017). This 16-element PSM system is composed of four foundational pillars: *Process Safety Leadership*, *Understanding Hazards and Risks*, *Risk Management* and *Review and Improve* (CSA, 2017) (Table 1).

Table 1 CSA Process Safety Management (PSM) Standard (CSA, 2017).
Process safety management elements

Process safety leadership	Understanding hazards and risks	Risk management	Review and improvement
Accountability	Process knowledge and documentation	Training and competency	Investigation
Regulations, codes, and standards	Project review and design procedures	Management of change	Audits process
Process safety culture	Process risk assessment and risk reduction	Process and equipment integrity	Enhancement of process safety knowledge
Conduct of operations—senior management responsibility	Human factors	Emergency management planning	Key performance indicators

CSA (2017) recommends that an organization's PSM program "should consider inherent safety in their policies, practices and procedures" (CSA, 2017). The current section therefore outlines how ISD can be incorporated within each of the 16 elements of the CSA PSM standard (CSA, 2017). In the following Section 4, the incorporation of ISD within the second pillar of *Understanding Hazards and Risks* (more specifically the element of *Process risk assessment and risk reduction*) using the bow tie methodology (a process hazard analysis, or PHA, technique) is discussed. Section 5 focuses on the incorporation of ISD within dust hazard analysis (DHA), which also falls primarily under the same *Process risk assessment and risk reduction* element.

3.1 PSM pillar 1: Process safety leadership

3.1.1 Element: Accountability

The incorporation of ISD within an organization is heavily dependent on the process safety culture, which is driven by company management and leadership. CSA (2017) states that senior management "shall provide direction to those responsible for design to consider inherently safer options." This quote presents the requirement for those in a leadership position to be champions of process safety and ISD within an organization, and to provide the tools needed by personnel to facilitate ISD consideration. Senior management can demonstrate their accountability for the PSM system by providing strong process safety leadership and direction in prioritizing process safety, providing sufficient resources, and establishing performance standards and enforcing them (Minerva, 2013). This proactive leadership role is integral to addressing an identified challenge with incorporating ISD within project design—there is a lack of a driver to move away from the status quo and try new inherently safer designs due to project timelines and budgets. Strong process safety leadership can drive this priority and provide the required resources. This is evident in work published by Edwards, Foster, Linwood, McBride-Wright, and Russell (2015) who outline an ISD workshop used within an organization to identify, assess and implement ISD. Such a tool deployed by management and project leadership demonstrates their commitment to process safety and ISD. It also places an emphasis on providing a conducive environment to personnel, thus demonstrating ultimate accountability for the PSM system.

Within this element of *Accountability*, CSA (2017) also states that senior management "shall ensure the integrity of equipment and processes are addressed through completing risk assessment and risk reduction throughout the facility's lifecycle." A hierarchy of hazards control analysis (HCA) is a tool that can be used to identify opportunities for ISD, as outlined in the

California Occupational Safety and Health Standards Board *Process Safety Management for Petroleum Refineries* regulation. (ISD regulations are discussed further in the *Regulations, codes and standards* element.) The HCA can be completed or revalidated every 5 years along with the standard 5-year process hazard analysis (PHA) revalidations. This is a tool that can be used by management to leverage ISD as a risk reduction barrier and demonstrate its commitment to the principles of inherently safer design.

3.1.2 Element: Regulations, codes and standards

The largely voluntary PSM system in Canada can be contrasted with other jurisdictions under different regulatory regimes in which there are prescriptive regulations for the use of the hierarchy of controls and consideration of ISD for risk management within facilities. These prescriptive regulations could be used by senior leadership to benchmark industry best-practices with which to improve their own internal standards and management systems. Jurisdictions with ISD regulations include Contra Costa County and the City of Richmond in the state of California. In Contra Costa County, where there is a high concentration of chemical process industry facilities (CCPS, 2019), the consideration of ISD has been legislated since 1998 with the enactment of the *Industrial Safety Ordinance* (*ISO*). The neighboring City of Richmond adopted the legislation (*Richmond Industrial Safety Ordinance*, or *RISO*), which became effective February 2002 (CCHS, 2019a; *Richmond Industrial Safety Ordinance*). Currently, eight facilities are covered under *ISO/RISO*, including three petroleum refineries and three chemical facilities in Contra Costa County, along with one petroleum refinery and one chemical facility in the City of Richmond. More information about these facilities can be found on the CCHS *ISO* Facilities website (CCHS, 2019b).

Also requiring the consideration of ISD is the state of California's *Process Safety Management for Petroleum Refineries* regulation implemented in 2017, as mentioned earlier under the PSM element *Accountability*. This regulation states that ISD is required to be implemented "to the greatest extent feasible," in addition to the following requirements (California Occupational Safety and Health Standards Board, 2016):

- Perform periodic safety culture assessments.
- Incorporate damage mechanism hazard reviews into process hazard analyses.
- Conduct root-cause analyses after significant accidents or releases.
- Account for human factors and organizational changes.

- Use structured methods, such as layer of protection analysis, to ensure adequate safeguards in process hazard analyses.

Again, these prescriptive PSM regulations can be used by management teams in other organizations to improve their PSM framework. The state of New Jersey has also established consideration of ISD within an Inherently Safer Technology (IST) Review under the *Toxic Catastrophe Prevention Act* (*TCPA*), which can be completed using one of the three methods outlined in CCPS (2009): (1) checklist PHA, (2) independent PHA, and (3) integral to PHA (ACUTech, 2019; State of New Jersey, 2017).

The state of Washington has also established a PSM Advisory Committee to improve process safety in the state's petroleum refining industry. This advisory committee outlines the following motivations for updating the PSM regulations: the current rule is outdated (it has not been updated in over 20 years), it does not reflect industry practices and improvements in processes and engineering, and other jurisdictions following the same PSM rule still have process safety incidents occurring (Washington State Department of Labor and Industries, 2017). The PSM Advisory Committee has proposed adopting California's PSM regulation because of the following reasons: the regulation has been developed and implemented, the regulation addresses recommendations made by the US Chemical Safety Board (CSB), and it is more cost effective for companies to follow consistent regulations in the different jurisdictions within which they operate (i.e., companies that have facilities in both Washington and California) (Washington State Department of Labor and Industries, 2018). These reasons and motivations should also be considered by management teams and leadership in other organizations when considering improvements and the incorporation of ISD within their PSM framework. A discussion draft of Washington state's proposed rule *Safety standards for process safety management of highly hazardous chemicals* is available at: https://lni.wa.gov/safety-health/grants-committees-partnerships/advisory committees/_psmdocs/PSM-DiscussionDraftRound2.pdf

3.1.3 Element: Process safety culture

CSA (2017) states "process safety leadership and competency shall be the core values of every manager from the CEO to the front line supervisors." Several US Chemical Safety Board investigation reports outline the need for both an improved safety culture and the prioritization of ISD implementation following incidents such as: the fatal explosion and fire from the catastrophic rupture of a heat exchanger at the Anacortes Tesoro Refinery in

2010, the Chevron Richmond Refinery pipe rupture and fire in 2012, the Williams Geismar Olefins Plant reboiler rupture and fire in 2013, and the toxic chemical release at the DuPont LaPorte Chemical Facility in 2014 (CSB, 2014, 2015a, 2016, 2017, 2019b). These CSB recommendations emphasize the correlation between the concepts of safety culture and uptake of inherently safer design.

Works by Andrew Hopkins in the area of safety culture are highly valuable resources (e.g., Hopkins, 2005, 2009). Hopkins (2005) describes the content of safety culture, which is comprised of four sets of practices: reporting culture, just culture, learning culture, and flexible culture. Hopkins (2005) also describes the processes of collective mindfulness: preoccupation with failure, reluctance to simplify, sensitivity to operations, and commitment to resilience and deference to expertise. The CCPS book "Essential practices for creating, strengthening, and sustaining process safety culture" (CCPS, 2018a) is also a useful reference for understanding the importance of process safety culture, the core principles and process safety leadership characteristics involved, and techniques for developing and maintaining process safety culture.

Investing resources to prioritize ISD within a process safety management system by leadership engrains it within the organization's safety culture (MKOCPSC/ISC, 2017). For example, the ISD workshop at Granherne described by Edwards et al. (2015) was found to help improve the organization's process safety culture. Safety culture assessments should be completed regularly to understand the state of safety culture, as is the practice in facilities in Contra Costa County, California that are regulated under the *Industrial Safety Ordinance* (*ISO*) (where, as previously indicated, the consideration of ISD by regulated facilities is mandatory) (CCHS, 2017a).

Expressed in a way process safety practitioners well understand: "A good safety culture is what everyone does when the managers aren't around."

3.1.4 Element: Conduct of operations: Senior management responsibility

CSA (2017) emphasizes that senior management is responsible for the "organization's overall process safety" and outlines aspects that need to be driven by management, including "ensuring personnel have required competencies and behaviors" and "maintaining the effectiveness of safety critical systems and equipment." For the incorporation of ISD within a facility, a continuous improvement cycle, well-represented by Forest (2012) in Fig. 2, must be followed. While ISD is directed by managers, they must also

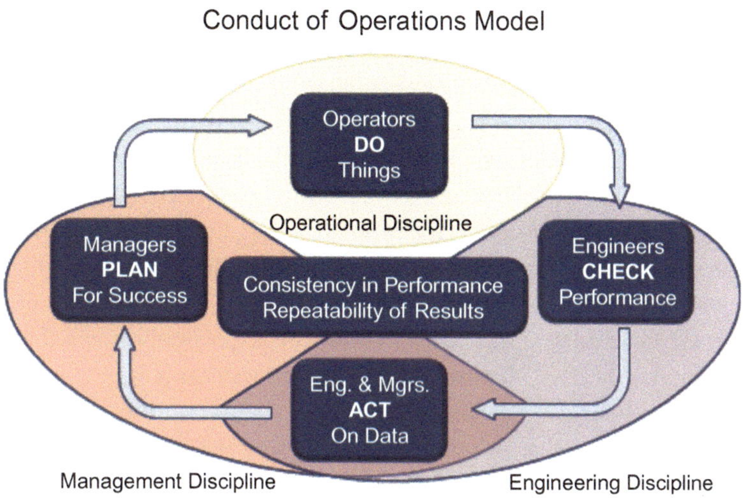

Fig. 2 Conduct of operations model (Forest, 2012) (with permission).

ensure operators understand the ISD features that are present as well as the operators' role in safety management. Engineers work to identify potential ISD opportunities to improve the effectiveness of safety systems and equipment. They make ISD recommendations to senior management, thus allowing the continuous improvement cycle to repeat.

The CSB investigation report following the DuPont La Porte incident outlines the need for management to ensure an inherently safer design review and a more robust PHA are completed to identify opportunities for implementation of ISD to eliminate harm to personnel and public (CSB, 2015b). Both CSB (2015b) and CSA (2017) identify that leadership should make this information available to personnel to create process safety risk awareness.

3.2 PSM pillar 2: Understanding hazards and risks

3.2.1 Element: Process knowledge and documentation

This PSM element encompasses the documentation and understanding of information related to the materials, process, and equipment design in order to mitigate hazards. The element is vital to the incorporation of ISD principles given that the basis of ISD is the underlying chemistry and physics of a process (Kletz & Amyotte, 2010). It is therefore crucial to have an intimate understanding of the process, equipment, and materials to identify opportunities for effective application of ISD principles, particularly within the element of *Process risk assessment and risk reduction* (described later in this section).

For example, in the DuPont La Porte incident described in CSB (2019b), the toxic and hazardous properties of chlorine and methyl mercaptan were understood, which presented opportunities to apply the ISD principle of substitution (although this had not been completed at the facility). This is contrasted with the incident at CTA Acoustics presented in CSB (2005). There was some awareness of combustible dust hazards within CTA Acoustics, but most personnel lacked an understanding of the explosion hazards presented by combustible dusts, specifically the phenolic resins being used within the facility. The phenolic resin safety data sheets were found to provide inadequate information about the explosion hazard. This lack of process knowledge documentation and understanding led to missed opportunities for ISD—specifically, minimizing horizontal flat surfaces to prevent dust accumulation (as recommended by the pertinent NFPA 654 standard) (Amyotte et al., 2009).

California Occupational Safety and Health Standards Board (2017) identifies requirements for written Process Safety Information (PSI) prior to completing a PHA, HCA, safeguard protection analysis (SPA) or damage mechanism review (DMR). This comprehensive documentation allows the employer and employee to identify and understand hazards associated with the process; such knowledge provides the basis for identifying ISD opportunities.

3.2.2 Element: Project review and design procedures

This is an important element for ISD consideration given that inherent safety is most effective at the design stage. Within the element of *Project review and design procedures*, CSA (2017) states that "siting decisions … shall consider inherently safer aspects" and the design process "should include items such as inherent safety."

The facility plot plan, specifically, must consider facility siting with respect to buffer zones, location of control rooms and location of storage facilities; this has strong connections with the ISD principle of moderation, in particular the sub-principle limitation of effects. There are two relevant API (American Petroleum Institute) standards that can be used address facility siting and in turn incorporate ISD through limitation of effects: RP (Recommended Practice) 752, Management of Hazards Associated With Location of Process Plant Permanent Buildings, and RP 753, Management of Hazards Associated With Location of Process Plant Portable Buildings (API, 2019a, 2019b). The Inherently Safer Design Review/Inherently Safer System Checklist outlined by CCHS (2011a,

2011b), based on CCPS (2009), is another tool that can be used by designers at the conceptual design, facility siting, and preliminary process design stages to avoid hazards.

3.2.3 Element: Process risk assessment and risk reduction

As previously mentioned under the PSM element *Regulations, codes and standards*, three Inherent Safety Analysis methods are outlined in CCPS (2009)—Checklist PHA, Independent PHA, and Integral to PHA—to explicitly consider ISD as a risk reduction barrier. An approach for incorporating ISD within the bow tie methodology is given in Section 4. The incorporation of ISD within PHAs also supports the element *Process knowledge and documentation* with respect to maintaining company memory (Amyotte, Goraya, et al., 2007). This formalization of ISD consideration within a PHA ensures that proper documentation and the justification of design decisions are maintained.

3.2.4 Element: Human factors

Within the *Human factors* element, CSA (2017) states that "human factors reviews shall be completed." The ISD principle of simplification has many links to human factors. Human factors reviews can identify opportunities for ISD through simplification as this principle primarily relates to design of the operator-process/equipment interface, written procedures, and the working environment. The ISD checklist in CCPS (2009) gives ISD considerations for human factors. These include simplified control displays, limited instrumentation complexity, clearly displayed information about process conditions, logical arrangement of controls and displays that match operator expectations, safety alarms that are easily distinguished from process alarms, correction of nuisance alarms, and elimination of redundant alarms to prevent complacency (CCPS, 2009). Kletz and Amyotte (2010) also discuss the consideration of human factors through making incorrect assembly impossible.

The Bow Ties in Risk Management concept book (CCPS/EI, 2018) is a useful resource for understanding the role of human factors in a PSM framework and barrier-based risk management. The chapter titled "Addressing Human Factors in Bow Tie Analysis" discusses the consideration of human factors within the bow tie methodology, the role human factors have on process safety barriers, and identifying how weaknesses can be addressed. As mentioned above, inherent safety can help reduce human error by means of human factors consideration. CCPS/EI (2018) touches on some barriers that capture ISD considerations to address human factors, including assessing job design, as well as the design of the ergonomics of safety critical elements and associated work systems. Other ISD barriers to address human factors can be found in the examples listed above in relation to simplification; for more examples and guidance, readers are referred to CCPS (2009) and Kletz and Amyotte (2010).

3.3 PSM pillar 3: Risk management

3.3.1 Element: Training and competency

There is a strong need for ISD-specific training and education within organizations for personnel to understand the fundamentals of ISD and the hierarchy of controls, and to ultimately become competent with the concepts in order to effectively apply them. In facilities regulated under *ISO* in Contra Costa County, as well as those regulated under the *Process Safety Management for Petroleum Refineries* regulation in California, personnel who are completing PHAs need to understand ISD. The previously described ISD workshop approach at Granherne helps to improve understanding of ISD concepts, allowing personnel to have a working knowledge of ISD and apply it to various project design phases (Edwards et al., 2015). There are also CCPS and IChemE training sessions available for this purpose (AIChE, 2019; IChemE, 2019).

3.3.2 Element: Management of change

CSA (2017) states "the primary focus of management of change (MOC) shall be to manage risks related to design changes and modifications to equipment, procedures and organization." CCPS (2011) details further the types of changes that will normally trigger the MOC process, some of which include changes to equipment, facilities and operating parameters; each of these changes includes opportunities to incorporate ISD.

The Chevron Richmond Refinery 2012 pipe rupture incident contains lessons learned regarding the incorporation of ISD within MOC. The root cause of the incident was the use of an inadequate material of construction. ISD could have been considered through "substitution of a higher chromium-content steel alloy which is less susceptible to sulfidation corrosion than the low-silicon carbon steel" that was being used for the 4-sidecut pipe at the time it ruptured (CSB, 2015a). With respect to management of change and material of construction, there are three primary changes that were not addressed, which are outlined by the interim investigation report (CSB, 2013) and final investigation report (CSB, 2015a, 2015b):

1. Equipment that removed hydrogen sulfide (H_2S) was itself removed in 1991, but an MOC procedure was not completed to assess what effect this change would have on sulfidation corrosion rates of the 4-sidecut. MOC should have captured both the use of low-silicon carbon steel and the increased sulfur content in the process stream.
2. Trend data for the 4-sidecut showed the sulfur content increased consistently over the course of 1984–2012. The subsequent changes in increased sulfidation corrosion rates were not assessed through the MOC process.
3. The temperature of the 4-sidecut draw also increased from 625 to 680 °F, further increasing sulfidation corrosion rates.

The Contra Costa County *ISO*, as well as California's *Process Safety Management for Petroleum Refineries* regulation, also state that ISD and the hierarchy of controls must be considered prior to implementing change (using one of the ISD PHA methods previously discussed or an ISD checklist, such as that in Amyotte, Goraya, et al., 2007 or CCPS, 2009). By considering ISD during the MOC process, opportunities to moderate risk with respect to building locations or substitute hazardous chemicals with less hazardous alternatives may be identified. Amyotte, Goraya, et al. (2007) detail the incorporation of ISD within MOC and give a protocol for the explicit consideration of inherent safety within the MOC process (as shown in Fig. 3).

Amyotte, Goraya, et al. (2007) and Amyotte, Pegg, et al. (2007) also provide an example of an inherent safety checklist corresponding to each ISD guideword (minimize, substitute, moderate, simplify). Excerpts of these questions include:

- Is the storage of all hazardous gases, liquids, and solids minimized?
- Can a less toxic, flammable, or reactive material be substituted for use?
- Are all hazardous gases, liquids, and solids stored as far away as possible to eliminate disruption to people, property, production, and environment in the event of an incident?
- Are machine controls located to prevent unintentional activation while allowing easy access for stopping the machine?

3.3.3 Element: Process equipment and integrity

CSA (2017) outlines end of service requirements—"the organization shall be accountable to dismantle, decommission, and dispose of equipment and related waste related to its operations"—which is connected to the ISD

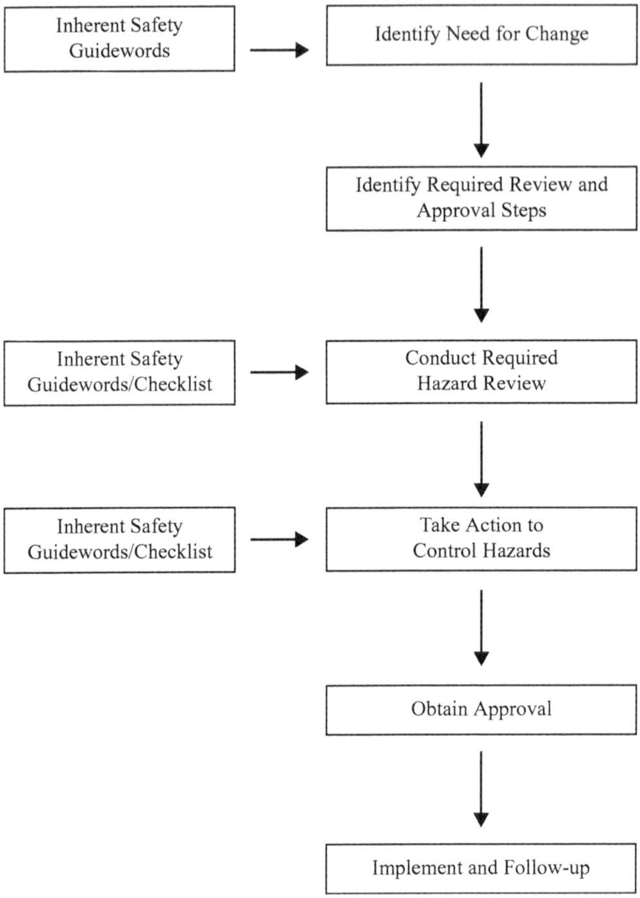

Fig. 3 An inherent safety-based management of change protocol (Amyotte, Goraya, et al. (2007) (with permission).

principle of minimization. Eliminating equipment that is decommissioned or no longer needed minimizes risk associated with this equipment.

CSA (2017) also states that in the pre-start-up safety review (PSSR) process, the organization must verify that a PHA has been completed before start-up. As outlined in California's *Process Safety Management for Petroleum Refineries* regulation, there is also the opportunity to consider the hierarchy of controls and ISD within the PSSR process by completing an HCA.

3.3.4 Element: Emergency management planning

An Emergency Management Program (EMP) is developed as a risk reduction barrier to manage the consequences of hazardous scenarios.

Considering ISD is an important and effective component of the measures available to lessen the effects of incidents. There is a strong connection between the element of *Process risk assessment and risk reduction* and the element of *Emergency management planning*; within the former element, consequence analysis is undertaken, including the consideration of any knock-on (domino) effects. This presents the opportunity for ISD to play an explicit role in emergency management through applying the principle of avoiding knock-on effects, a sub-principle of moderation. Examples of avoiding domino events include siting process units to minimize impacts to plant control and off-site facilities, along with designing fire breaks and corridors between units.

3.4 PSM pillar 4: Review and improve

3.4.1 Element: Investigation

CSA (2017) states that incident investigations "shall include recommendations to prevent recurrence." These recommendations and lessons learned from incident investigations are an opportunity to incorporate ISD. Corrective actions based on ISD solutions represent the most effective and reliable opportunities for preventing the incident from occurring again by eliminating or reducing the hazard. Amyotte, MacDonald, and Khan (2011) and Amyotte, Irvine, and Khan (2018) describe the analysis of CSB recommendations with respect to the hierarchy of controls, as well as applicable PSM elements. Examples of ISD recommendations include substituting natural gas with less hazardous air for cleaning fuel gas pipe, managing deadleg piping, substituting chlorine with sodium hypochlorite, using limitation of effects (moderation) for process unit segregation to avoid domino effects, and minimizing beams, ledges and overhead fixtures that can collect combustible dusts (Amyotte et al., 2011, 2018).

A protocol to explicitly incorporate ISD considerations in an incident investigation is detailed by Goraya, Amyotte, and Khan (2004) (as shown in Fig. 4). This protocol incorporates inherent safety guidewords, not only at the recommendations stage, but also within the onsite investigation and root-cause identification stages. Goraya et al. (2004) also provide an example of an inherent safety checklist corresponding to each ISD guideword (minimize, substitute, moderate, simplify).

3.4.2 Element: Audits process

Audits should be completed by trained personnel to assess an organization's PSM system and the extent of incorporation of ISD. CCPS (2009) states that

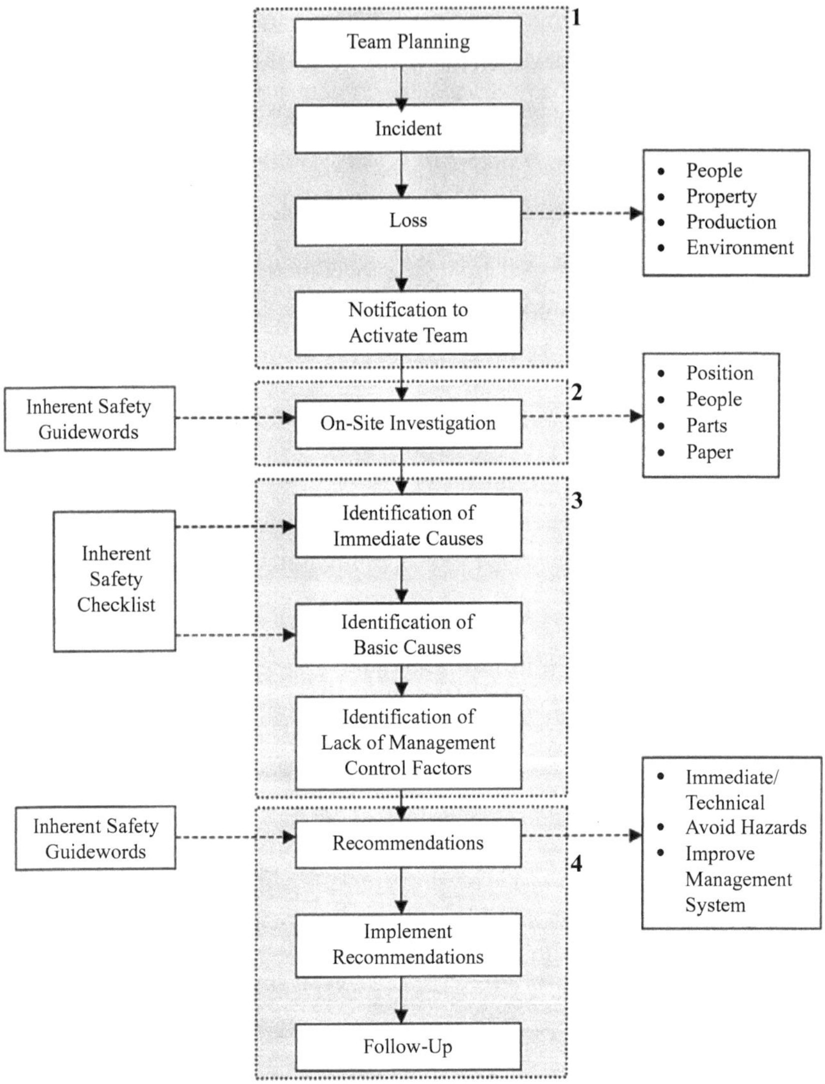

Fig. 4 An inherent safety-based incident investigation methodology (Goraya et al., 2004) (with permission).

audits help develop methods to measure effectiveness of ISD efforts and provide feedback to personnel to improve performance. Audits can help identify corrective actions and must be repeated if the system has changed. Regulated facilities within the Contra Costa County *ISO* complete internal audits once every 3 years (CalARP; CCHS, 2014). The facilities also undergo external audits conducted by the regulator (CCHS), the results of which are publicly available on the CCHS *ISO* website (CCHS, 2019c). These safety audit summaries outline opportunities for ISD improvement, such as incorporating ISD within a facility's PHA review/revalidation process, establishing more expertise in ISD, and improving ISD consideration and documentation with respect to facility siting (CCHS, 2011b, 2017b, 2019d).

3.4.3 Element: Enhancement of process safety knowledge

Process safety data related to "what went right," or ISD success stories, is needed in order to provide personnel and organizations with example-based guidance to increase the implementation of ISD within their facilities. These lessons may come from outside the facility or learned from within. This process safety data should be maintained in a library or database within the organization that can be easily accessed. Personnel should also be made aware of other types of process safety resources. Resources like the CCPS Process Safety Beacon, a monthly newsletter, often contain links to ISD; for example, as described in CCPS (2010a), the inherently safer solution to hazards associated with long-bolt flangeless valves would be to replace with a fully flanged valve (CCPS, 2010a). Personnel should also be encouraged to watch relevant US Chemical Safety Board incident videos (https://www.csb.gov), which are valuable resources, and as previously discussed, identify many ISD opportunities based on lessons learned from incidents. Another, less-traditional, media format available for personnel to enhance their process safety knowledge is podcasts. The Dust Safety Science Podcast, produced by DustEx Research Limited, is an excellent resource for learning about hazards associated with combustible dusts (Dust Safety Science, 2019). The podcast touches on the topic of ISD and the hierarchy of controls in several episodes, as well as dust hazard analyses (which are discussed in Section 5).

3.4.4 Element: Key performance indicators

As CCPS (2011) (updated in CCPS, 2018b) states: "you don't improve what you don't measure." The same holds for the incorporation of ISD within an organization. CSA (2017) states that "performance indicators for process safety shall be developed and monitored and improvements identified. Both leading and lagging indicators shall be used." Leading indicators include proactive monitoring of safety systems that play a role in preventing incidents, while lagging indicators refer to reactive monitoring of incidents (CCPS, 2018a, 2018b). API RP 754, *Process Safety Indicators for the Refining and Petrochemical Industries* (API, 2010), defines a key performance indicators (KPIs) and metrics program as one that "provides useful information for driving improvement and when acted upon, contributes to reducing risks of major hazards by identifying the underlying causes and taking action to prevent recurrence."

Using the tier-based, pyramid approach for leading and lagging indicators described by Azizi (2016) (based on API, 2010), a program to measure leading and lagging indicators related to the use of ISD within an organization should be developed. Examples of leading indicators (at the base of the pyramid) related to operating discipline and management system performance (Tier 4) include:

- Measurement of process safety culture, as recommended by CCPS (2018a, 2018b). Measuring safety culture assessments completed can be used as a leading indicator. Using a safety culture survey prior to ISD workshop implementation and afterward can generate metrics to track changes in the safety culture.
- Measurement of "barriers related to the hazards inherent in operations managing hazardous materials and energies," again as recommended by CCPS (2018a, 2018b). Measuring ISD opportunities and implementation of ISD for hazard elimination or reduction could be captured as a leading metric.

An example of Tier 1 or Tier 2 lagging indicators (process safety incidents at the top of the tier-based safety pyramid) is found in CCHS (2019a), which contains data for major catastrophic accidental releases (MCAR) in Contra Costa County from 1998 to 2018. CCHS (2019a) plots the number of MCARs over the 20 years since the *ISO* was implemented. As a point of discussion, as stated in CCPS (2019), while it is difficult to "establish a direct causal relationship" due to the relatively low number of annual MCARs, the observed trend is decreasing MCARs since the *ISO* was established. The measurement of ISD incorporation as it relates to process safety incidents is valuable data to collect.

Other useful resources for KPIs include Amyotte and Lupien (2017), which provides information related to KPIs and metrics within a PSM framework. Additionally, the evergreen webpage created by CCPS (https://www.aiche.org/ccps/process-safety-metrics) for information and documents regarding process safety metrics is also a valuable resource.

In closing Section 3, it is hopefully clear that ISD can be incorporated into a PSM framework within each of the system elements. To facilitate this incorporation, however, specific tools need to be developed (Amyotte et al.,

2003; Amyotte, Goraya, et al., 2007). In the following section, one such tool—a protocol for incorporating ISD into the bow tie methodology—is discussed.

Having concluded the current section, readers are invited to consider the second *Opportunity for Reflection* exercise in this chapter. With respect to the incorporation of ISD in process safety management systems:

- What would you like to *start* doing?
- What would you like to *stop* doing?
- What would you like to *continue* doing?

4. ISD application to the bow tie methodology for process hazard analysis

As stated in Section 2, ISD must be integrated within the framework of a process safety management system to be explicitly considered. The need for a tool or protocol that facilitates this integration has been identified. The current section focusses on incorporating ISD specifically within process hazard analysis (PHA). PHA refers to hazard evaluation methods that help identify and analyze dangers associated with processes (CCPS, 2008); as such, with respect to the PSM framework, a PHA protocol falls primarily within the element *Process risk assessment and risk reduction* in Pillar 2 *Understanding Hazards and Risks*.

The incorporation of ISD into PHA is a promising way to explicitly use the principles of inherent safety. As previously described, CCPS (Center for Chemical Process Safety) (2009) outlines three ISD review methods that incorporate inherent safety into PHAs: (1) checklist PHA, (2) independent PHA, and (3) integral to PHA. Within the integral to PHA method, ISD guidewords and questions are included in a PHA using techniques such as HAZOP or What-If? analysis. Such ISD review methods are beneficial at both the initial design and operational life cycle stages. At the design stage, the proposed design and safeguards can be examined to identify any ISD opportunities. At the operational stage, completing an ISD review on existing designs and processes can identify opportunities to use ISD to improve the current state.

Another PHA technique in which ISD can be explicitly incorporated is the bow tie methodology. The remainder of this section outlines a protocol that can be used at both the design and operational life cycle stages to identify opportunities for ISD. First, an overview of bow tie diagrams as a PHA tool is given. Second, the concept of example-based guidance and its importance and relevance to a PSM framework are presented. Finally, a high-level overview of the protocol is provided and its application to the Bhopal gas tragedy is outlined.

4.1 Bow tie methodology

Bow tie diagrams are a barrier-based risk management tool that demonstrate and communicate how various factors can cause loss of control of a hazard and lead to undesirable consequences. The generic bow tie structure is shown in Fig. 5. The elements of a bow tie diagram are as follows: hazard, top event, threats, consequences, barriers, degradation factors, and degradation controls. The definitions of these bow tie elements are shown in Table 2 (CCPS/EI, 2018; CGE, 2017). Specialized software for drawing bow tie diagrams and completing related tasks is commercially available. An example of this software is BowTieXP, developed by CGE Risk Management Solutions; the bow tie diagrams, like the one in Fig. 5 and others within this section, have been drawn using BowTieXP.

One of the greatest strengths and benefits of bow tie diagrams is that they are a visual tool able to communicate hazardous scenarios to a wide range of audiences. Bow ties show direct cause and effect lines, which can make it easier to understand how hazardous events and consequences can occur. Hatch, McCulloch, and Travers (2017) outline the strengths of the bow tie approach compared with the HAZOP method for completing a PHA. Bow tie diagrams allow barrier weaknesses and degradation factors to be clearly displayed. This is compared with the single "safeguard" column used in standard HAZOP formats, which can make it challenging to understand the efficacy and criticality of safeguards. This a significant benefit, specifically for incorporation of ISD as a type of barrier. Additionally, when bow tie software (e.g. BowTieXP) is used, metadata about the barriers, such as barrier type (which can be directly related to the hierarchy of controls), can be captured and displayed to help a user understand the types of barriers that are being deployed and whether they are diverse (CGE, 2019a). Based on CCPS/EI (2018), the process outlined in Fig. 6 is followed to build a bow tie.

For more information about the bow tie methodology, readers are referred to the following valuable resources: CCPS/EI (2018) *Bow Ties in Risk Management: A Concept Book for Process Safety*; Vaughen and Bloch (2016) *Use the Bow Tie Diagram to Help Reduce Process Safety Risks*; Klein and Vaughen (2017) *Process Safety: Key Concepts and Practical Approaches*;

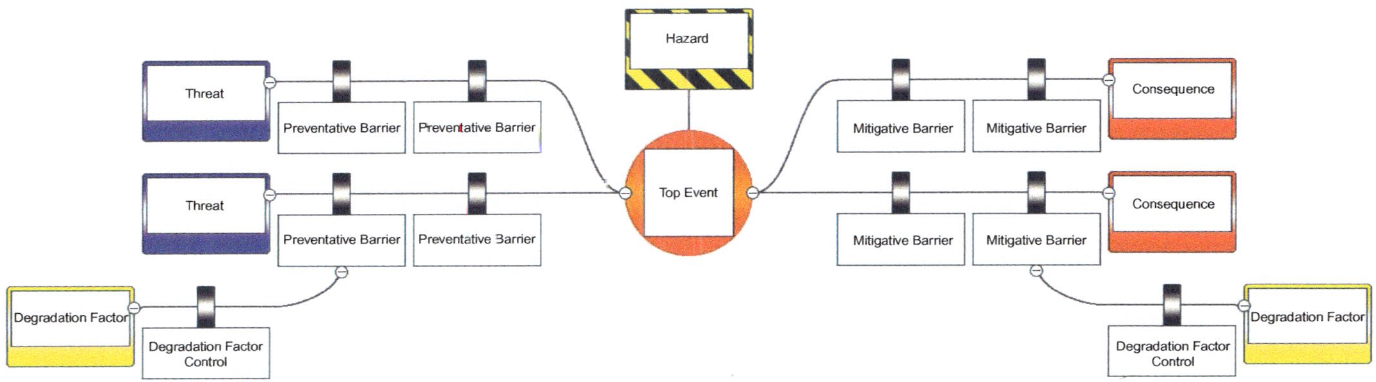

Fig. 5 Bow tie diagram and the elements—hazard, top event, threats, consequences, barriers, degradation factors and degradation controls.

Table 2 Bow tie element definitions (CCPS/EI, 2018; CGE, 2017).

Hazard	An operation, activity or material that poses a potential harm to people, property, the environment or business
Top event	The loss of containment or loss of control of a hazard that is preceded by threats and followed by consequences; the central "knot" of the bow tie diagram
Threats	Events or conditions that can cause the top event (can initiate the loss of containment or loss of control of a hazard). Threats are located on the left side of the top event
Consequences	Events or conditions that can result from the top event (outcomes of the loss of containment or loss of control of a hazard). Consequences are located on the right side of the top event. Consequences are typically negative effects to people, property, the environment or business (e.g., health and safety impacts, property loss, business interruption)
Barriers	Physical and non-physical controls and measures that can prevent threats from leading to the top event or mitigate consequences after the top event has occurred. Preventative barriers are to the left of the top event and mitigative barriers are to the right of the top event. Barriers are effective, independent and auditable
Degradation factors	Conditions or errors that defeat or degrade the effectiveness of barriers and compromise their function
Degradation controls	Physical and non-physical controls and measures that prevent degradation factors from compromising barriers

CGE (2019b) *CGE Knowledge Base Website*; Hatch et al. (2017) *Visual HAZOP: Exploiting the Power of Bowties to Improve Study Effectiveness and Enhance Engagement*; and Hatch, McCulloch, and Travers (2019) *Enhancing PHAs: The Power of Bowties.*

4.2 Example-based guidance

Example-based guidance (EBG) is the phrase used to encompass specific, practical applications of ISD that can be used to guide and inform other ISD opportunities (Amyotte et al., 2003; Amyotte, Goraya, et al., 2007). The need for example-based guidance, as a constituent of the PSM element *Enhancing PSM knowledge*, has been identified previously in Section 3. A listing of example-based guidance is included in this section (Table 3),

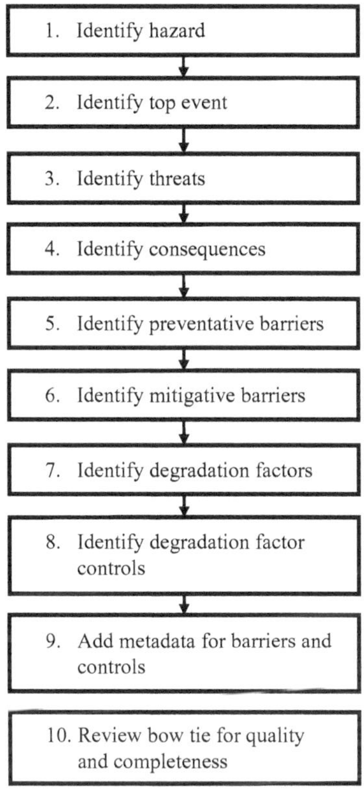

Fig. 6 Process for building bow tie diagram (CCPS/EI, 2018; CGE Risk Management Solutions, 2017).

which had been developed from four primary sources using relevant excerpts—CCPS (Center for Chemical Process Safety) (2009), Kletz and Amyotte (2010, 2019), and Mannan and Lees (2005). Table 3 also contains example-based guidance found in journal articles, CSB incident investigation reports, and Process Safety Beacons. As a sample of example-based guidance, users can refer to Table 3 in order to find examples that can be directly used within their facility, as well as use it as the basis for "mind-triggers" to help them recognize opportunities specific to their facility and unique application. Any known industrial applications of ISD, such as those undertaken in jurisdictions where ISD is legislated (e.g. Contra Costa County in the state of California) can also serve as valuable sources of example-based guidance. Research in this regard is currently underway at Dalhousie University.

Table 3 Inherently safer design example-based guidance.

ISD principle	Hazard, threat or consequence category	ISD example-based guidance	References
Substitution	Hazardous material	Substitute flammable with non-combustible	CCPS (Center for Chemical Process Safety), 2009
		Substitute with less volatile	CCPS (Center for Chemical Process Safety), 2009
		Substitute with less reactive	CCPS (Center for Chemical Process Safety), 2009
		Substitute with more stable	CCPS (Center for Chemical Process Safety), 2009
		Substitute with less toxic (e.g., sodium hypochlorite instead of elemental chlorine)	CCPS (Center for Chemical Process Safety), 2009
		Substitute flammable heat transfer fluid (i.e., operated above flash point) with low pressure steam	CCPS (Center for Chemical Process Safety), 2009
		Substitute hydrofluoric acid (HF) with modified hydrofluoric acid (MHF), which contains a vapor suppressant	Blais et al. (2012)
		Substitute HF-based alkylation process with new HF-free alkylation technology (ionic liquid and solid-state alkylation technologies are emerging alternatives)	CCPS (Center for Chemical Process Safety), 2009, CSB (2019c)
		Replace flammable solvents with alternative solvents with a high flash point for many formulated products such as agricultural chemicals, dyes, adhesives, paints and coatings, and paint removers	Mannan and Lees (2005)

	Reaction chemistry	Substitute manufacturing process with alternative process that uses less toxic raw materials (e.g., for manufacturing acrylonitrile, use new ammoxidation process that uses propylene and ammonia, rather than the conventional process that uses hydrogen cyanide)	CCPS (Center for Chemical Process Safety), 2009
		Substitute manufacturing process with alternative process that uses less toxic raw materials (e.g., for manufacturing acrylic esters, use propylene oxidation process instead of the Reppe process that uses a toxic nickel carbonyl catalyst)	CCPS (Center for Chemical Process Safety), 2009
		Substitute an alternative chemical that prevents the production of hazardous side reactions and products (e.g., vinyl chloride production in a reaction between sodium hydroxide and ethylene dichloride)	CCPS (Center for Chemical Process Safety), 2009
		Substitute conventional processes with alternative electrochemical techniques that may eliminate the hazardous materials or chemical intermediates or allow less severe operating conditions (e.g., naphthoquinone, anisaldehyde, benzaldehyde)	CCPS (Center for Chemical Process Safety), 2009
	Combustible solids	Identify a non-combustible solid that can replace the combustible dust in the process	Amyotte et al. (2009), Mannan and Lees (2005)
		Use explosion-proof vacuum in place of a broom for removing dust accumulations	Amyotte et al. (2009)

Continued

Table 3 Inherently safer design example-based guidance.—cont'd

ISD principle	Hazard, threat or consequence category	ISD example-based guidance	References
		Replace bucket elevators and other mechanical conveying systems with dense-phase pneumatic transport	Amyotte et al. (2009)
		Substitute process route (e.g., for preparation of powdered metals for refractory use (explosible aluminum/magnesium alloys); inert dust could be introduced at earlier stage of process to render the aluminum/magnesium alloys non-explosible throughout manufacturing)	Amyotte et al. (2009)
	Pipe integrity	Use higher chromium content steel to prevent sulfidation corrosion	CSB (2015a)
		Replace any piping to remove the bimetallic welds (e.g., piping that is a mixture of carbon steel and chrome should be replaced with only chrome piping)	CCPS (2012a)
Minimization	Reactors	Reduce inventory by using tubular reactors, which offer great potential as a reaction device; they can be used for many different applications (e.g., polymerization, carbonylation, dehydrogenation, hydrogenation, hydrocracking, hydroformulation, oxidative decomposition, partial oxidation, reforming)	CCPS (Center for Chemical Process Safety), 2009, Mannan and Lees (2005), Parr Instrument Company (2019)
		Use reactive distillation, a form of Process Intensification (PI), that combines the unit operations of chemical reaction and distillation into a single piece of equipment. For example, a reactive distillation process for the manufacture of methyl acetate from methanol and acetic acid uses a single reactive distillation column and two auxiliary columns in place of a continuous reactor, an extractor and eight distillation columns	Mannan and Lees (2005)

	Distillation operations	Modify and optimize distillation column packing and tray geometry to minimize hazardous liquid inventory	Mannan and Lees (2005)
		Modify and optimize column base geometry to minimize hold-up of hazardous material	Mannan and Lees (2005)
		Consider using centrifugal distillation devices, which have much smaller inventory	Mannan and Lees (2005)
	Extraction operations	Consider using centrifugal extractors, which have much smaller volume than traditional extraction columns	Mannan and Lees (2005)
	Heat exchangers	Use the most efficient heat exchanger possible. Various heat exchanger types have vastly different efficiency with respect to heat transfer area per unit volume of material inventory. For example, plate heat exchangers have a higher surface compactness than shell and tube exchangers; examine available design options to select the best for the application	Mannan and Lees (2005)
	Material inventory	Eliminate or reduce size of hazardous in-process storage inventory	CCPS (Center for Chemical Process Safety), 2009
		Reduce hazardous in-process storage and inventory by designing process equipment involving hazardous material with the smallest feasible inventory	CCPS (Center for Chemical Process Safety), 2009

Continued

Table 3 Inherently safer design example-based guidance.—cont'd

ISD principle	Hazard, threat or consequence category	ISD example-based guidance	References
	Equipment and piping inventory	Minimize the length of hazardous material piping	Mannan and Lees (2005)
		Design hazardous material piping for minimum piping diameter	Mannan and Lees (2005)
		Remove deadleg piping	CCPS (Center for Chemical Process Safety), 2009
		Remove out of service, abandoned, or unnecessary equipment, such as pipe	CCPS (Center for Chemical Process Safety), 2009
	Mixing operations	Use intensive mixers—form of Process Intensification (PI)—to minimize size of mixing vessel or reactor	CCPS (Center for Chemical Process Safety), 2009
	Combustible solids	Minimize fuel loadings with respect to dust layers—remove dust deposits from workplace to limit formation of dust suspensions	Amyotte et al. (2009)
		Minimize fuel loadings with respect to dust layers; minimize flat surfaces to prevent accumulation of combustible dusts	Amyotte et al. (2009)
Moderation	Operating conditions	Make operating conditions less severe (for hazardous reactants or products) by considering operating at lower pressure to limit potential release rate	CCPS (Center for Chemical Process Safety), 2009
		Make reaction conditions for hazardous reactants or products (temperature, pressure) less severe by using a catalyst	CCPS (Center for Chemical Process Safety), 2009

	Material concentration	Use less concentrated hazardous raw materials to reduce the hazard potential; use benzoyl peroxide as water wetted powder instead of pure, dry material	CCPS (Center for Chemical Process Safety), 2009, Mannan and Lees (2005)
		Use less concentrated hazardous raw materials to reduce the hazard potential; use aqueous ammonia or methylamine instead of the anhydrous material	CCPS (Center for Chemical Process Safety), 2009
	Material form	Convert materials to a less hazardous form; in manufacturing/metal grinding applications, use briquetter to turn flammable metal turnings (e.g., magnesium) into briquettes instead of turnings	ATM Recycling Systems (2017), PRAB (2020), RUF US Inc. (2016)
	Process design	Design the layout for processes handling flammable materials to minimize the number and size of confined areas and limit the potential for serious overpressure in the event of a loss of containment and subsequent ignition	CCPS (Center for Chemical Process Safety), 2009
		Ensure process materials (e.g., heating/cooling media) are compatible with process materials in event of inadvertent contamination (e.g., due to a tank coil or heat exchanger tube failure)	CCPS (Center for Chemical Process Safety) (2009)
		Design process units for hazardous materials to limit the magnitude of process deviations; select pumps with maximum capacity lower than safe rate of addition for the process	CCPS (Center for Chemical Process Safety) (2009)
		Design process units for hazardous materials to limit the magnitude of process deviations; use pipe size or fixed orifice to limit the maximum feed rate of gravity-fed systems within safe limits	CCPS (Center for Chemical Process Safety) (2009)

Continued

Table 3 Inherently safer design example-based guidance.—cont'd

ISD principle	Hazard, threat or consequence category	ISD example-based guidance	References
	Spills	Direct spills of flammable materials away from storage vessels to reduce the risk of a boiling liquid expanding vapor explosion (BLEVE) in the event of a fire	CCPS (Center for Chemical Process Safety) (2009)
		Prevent hazardous material liquid spills from entering drainage system/sewer (if potential for fire or hazardous reaction exists; e.g., water reactive material)	CCPS (Center for Chemical Process Safety) (2009)
	Combustible solids	Handle solids in the form of larger particle size granules, pellets, beads, or prills, rather than a fine powder; reduces the potential for worker exposure and also increases the minimum explosible concentration, thus reducing the hazard of dust explosions	CCPS (Center for Chemical Process Safety), 2009
		Reduce worker exposure and dust explosion hazards by formulating dyes as liquids, wet pastes or slurries rather than dry solids or powders	CCPS (Center for Chemical Process Safety) (2009)
	Materials storage	Design equipment to contain the materials that might be present inside at ambient temperature or the maximum attainable process temperature; design for higher than the maximum allowable working temperature to accommodate loss of cooling (reduce the reliance on external systems like refrigeration to control temperature such that vapor pressure is less than equipment design pressure)	CCPS (Center for Chemical Process Safety) (2009)
	Hazardous material transportation	Locate the plant to minimize the need for transportation of hazardous materials; co-locate with supplier/customer	CCPS (Center for Chemical Process Safety) (2009)
		Locate the plant to minimize the need for transportation of hazardous materials; on-site production of hazardous raw materials	CCPS (Center for Chemical Process Safety) (2009)

	Storage tanks	Ensure sufficient space between tank and process areas	Kletz and Amyotte (2019)
	Hazardous equipment	Use spectacle (or figure-eight) blinds instead of slip-plates	CCPS (Center for Chemical Process Safety) (2009), CCPS (2012a, 2012b)
		Reduce potential error by minimizing the number of components, such as gaskets, nuts, bolts and flanges used	CCPS (Center for Chemical Process Safety) (2009), Kletz and Amyotte (2010)
	Process unit siting	Consider limitation of effects; locate hazardous process units to eliminate or minimize on-site impacts on employees and other plant facilities including control rooms, fire protection systems, emergency response and communication facilities, and maintenance and administrative facilities	Kletz and Amyotte (2010)
Simplification	Pipe integrity	Eliminate insulation on piping to minimize chance for corrosion under insulation (CUI)	Kletz and Amyotte (2010)
		Use fewer pipe seams and joints	CCPS (Center for Chemical Process Safety), 2009
		Select appropriate pipe thickness/schedule that will provide extra corrosion/erosion allowance (e.g., Sch 80 is thicker than Sch 40 and may provide more corrosion resistance)	CCPS (Center for Chemical Process Safety), 2009
	Valves	Use check valves with easy to identify direction of flow	CCPS (Center for Chemical Process Safety) (2009)
		Use gate valves with rising spindles to clearly indicate open or closed position	CCPS (Center for Chemical Process Safety) (2009)

Continued

Table 3 Inherently safer design example-based guidance.—cont'd

ISD principle	Hazard, threat or consequence category	ISD example-based guidance	References
	Operation and maintenance	Outfit equipment (e.g., compressor) with double block and bleeds to avoid rolling feedgas blinds during compressor isolation	CSB (Chemical Safety Board) (2019d)
		Use dedicated hoses and compatible couplings for reactants where hose connections are used	CCPS (Center for Chemical Process Safety) (2009), CSB (2017)
	Robustness of equipment and materials of construction	Design equipment with MAWP (maximum allowable working pressure) to contain the maximum pressure generated without reliance on pressure relief systems, even if the "worst credible event" occurs	CCPS (Center for Chemical Process Safety) (2009)
		Use mixing feed nozzle instead of agitator or mechanical mixer for vessel mixing; these mixing nozzles have no moving parts, which makes them simple and reliable	CCPS (Center for Chemical Process Safety) (2009)
		Use simple flare designs; avoid obstructions such as water seals, molecular seals, filters and flame arrestors, as these devices are prone to choke or freeze	Kletz and Amyotte (2010)
		Use corrosion and creep resistant materials for furnace tubes	CCPS (Center for Chemical Process Safety) (2009)
		Design or select equipment to be impervious to the corrosion mechanisms that are present given the materials of construction and within the process, and the operating conditions (i.e., temperature, pH, concentration, viscosity)	CCPS (Center for Chemical Process Safety) (2009)
		Eliminate use of glass, plastic or other brittle materials as material of construction	CCPS (Center for Chemical Process Safety) (2009)

	Human factors	Use logical arrangement of controls and displays that match operator expectations	CCPS (Center for Chemical Process Safety) (2009)
		Use safety alarms that are easily distinguished from process alarms	CCPS (Center for Chemical Process Safety) (2009)
	Heat exchangers	Design the shell and tube sides of liquid-to-liquid heat exchangers, and the tubes of air-cooled heat exchangers to contain the maximum attainable pressure on either side. Such a design eliminates reliance on pressure relief devices or other active safeguards	CCPS (Center for Chemical Process Safety) (2009)
	Equipment utilities	Use independent power busses for redundant equipment to minimize consequences of partial power failures	CCPS (Center for Chemical Process Safety) (2009)
	Pumps	Use seal-less pumps (e.g., canned, magnetic drive)	CCPS (Center for Chemical Process Safety) (2009)
	Storage tanks	Ensure the design of drain valves system installed on dikes for rainwater removal allows for easy determination of valve status/position (open/closed); if valves are left open, spillage will be lost to drain	Kletz and Amyotte (2019)
		Ensure tanks containing incompatible materials are not sited within the same diking system	Kletz and Amyotte (2019)
		Ensure the material of construction of the tank, the stored material and any impurities are compatible	Kletz and Amyotte (2019)
		Use sloped tank bottom design and piping configured for total drainage	Kletz and Amyotte (2019)

Continued

Table 3 Inherently safer design example-based guidance.—cont'd

ISD principle	Hazard, threat or consequence category	ISD example-based guidance	References
	Equipment assembly	Design equipment that does not allow for incorrect assembly or at least allows an incorrect assembly to be apparent/determined (e.g., design compressor valves with inlet and exit valves that are not interchangeable)	CCPS (Center for Chemical Process Safety) (2009), Kletz and Amyotte (2010)
		Design equipment that does not allow for incorrect assembly or at least allows an incorrect assembly to be apparent/determined (e.g., use flanged or welded pipes instead of screwed couplings and compression couplings)	Kletz and Amyotte (2019)
	Equipment status	Use figure-eight/spectacle blinds that make it easier to see the position than blinds (spades or slip plates)	Kletz and Amyotte (2010)
		Use gate valves with rising spindles that make it easier than non-rising spindles to show if the valve is open or closed	Kletz and Amyotte (2010)
	Equipment tolerance	Use expansion loops in piping instead of bellows for thermal expansion; bellows are not tolerant of incorrect installation	CCPS (Center for Chemical Process Safety) (2009), Kletz and Amyotte (2010)
		Use articulated arms or fixed pipes instead of hoses for loading/unloading of hazardous materials	CCPS (Center for Chemical Process Safety) (2009), Kletz and Amyotte (2010)
	Process operation control	Use robust processes that will still operate even though there are approximations in the model of the process on which the control system is based	Kletz and Amyotte (2010)
		Ensure process control limits are not close to maximum safe operating limits, and margin between control limit and maximum safe operating limits is wide to allow for recovery if needed	Kletz and Amyotte (2010)

As it relates to the current protocol for incorporating ISD into bow tie diagrams, example-based guidance serves to identify ways a hazard may be eliminated, as well as to identify ISD barriers for threats and consequences. Thus, the ISD example-based guidance library represented by Table 3 can be organized by category of hazards, threats and consequences to help the user determine where to look for a suitable or relevant ISD barrier.

4.3 Protocol for incorporating ISD into bow tie diagrams

Currently, the barriers emphasized in typical bow tie diagrams are primarily administrative, active and passive. For example, the barrier types outlined in CCPS/EI (2018) are as follows: passive hardware, active hardware, active hardware and human, active human, and continuous hardware. The previously stated objective of the protocol described in this section is to identify opportunities for ISD barriers during both process design and operational life cycle stages. The protocol, shown in Fig. 7, is a tool that allows ISD to be explicitly considered within bow tie diagrams to manage risk.

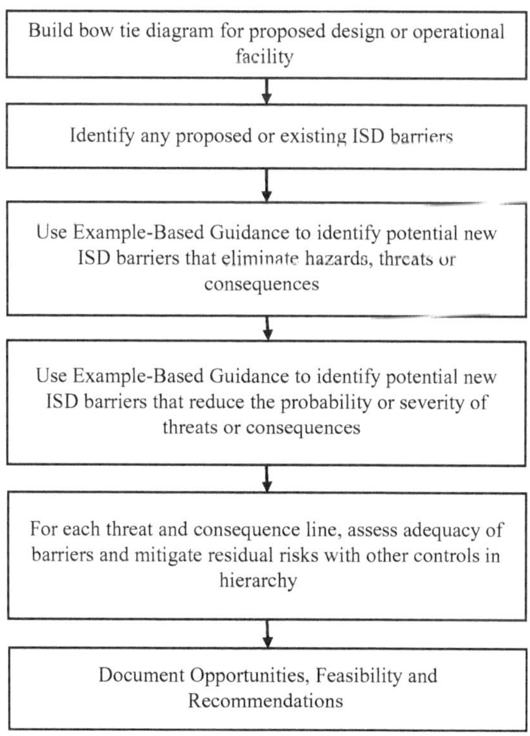

Fig. 7 Protocol overview for incorporating ISD into bow tie diagrams at design and operational life cycle stages.

The protocol follows a structure and approach to risk management similar to that outlined by Amyotte (2013) in Fig. 8. The protocol is also aligned with a number of initiatives identified by CCPS (2019) as current gaps in the field of inherent safety—incorporation of ISD into PSM, developing ISD databases and libraries, and developing a tool to apply ISD.

The following paragraphs give a description and overview of the protocol. First, a bow tie diagram is drawn for the hazardous scenario that is being analyzed. The process for drawing bow tie diagrams shown in Fig. 6 can be referenced. CCPS/EI (2018), which provides guidelines on properly wording the bow tie elements to produce an effective diagram, can also be consulted.

In the second stage of the protocol, any existing or proposed ISD barriers are identified using the definition of the hierarchy of controls (Section 2) to determine the barrier type with respect to the method of control (administrative, active engineered, passive engineered, and ISD). If software (e.g., BowTieXP) is being used to complete the bow tie, the "barrier type" metadata field can be used to label and display the barrier type. In contrast to the barrier types outlined by CCPS/EI (2018), the current protocol encourages the user to categorize barrier types based on the hierarchy of controls convention—ISD, passive, active and administrative. Labeling the barriers helps ensure diverse barriers are being deployed and all levels of control in the hierarchy are being considered.

In the third stage of the protocol, example-based guidance is used to identify potential ISD barriers based on the principles of substitution, minimization, moderation, and simplification. These ISD barriers could eliminate a hazard, or threats and consequences that could lead to or be caused by a top event, respectively (given that these threats and consequences are hazard specific). Barriers that could serve as degradation factor controls may also be identified (such as those related to human factors, as discussed in Section 3).

Next in the fourth stage of the protocol, example-based guidance is used again to identify potential ISD barriers based on the principles of substitution, minimization, moderation, and simplification. Within this step, however, ISD barriers that could reduce the likelihood or severity of a top event to arise from threats, as well as reduce the likelihood or severity of potential consequences, are identified. It can be noted that barriers that eliminate a hazard may be referred to as *first-order ISD* and those that reduce likelihood or severity may be referred to as *second-order ISD* (CCPS (Center for Chemical Process Safety), 2009).

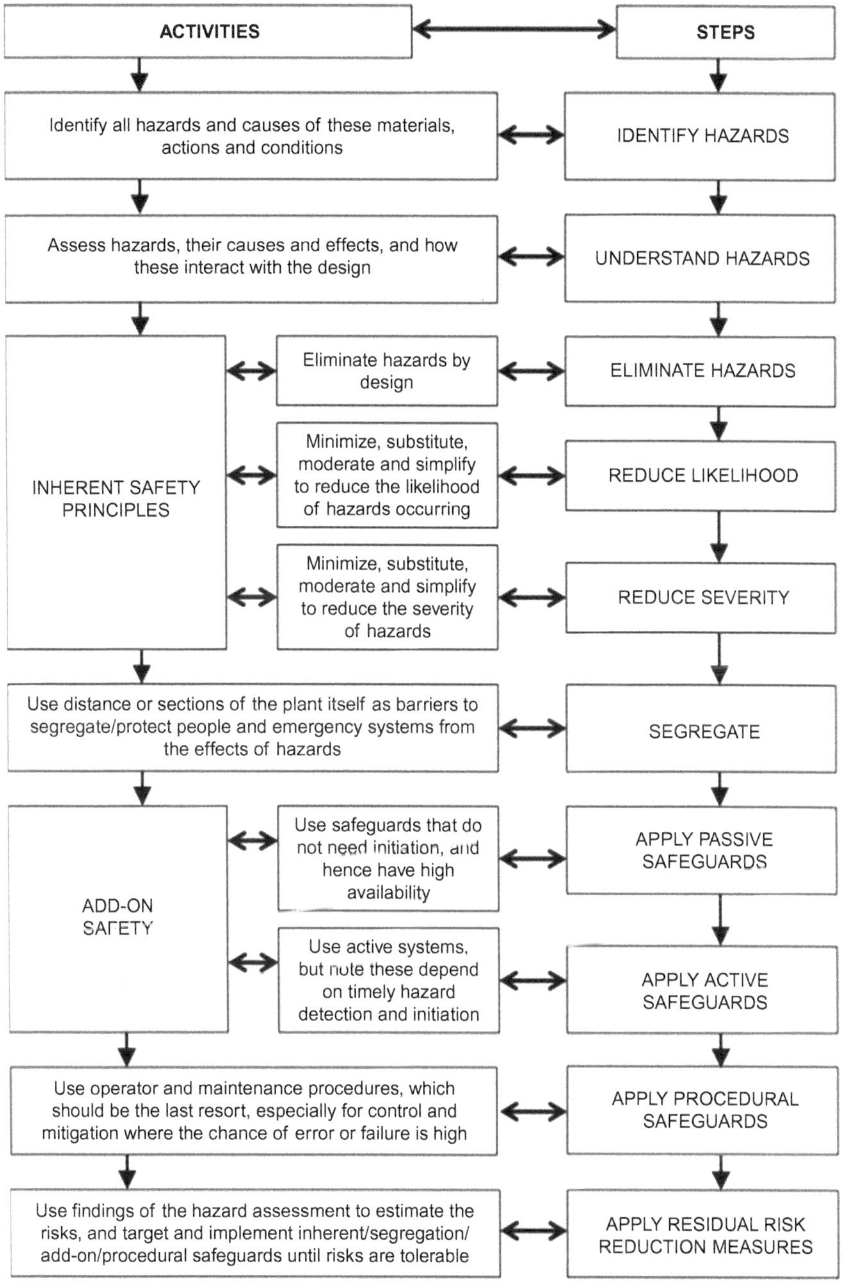

Fig. 8 From Amyotte (2013)—a systematic approach to loss prevention incorporating the hierarchy of controls. *Adapted from Kletz, T.A., & Amyotte, P.R. (2010).* Process plants: A handbook for inherently safer design *(2nd ed.). Boca Raton, FL: CRC Press/Taylor & Francis with the modification that the inherent safety principles of minimization, substitution, moderation and simplification can all be effective for both prevention and mitigation purposes (Amyotte et al., 2011) (with permission).*

Use of example-based guidance during the protocol to identify potential ISD opportunities may result in elimination of a threat or consequence line (such as using alternate reaction chemistry or eliminating the storage of a hazardous chemical), or reduction of the likelihood or magnitude of the threat or consequence (such as reducing the stored inventory of a hazardous chemical or designing an exclusion zone). All different ISD opportunities should be documented. These different potential barrier approaches can either be documented on the same bow tie or separate bow ties depending on the complexity of the diagram and the barriers identified. This allows the different layers of protection schemes to be compared with each other (as identified by CCPS (Center for Chemical Process Safety), 2009) in order to properly inform recommendations and decision-making. If there are different ISD barrier approaches that could be used for the same threat or consequence lines, these could be displayed on the same bow tie with the threat or consequence repeated for each of the different approaches for the interim decision-making step to qualitatively compare the approaches. Otherwise, bow tie diagrams require threats and consequences to be unique (CCPS/EI, 2018).

In the fifth stage of the protocol, after ISD opportunities have been considered for each of the threat and consequence lines, any remaining risk that could not be addressed using ISD may be mitigated using other add-on barriers from the hierarchy of controls (passive, active, administrative).

Last in the protocol, all ISD considerations are documented. ISD opportunities and recommendations can be assessed for feasibility and subsequently used in design decisions; as mentioned previously, it is important to document these ISD considerations. This documentation stage is also aligned with the ISD review checklist procedure outlined by CCPS (Center for Chemical Process Safety) (2009).

4.4 Example of protocol application: Bhopal gas tragedy

The Bhopal gas tragedy can be used as a case study for the application of the protocol, given that numerous reviews of the disaster have found there were opportunities for ISD application (Khan & Amyotte, 2003; Kletz & Amyotte, 2019; NRC (National Research Council), 2012). The following series of bow ties demonstrate the progression of the protocol application. As per the first stage of the protocol, a bow tie diagram is drawn to represent the scenario associated with the hazard of MIC storage tanks and the loss of containment (LOC) of MIC as the top event (Fig. 9). This bow tie provides a

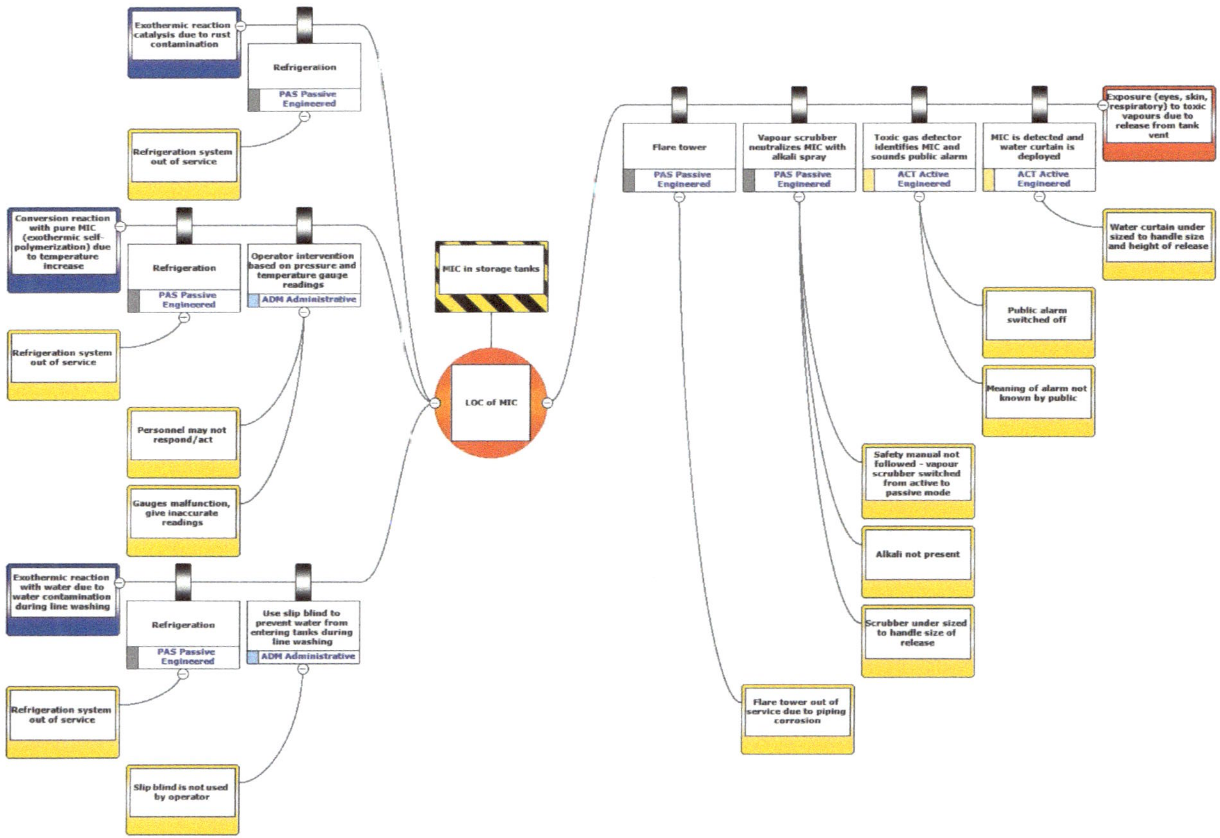

Fig. 9 High level bow tie diagram of Bhopal gas tragedy with hazard as MIC in storage tanks and top event as loss of containment of MIC.

high-level overview of some of the risks associated with the storage of methyl isocyanate (MIC) for pesticide manufacturing at the Bhopal facility at the time of the incident (Bowonder, 1987; NRC (National Research Council), 2012; Vaughen & Bloch, 2016). The references identified in the first sentence of this paragraph provide a more comprehensive description of the manufacturing process and other technical issues and risks involved.

The second stage of the protocol builds on the bow tie drawn previously (Fig. 9). Each barrier is categorized with respect to the hierarchy of controls and existing or proposed ISD barriers are identified. As seen in Fig. 9, barriers can be labeled and color-coded based on the barrier type, which helps to easily communicate the different types of barriers being deployed.

The bow tie diagram in Fig. 9 demonstrates a number of things. First, the existing barriers are add-on passive and active engineered safety devices and administrative controls (procedures), which are known to be less effective and reliable based on the hierarchy of controls. The reduced effectiveness and reliability of these passive engineered, active engineered and administrative barriers is also evident by their associated degradation factors displayed in the diagram. Many of these degradation factors are associated with deficiencies in maintenance, which is an issue that has been well-documented (NRC, 2012). It can also be noted that there is no barrier against rust contamination; the design specification stated that stainless-steel piping should be used, but this design specification was not followed, and carbon steel was used instead.

The third stage of the protocol can be completed next, where example-based guidance is used to identify new ISD barriers that could be used to eliminate the hazard, threats or consequences; these are shown in Fig. 10 with green labels reading "ISD Inherently Safer Design." Barrier categories are also included based on the ISD principles (minimization, substitution, moderation, and simplification) and are color-coded (orange, red, blue, and purple, respectively). Note that degradation factors in the subsequent bow tie diagrams are still present, but are hidden due to space considerations. Referring to the example-based guidance in Table 3, potential ISD barriers are identified that could eliminate hazards, threats or consequences. The barriers added to the bow tie include substituting alternate reaction chemistry to manufacture the pesticide or eliminating the MIC storage.

The fourth stage of the protocol is completed next, where example-based guidance is employed to identify new ISD barriers that could be used to reduce the likelihood or magnitude of threats or consequences. Referring

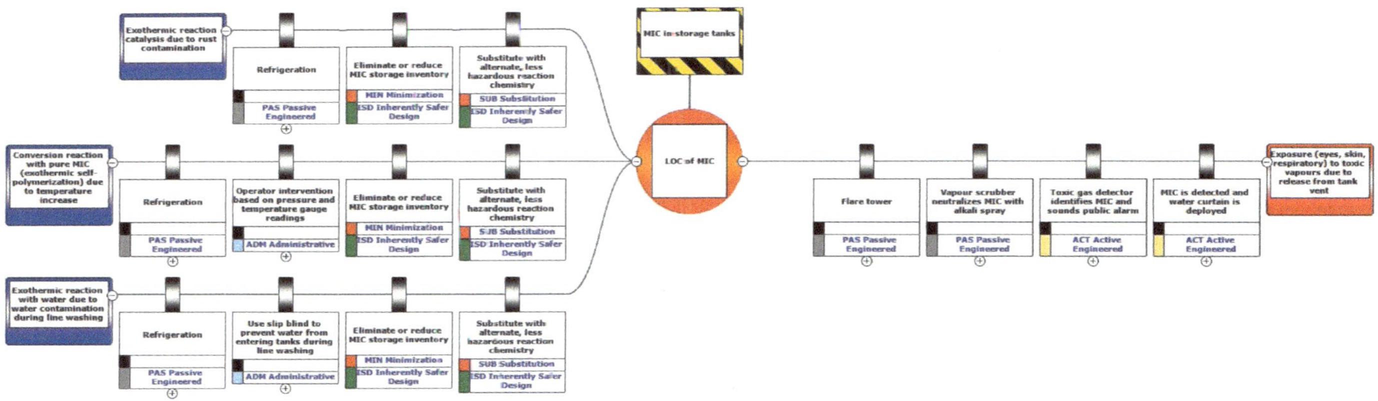

Fig. 10 High level bow tie diagram of Bhopal gas tragedy with hazard as MIC in storage tanks and top event as loss of containment of MIC with first-order ISD opportunities identified.

to the example-based guidance in Table 3 again, potential ISD barriers are identified, including using an alternate material of construction that is more corrosion resistant, reducing the MIC storage, or designing an exclusion zone around the facility to segregate the hazard (shown with green ISD labels in Fig. 11). For simplicity, the barriers "eliminate MIC storage" and "minimize MIC storage" have been combined in this bow tie. With respect to using more corrosion resistant materials of construction, in order to address the risk of a loss of containment event associated with rust contamination as previously mentioned, an inherently safer design barrier is to substitute the carbon steel pipe with more corrosion resistant pipe (minimum stainless steel).

In the fifth stage of the protocol, additional barrier types from the hierarchy of controls can be used to further reduce risk. In the Bhopal gas tragedy, it has been documented that an established shelter in place procedure (administrative barrier) for the surrounding residents could have mitigated the impacts of an MIC release. This mitigative barrier is shown in Fig. 12 with a light blue barrier type label "ADM Administrative."

In closing, the technical literature describes several ways in which ISD could have played a role in preventing and mitigating the Bhopal gas tragedy. Bow tie diagrams help to communicate hazardous scenarios like that at Bhopal; the current protocol further helps to identify potential ISD opportunities to manage risk.

4.5 ISD bow tie workshop for protocol implementation within an organization

To adopt the protocol as a viable tool within an organization, a training workshop should be provided as part of the PSM element *Training and competency*. Here, we describe an example ISD bow tie workshop having the following high-level structure:

1. Introduction to the bow tie methodology.
2. Overview of the hierarchy of controls and how it aligns with bow tie diagrams.
3. Completion of ISD/bow tie methodology exercises (could be completed using bow tie modeling software if available in the organization).

The workshop encompasses the following material:

- Overview of the bow tie method, including definitions of the bow tie elements and formulating these elements (as described by CCPS/EI, 2018).

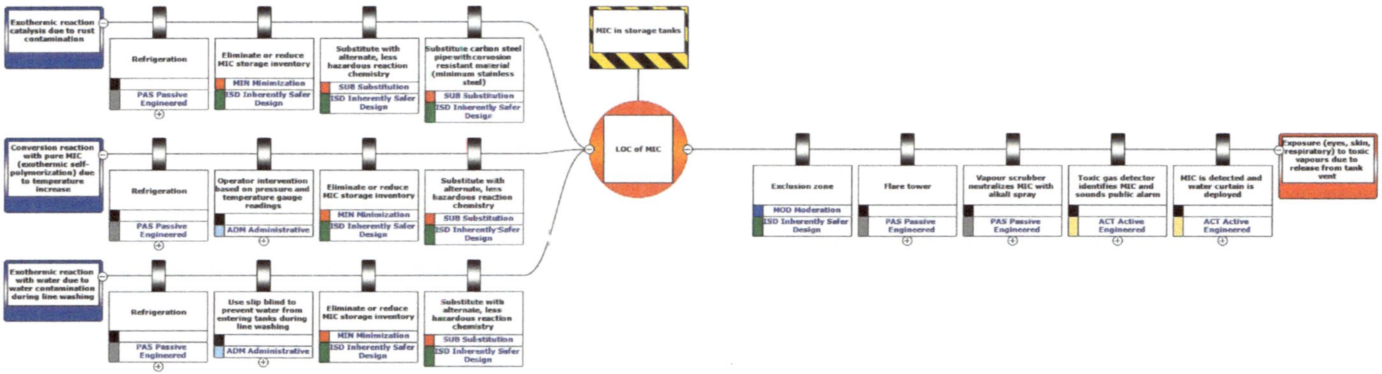

Fig. 11 High level bow tie diagram of Bhopal gas tragedy with hazard as MIC in storage tanks and top event as loss of containment of MIC with first- and second-order ISD opportunities identified.

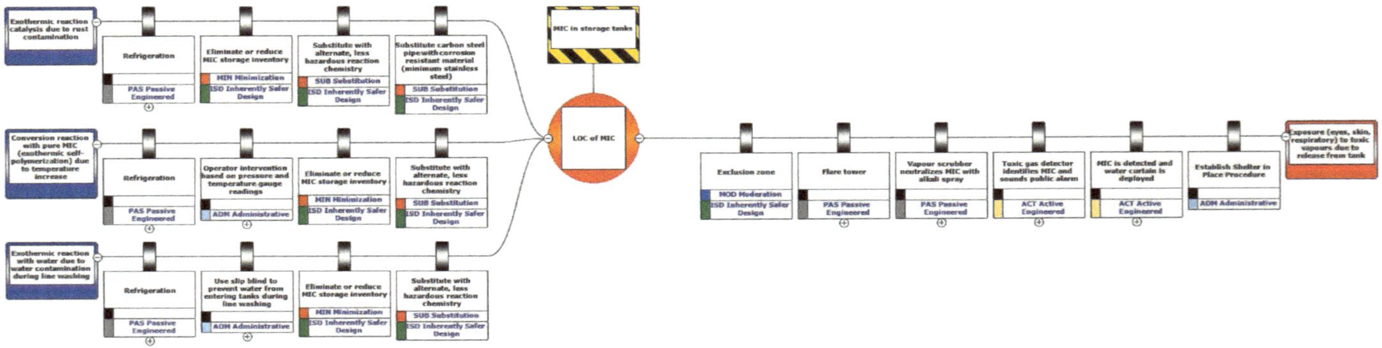

Fig. 12 High level bow tie diagram of Bhopal gas tragedy with hazard as MIC in storage tanks and top event as loss of containment of MIC after protocol application.

- Fundamentals of hierarchy of controls and ISD, with an emphasis on ISD principles, example-based guidance and ISD barriers in bow ties.
- Group exercises for workshop participants to build bow tie diagrams together (exercises developed by workshop facilitators could be based on worked examples found in CCPS/EI, 2018):
  - Workshop participants identify barrier types and barrier categories (where barrier types relate to the hierarchy of controls and barrier categories are the ISD principles).
  - Workshop participants identify any current ISD barriers/considerations and propose any potential ISD considerations.
- Completion of a bow tie example based on a word problem to help encourage brainstorming. This exercise entails a written description of a scenario and design details where participants draw a bow tie based on the description and identify potential opportunities for ISD considerations.

Organizers and facilitators should prepare a comprehensive package for distribution prior to conducting the workshop. The workshop package and structure would benefit from consideration of *Bloom's Taxonomy of Educational Objectives* to help elucidate the instructional objectives and learning outcomes. Kletz and Amyotte (2010) include instructional objectives specifically for ISD. Examples of workshop learning outcomes include:

- Knowledge Base: Explain the bow tie methodology for process hazard analysis.
- Use of Engineering Tools, Problem Analysis: Use the bow tie methodology and the full hierarchy of controls to model hazardous scenarios related to process safety.
- Teamwork, Communication Skills: Collaborate with team members to develop bow tie diagrams for risk communication.

The workshop environment is an important consideration. Facilitators should use an appropriate space that promotes brainstorming, teamwork and creative thinking, as emphasized by Edwards et al. (2015). Facilitators should also collect participant feedback for continuous improvement. It is important to understand the best way for the workshop to be administered in the organization to ensure its effectiveness. At the beginning and end of the workshop, facilitators should encourage participants to consider what they hope to learn from the session. Upon completion, facilitators should ask participants to reflect to see if they learned and accomplished what they hoped to, and share feedback on how the workshop could be improved. Following this educational training workshop to help personnel understand

ISD and the bow tie methodology, an ISD workshop like that detailed in Edwards et al. (2015) can be used in the organization to apply the ISD/bow tie protocol to new scenarios.

Having concluded the current section, readers are invited to consider the third *Opportunity for Reflection* exercise in this chapter. With respect to ISD application to the bow tie methodology for process hazard analysis:

- What would you like to *start* doing?
- What would you like to *stop* doing?
- What would you like to *continue* doing?

5. ISD application to dust hazard analysis

This section provides a brief background to dust hazard analysis (DHA), the elements of a typical DHA process, and the three most used types of DHA. The section also outlines how the ISD philosophy can be incorporated into conducting a DHA to address (prevent or mitigate) dust explosion scenarios in process equipment, including an example for a cyclone separator.

5.1 Dust hazard analysis (DHA)

According to NFPA 652 (2019), a Dust Hazard Analysis (DHA) is defined as "a systematic review to identify and evaluate the potential fire, flash fire, or explosion hazards associated with the presence of one or more combustible particulate solids in a process or facility." Until the introduction of the term DHA, hazard analysis of combustible dust-handling operations was conducted using a Process Hazard Analysis (PHA) (Perry, Ozog, Murphy, & Stickles, 2009).

The main purpose in conducting a DHA is to evaluate existing controls and develop recommendations for additional protections as appropriate. Although a DHA is a relatively new strategy, it has become very important in ensuring the safe operation of facilities where combustible dusts are processed or handled. Due to its importance, it is required under NFPA 652 that a dust hazard analysis be conducted for all existing process plants by September 7, 2020 (NFPA, 2019), which is a 2-year extension from the NFPA 652 (2016) deadline date of September 7, 2018. The standard also applies retroactively to all facilities and requires that upgrades or new installations must perform a DHA within 3 years to comply with this standard. In the United States, process plants (involving combustible dusts) that do not take these hazards seriously and perform a DHA are likely to receive citations and risk penalties (including fines and unrealistic dates to achieve compliance) from the authority having jurisdiction (AHJ).

Although DHAs can be a time-consuming, expensive endeavor, the impacts and benefits are profound. When the correct approach is employed, valuable resources can be conserved, regulatory requirements can be met, and most importantly, risk can be reduced. Also, combining the requirements of NFPA 652 (2019) with the established hazard analysis process will make implementation simpler and consistent, and the process renewable (Murphy & Borene, 2019).

A DHA involves detailed consideration of the dust present within the complete framework of the work area, the equipment used, the types of operations, and the buildings housing these equipment and processes. Regardless of the type of DHA being conducted, there are several common features that constitute the minimum requirements of any DHA. The basic steps for performing a DHA are as follows (Frank, 2019):

1. Identify hazards associated with operations.
2. Identify credible scenarios through which the potential loss associated with the hazards could be manifested.
3. Determine the potential harm (consequences) associated with these scenarios.
4. Determine the likelihood of the scenarios occurring.
5. Identify existing engineered and administrative safeguards that could effectively prevent or mitigate the scenarios.
6. Evaluate the risk of operations and determine if additional safeguards are warranted.
7. Propose additional appropriate safeguards if warranted; then re-evaluate the risk.

NFPA 652 (2019) also specifies several considerations to help identify and evaluate the fire, deflagration and explosion hazards of each part of the process and each building or compartment:

1. Apply dust characteristics of materials representative of the dust present.
2. Identify safe operating ranges.
3. Address potential transport of materials and/or deflagration propagation:
   - Intended and unintended dispersion of dust between parts of the process,
   - Fugitive dust emissions into the building/compartment, and
   - Deflagration propagation between parts of the system.
   - Use a qualified person to lead the DHA.

4. Document as a fire hazard each part of the process and buildings that contains a combustible particulate solid and can potentially include an oxidizing atmosphere and a credible ignition source.
5. Document as a dust deflagration hazard each part of the process and buildings that contains a combustible particulate solid and can potentially include an oxidizing atmosphere, a credible ignition source, and a credible suspension mechanism.
6. Evaluate buildings/compartments where combustible dust is present:
   - Address potential combustible dust migration between buildings or compartments, and
   - Address potential deflagration propagation between buildings or compartments.
7. Compare actual or intended dust accumulation to the threshold housekeeping dust accumulation that presents a potential for flash-fire or explosion.
8. Identify safeguards.
9. Recommend additional safeguards, as necessary.
10. Document the results, including action items and a plan for implementation.

Methods of performing a DHA are mainly categorized in 3 different ways depending on the intended outcome of the entire process or what exactly the DHA team or consultant seeks to achieve. These are explained as follows:

1. *Prescriptive*: This type of DHA depends on experienced-based codes and standards to define process safety and loss prevention efforts (CCPS, 2010b). The method is suitable for straightforward processes where there is full knowledge about the associated dust hazards. The prescriptive type DHA tends to focus less on assessing the effect of material-specific explosibility parameters on the hazards, and more on whether the dust being handled is considered explosible, to suggest recommendations.
2. *Performance-based design*: This type of DHA basically mirrors available fire codes and the various fire protection standards (CCPS, 2010b). The method can be used where prescriptive protection measures do not directly meet operational goals of the process plant. Performance-based DHAs consider material characteristics to show that even though the materials being processed are explosible, there is no risk of deflagration or explosion under certain operating conditions, and then propose appropriate safeguards.
3. *Risk-based*: This type of DHA is used to estimate the risk present in a process and decide if it is in agreement with acceptable levels of exposure (CCPS, 2010b). A risk-based approach assesses the risk present within a

piece of equipment using material characteristics and specific operating conditions and compares it to what is considered a tolerable risk level before making recommendations on appropriate safeguards. For process facilities with similar types of process equipment working under comparable conditions, a risk-based DHA is recommended for risk identification and evaluation due to its detailed and comprehensive nature. It should be noted, however, that more effort and skill are required compared to the other two types of DHA. A distinct feature of a risk-based DHA is the production of a risk ranking system whereby a value is assigned to the likelihood of occurrence and degree of consequence. This ranking system prioritizes the identified risks so they can be addressed from highest to moderate and finally to the lowest risk. This allows for investment in higher risk processes and activities first before subsequently addressing other levels of risk associated with the various scenarios and conditions that may be present in the process.

5.2 Incorporating ISD into DHA

This section illustrates how the ISD philosophy can be incorporated into conducting a DHA to prevent or mitigate a dust explosion hazard. In the ensuing section, a DHA-ISD strategy is applied to dust explosion scenarios in a cyclone separator as an example of how process equipment combustible dust hazards can be represented in a DHA-ISD worksheet.

The DHA approach suggested here is tailored according to one of the commonly used types of DHA procedures (i.e. the prescriptive type) and the examples provided in NFPA 652 (2019). Considerations fundamental to the development of a comprehensive DHA worksheet include the following:

- Identify parts, equipment, processes, and building compartments of the process system.
- Identify materials posing combustible dust hazards that may be present in the part of the process.
- Identify conditions that could lead to fire, deflagration, or explosion hazards under normal as well as foreseeable upset operating conditions.
- Consider deflagration hazards in the particular process section under consideration.
- Identify and document hazards.
- Identify and document any existing safeguards.
- Identify and document any recommendations.

Table 4 List of ISD recommendations to be incorporated into DHA process.

ISD principle	ISD recommendations for DHA
Minimization	• Reduce amount of combustible material being processed • Avoid as much as possible identified sources of ignition in dust generating areas • Minimize as much as possible operations or procedures that generate dust • Reduce the frequency of operations that disperse dust into the vicinity • Keep dust concentration outside explosible range
Substitution	• Convey materials in an inert (e.g., N_2 or Ar) atmosphere instead of air • Replace dilute-phase transport with dense-phase pneumatic transport • Use mass flow silos instead of funnel flow silos • Replace a work procedure with another that reduces dust production and accumulation • Replace a process route that produces large amounts of dust with one that produces lesser amount of dust • Replace a more combustible, reactive or hazardous material with a less combustible, reactive or hazardous material
Moderation	• Increase particle size and moisture content to reduce reactivity • Alter the composition of a combustible dust by admixture with solid inertant to reduce its hazardous nature • Separate possible ignition sources from dust generating areas and processes • Isolate dust generating processes from flammable gas units to prevent the formation of explosible hybrid mixtures
Simplification	• Reduce number of size reduction equipment and processes • Use equipment constructed to withstand explosion pressures • Ensure information on the hazardous properties of combustible dusts is clear and easily understandable by all • Use dust-tight equipment

Partly extracted from Amyotte, P. R. (2013). *An introduction to dust explosions: Understanding the myths and realities of dust explosions for a safer workplace*. Oxford, UK: Butterworth-Heinemann.

Integration of inherent safety principles to achieve a DHA-ISD worksheet can be accomplished by incorporating the set of dust-specific ISD measures given in Table 4. This list serves as a pool of ISD remedies from which risk reduction recommendations can be selected for a given scenario.

As emphasized throughout this chapter, the first point to consider before an attempt to integrate ISD into any risk reduction strategy is to cultivate a mindset of risk elimination and prevention rather than risk control. With this in mind, integration of ISD into DHA should improve the effectiveness and efficiency of application. Inherently safer design principles have been successfully incorporated into other risk assessment tools such as Checklist, What-if analysis, and Bow Tie analysis (as per Section 4). In all these cases, the resultant strategy is typically more efficient and cost effective as well as being simpler.

5.3 Typical case scenario—Cyclone separator

Cyclone separators are connected to solid particulate matter producing equipment to provide a way of removing fine material from air or other gas streams at low price and with minimal maintenance. In a combustible dust handling facility, they are affixed to extract particles from other process equipment. Cyclones are principally centrifugal separators that comprise an upper cylindrical part called the barrel and a lower conical part referred to as the cone (Fig. 13 from He et al., 2017). The particle-laden air stream enters tangentially at the top of the barrel and is forced to follow the curved geometry of the cyclone. The inertia of the particles in the flow causes them to move toward the cyclone wall where they collide and then drop due to gravity. The particles travel downward into the cone forming an outer vortex. The increasing air flow rate within the outer vortex leads to a centrifugal

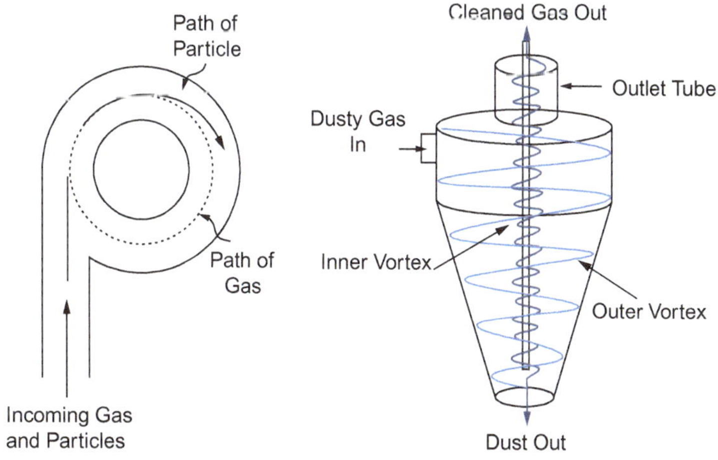

Fig. 13 Schematic for a cyclone separator in a dust collection system (He, Chen, & Xu, 2017) (with permission).

Table 5 ISD principles incorporated into DHA for cyclone separator.

Equipment	Particulate deflagrable?	Dust suspended in air?	Suspended dust concentration	Potential ignition source	Hazard	Existing hazard management	Recommendations
Cyclone separator	Yes	Yes	>Minimum explosible concentration (MEC) of combustible dust material	Electrostatic discharge from ungrounded component or tramp metal; friction; chemical reaction	Explosion	Venting; isolation valve	Minimize as much as possible operations or procedures that generate dust Replace a process route that produces large amounts of dust with one that produces lesser amount of dust Replace a more combustible, reactive or hazardous material with a less combustible, reactive, or hazardous material Separate possible ignition sources from dust generating areas and processes Reduce number of size reduction equipment and processes

force on the particles, separating them from the air stream. When the air reaches the bottom of the cone, it begins to flow radially inwards and goes out through the top as clean air or gas while the particulates fall into the dust collection chamber connected to the lowest part of the cyclone.

The operations in a cyclone generate fines that can form combustible clouds with explosible concentrations. In fact, most dust collector explosions are recorded to have started from the cyclone unit. In the cyclone, there could be a number of effective ignition scenarios including tramp metal hitting the walls of the cyclone, electrostatic discharges, friction, and chemical reactions. Table 5 shows a DHA worksheet that incorporates the principles of ISD at the recommendations stage to prevent and mitigate the dust explosion hazard identified in the cyclone separator. At least one ISD remedy from each of the four principles (as seen in Table 4) has been applied to address the hazard. Hence Table 5 represents a DHA-ISD worksheet.

Having concluded the current section, readers are invited to consider the fourth *Opportunity for Reflection* exercise in this chapter. With respect to ISD application to dust hazard analysis:

- What would you like to *start* doing?
- What would you like to *stop* doing?
- What would you like to *continue* doing?

6. Concluding remarks

The current chapter has provided an overview of inherently safer design, the fundamental principles on which it is based, its benefits, and the challenges associated with its implementation. The explicit and systematic consideration of inherently safer design within a process safety management framework was described. A protocol for the incorporation of inherently safer design within process hazard analysis, specifically within bow tie analysis, was presented. Lastly, an overview of dust hazard analysis and the incorporation of inherently safer design within dust hazard analysis was outlined. Our objective has been to enable readers to gain a better understanding of the benefits of inherently safer design and some of the ways ISD can be incorporated in their organizations.

Acknowledgment

The authors gratefully acknowledge the funding assistance of the Natural Sciences and Engineering Research Council of Canada (NSERC).

References

ACUTech. (2019). *Inherently safer design analysis approaches*. ACUTech. Retrieved December 14, 2019, from. State of New Jersey https://www.nj.gov/dep/enforcement/tcpa/downloads/IST_workshop_sample_worksheets.pdf.

AIChE (American Institute of Chemical Engineers). (2019). *SAChE certificate program—Inherently safer design*. Retrieved November, 27, 2019 from https://www.aiche.org/academy/courses/ela984/sacher-certificate-program-inherently-safer-design.

Amyotte, P. R. (2013). *An introduction to dust explosions: Understanding the myths and realities of dust explosions for a safer workplace*. Oxford, UK: Butterworth-Heinemann.

Amyotte, P. R. (2019). What went right. In *Proceedings of hazards 29* (pp. 314–323). Birmingham, UK: IChemE. May 22-24 2019.

Amyotte, P. R., Berger, S., Edwards, D. W., Gupta, J. P., Hendershot, D. C., Khan, F. I., et al. (2016). Why major accidents are still occurring. *Process Safety Progress, 14*, 1–8.

Amyotte, P. R., Goraya, A. U., Hendershot, D. C., & Khan, F. I. (2007). Incorporation of inherent safety principles in process safety management. *Process Safety Progress, 26*, 333–346.

Amyotte, P. R., Khan, F. I., & Dastidar, A. G. (2003). Reduce dust explosions the inherently safer way. *Chemical Engineering Progress, 99*, 36–43.

Amyotte, P. R., & Lupien, C. S. (2017). Elements of process safety management. In F. Khan (Ed.), *Vol. 1. Methods in chemical process safety* (pp. 87–148). Cambridge, MA: Elsevier/Academic Press.

Amyotte, P., Irvine, Y., & Khan, F. (2018). Chemical safety board investigation reports and the hierarchy of controls: Round 2. *Process Safety Progress, 37*, 459–466.

Amyotte, P. R., MacDonald, D. K., & Khan, F. I. (2011). An analysis of CSB investigation reports concerning the hierarchy of controls. *Process Safety Progress, 30*, 261–265.

Amyotte, P. R., Pegg, M. J., & Khan, F. I. (2009). Application of inherent safety principles to dust explosion prevention and mitigation. *Process Safety and Environmental Protection, 87*, 35–39.

Amyotte, P. R., Pegg, M. J., Khan, F. I., Masaharu, N., & Yingxin, T. (2007). Moderation of dust explosions. *Journal of Loss Prevention in the Process Industries, 20*, 675–687.

API (American Petroleum Institute). (2010). *Fact sheet: Process safety performance indicators for the refining and petrochemical industries. ANSI/API Recommended Practice 754* (1st ed.). Washington, DC: American Petroleum Institute. Retrieved December 15, 2019, from https://www.api.org/~/media/Files/Oil-and-Natural-Gas/Refining/Process%20Safety/RP-754-Fact-Sheet.pdf.

API (American Petroleum Institute). (2019a). *Management of hazards associated with location of process plant permanent buildings. ANSI/API recommended practice 752* (3rd ed.). Washington, DC: American Petroleum Institute.

API (American Petroleum Institute). (2019b). *Management of hazards associated with location of process plant portable buildings. ANSI/API recommended practice 753* (1st ed.). Washington, DC: American Petroleum Institute.

ATM Recycling Systems. (2017). *Recycling system of swarf processing from secondary raw materials to reusable briquettes*. Retrieved January 10, 2020, from https://www.atm-recyclingsystems.com/systems/recycling-system-of-swarf-processing.html.

Azizi, W. (2016). Predict incidents with process safety performance indicators. *Chemical Engineering Progress, 112*(2), 22–25.

Blais, D.H.J., Bodeux, D.F., Burgwin, S.L., Chan, A.D., Locke, G.S., Squires, J.H., Virtue, S.J., 2012. HF alkylation process with internal acid regeneration. US 8.212,098 B2.

Bowonder, B. (1987). An analysis of the Bhopal accident. *Project Appraisal, 2*(3), 157–168.

California Code of Regulations, California Office of Emergency Services. California accidental release prevention (CalARP) regulations, Title 19, Division 2, Chapter 4.5.

California Occupational Safety and Health Standards Board. (2016). *Informative digest.* In *Process safety management for petroleum refineries.* §5189.1. Retrieved December 14, 2019, from https://www.dir.ca.gov/OSHSB/documents/noticeSep2016-Process-Safety-Management-for-Petroleum-Refineriess.pdf.

CCHS (Contra Costa County Health Services). (2011a). *Industrial Safety Ordinance (ISO) guidance document, attachment C inherently safer systems checklist.* CCHS. Retrieved December 15, 2019, from https://cchealth.org/hazmat/pdf/iso/attachment_c.pdf.

CCHS (Contra Costa County Health Services). (2011b). *Safety audit summary.* Retrieved November, 27, 2019 from https://cchealth.org/hazmat/facilities/pdf/shell-chemical-audit-summary.pdf.

CCHS (Contra Costa County Health Services). (2014). *Industrial safety ordinance. Section A: Safety program management:* CCHS. Retrieved November, 27, 2019, from https://cchealth.org/hazmat/pdf/iso/section_a.pdf.

CCHS (Contra Costa County Health Services). (2017a). *Industrial safety ordinance (ISO) guidance document.* In *Section F: Safety culture assessments.* CCHS Retrieved November, 27, 2019, from https://cchealth.org/hazmat/pdf/iso/section_f.pdf.

CCHS (Contra Costa County Health Services). (2017b). *Safety audit summary.* CCHS. Retrieved November, 27, 2019, from https://cchealth.org/hazmat/facilities/pdf/air_liquide_rodeo_audit_summary.pdf.

CCHS (Contra Costa County Health Services). (2019a). *Twentieth anniversary of the industrial safety ordinance.* Retrieved February 14, 2019, from https://cchealth.org/hazmat/pdf/iso/Report-20th-Anniversary.pdf.

CCHS (Contra Costa County Health Services). (2019b). *ISO Facilities.* CCHS. Retrieved December 13, 2019, from https://cchealth.org/hazmat/iso/facilities.php.

CCHS (Contra Costa County Health Services). (2019c). *Risk management plans—Facility safety audit summaries.* CCHS. Retrieved November 27, 2019, from https://cchealth.org/hazmat/rmp/.

CCHS (Contra Costa County Health Services). (2019d). *Safety audit summary.* CCHS. Retrieved November, 27, 2019, from https://cchealth.org/hazmat/facilities/pdf/general-chemical-richmond-audit-summary.pdf.

CCPS (Center for Chemical Process Safety). (2008). *Guidelines for hazard evaluation procedures* (3rd ed.). Hoboken, NJ: John Wiley & Sons, Inc.

CCPS (Center for Chemical Process Safety). (2009). *Inherently safer chemical processes: A life cycle approach* (2nd ed.). Hoboken, NJ: John Wiley & Sons, Inc.

CCPS (Center for Chemical Process Safety). (2010a). *Process safety beacon: Fire protection—Long-bolt flangeless valves.* CCPS. Retrieved November, 27, 2019 from https://www.aiche.org/sites/default/files/cep/20100424.pdf.

CCPS (Center for Chemical Process Safety). (2010b). *Guidelines for risk based process safety.* Hoboken, NJ: John Wiley & Sons.

CCPS (Center for Chemical Process Safety). (2011). *Process safety leading and lagging metrics… you don't improve what you don't measure.* New York, NY: Center for Chemical Process Safety.

CCPS (Center for Chemical Process Safety). (2012a). *Guidelines for engineering design for process safety* (2nd ed.). Hoboken, NJ: John Wiley & Sons, Inc.

CCPS (Center for Chemical Process Safety). (2012b). *March 2012 process safety beacon: Double block and bleed.* CCPS. Retrieved December 16, 2019, from https://www.aiche.org/sites/default/files/cep/20120325.pdf.

CCPS (Center for Chemical Process Safety). (2018a). *Essential practices for creating, strengthening, and sustaining process safety culture.* Hoboken, NJ: John Wiley & Sons, Inc.

CCPS (Center for Chemical Process Safety). (2018b). *Process safety metrics—Guide for selecting leading and lagging metrics.* New York, NY: Center for Chemical Process Safety. Retrieved November 27, 2019, from https://www.aiche.org/sites/default/files/docs/pages/ccps_process_safety_metrics_-_v3.2.pdf.

CCPS (Center for Chemical Process Safety). (2019). *Inherently safer chemical processes: A life cycle approach* (3rd ed.). Hoboken, NJ: John Wiley & Sons, Inc.

CCPS/EI (Center for Chemical Process Safety/Energy Institute). (2018). *Bow ties in risk management: A concept book for process safety*. New York, NY/London, UK: Wiley.

CGE Risk Management Solutions. (2017). *BowTieXP methodology manual*. Revision 16 (July 2017).

CGE Risk Management Solutions. (2019a). *BowTieXP software manual*. Revision 39 (March 8, 2019).

CGE Risk Management Solutions. (2019b). *Knowledge base*. CGE. Retrieved November 27, 2019, from https://www.cgerisk.com/knowledgebase/Main_Page.

CSA. (2017). *CAN/CSA-Z767-17, Process safety management*. Toronto, ON: CSA Group.

CSB (Chemical Safety Board). (2005). *Investigation report—Combustible dust fire and explosions*. Washington, DC: CTA Acoustics Inc. Report No. 2003-09-I-KY.

CSB (Chemical Safety Board). (2013). *Interim investigation report—Chevron Richmond refinery pipe rupture and fire*. Report No. 2012-03-I-CA. Washington, DC.

CSB (Chemical Safety Board). (2014). *Final investigation report—Catastrophic rupture of heat exchanger, Tesoro Anacortes refinery, report no*. 2010-08-I-WA, DC: Washington.

CSB (Chemical Safety Board). (2015a). *Final investigation report—Chevron Richmond refinery pipe rupture and fire*. Report No. 2012-03-I-CA, DC: Washington.

CSB (Chemical Safety Board). (2015b). *Interim recommendations report—Toxic chemical release at the DuPont La Porte Chemical facility*. Report No. 2015-01-I-TX DC: Washington.

CSB (Chemical Safety Board). (2016). *Case study—Williams Geismar olefins plant reboiler rupture and fire*. Report No. 2013-03-I-LA, DC: Washington.

CSB (Chemical Safety Board). (2017). *Case study—Key lessons for preventing inadvertent mixing during chemical unloading operations —Chemical reaction and release in Atchison, Kansas at MGPI processing Inc.*. Report No. 2017-01-I-KS, DC: Washington.

CSB (Chemical Safety Board). (2019a). *Factual statement—Fire and explosions at Philadelphia energy solutions refinery hydrofluoric acid alkylation unit*. Report No. 2019-06-I-PA DC: Washington.

CSB (Chemical Safety Board). (2019b). *Final investigation report—Toxic chemical release at the DuPont La Porte chemical facility*. Report No. 2015-01-I-TX DC: Washington.

CSB (Chemical Safety Board). (2019c). *CSB Letter to EPA regarding HF (April 23, 2019)*. Retrieved December 14, 2019, CSB. from https://www.csb.gov/assets/1/6/letter_to_epa_4.23.2019.pdf.

CSB (Chemical Safety Board). (2019d). *Inherently safer systems review*. CSB. Retrieved January 18, 2019, from https://www.csb.gov/assets/1/20/ist_document.pdf?14935.

CSChE (Canadian Society for Chemical Engineering). (2012). *Process safety management guide* (4th ed.). Ottawa, ON: CSChE.

Dust Safety Science. (2019). *Dust Safety Science Podcast*. Dust Safety Science. Retrieved November 27, 2019, from https://dustsafetyscience.com/dust-safety-science-podcast/.

Edwards, D., Foster, J., Linwood, D., McBride-Wright, M., & Russell, P. (2015). Inherent Safety: It's common sense, now for common practice!. In *Proceedings of hazards 25* (pp. 1–14). Edinburgh, UK: IChemE. (May 13-15 2015).

Forest, J. (2012). Management discipline. *Process Safety Progress*, *31*, 334–336.

Frank, W. (2019). Dust explosions: Hazard identification. In P. Amyotte & F. Khan (Eds.), *Vol. 3. Methods in chemical process safety* (pp. 123–165). Cambridge, MA: Elsevier/Academic Press.

Goraya, A., Amyotte, P. R., & Khan, F. I. (2004). An inherent safety-based incident investigation methodology. *Process Safety Progress*, *23*, 197–205.

Hatch, D., McCulloch, P., & Travers, I. (2017). Visual HAZOP: Exploiting the power of bowties to improve study effectiveness and enhance engagement. *Chemical Engineer*, *917*, 27–32.

Hatch, D., McCulloch, P., & Travers, I. (2019). Enhancing PHAs: The power of bowties. *Chemical Engineering Progress*, *115*, 20–26.

He, J., Chen, K., & Xu, J. (2017). Urban air pollution and control. In M. A. Abraham (Ed.), *Encyclopedia of sustainable technologies* (pp. 243–257). Elsevier.

Hendershot, D. C. (2011). Inherently safer design: An overview of key elements. *Professional Safety, 56*(2), 48–55 American Society of Safety Engineers. (February 2011).

Hopkins, A. (2005). *Safety, culture and risk.* Sydney, Australia: CCH Australia Limited.

Hopkins, A. (Ed.), (2009). *Learning from high reliability organisations.* Sydney, Australia: CCH Australia Limited.

IChemE (Institution of Chemical Engineers). (2019). *Inherent safety in design and operation development.* IChemE. Retrieved November 27, 2019, from https://www.icheme.org/career/training/courses/inherent-safety-in-design-and-operation-development/.

Jafari, M. J., Nourai, F., Pouyakian, M., Torabi, S. A., Rafiee Miandashti, M., & Mohammadi, H. (2018). Barriers to adopting inherently safer design philosophy in Iran. *Process Safety Progress, 37*(2), 221–229.

Khan, F. I., & Amyotte, P. R. (2003). How to make inherent safety practice a reality. *Canadian Journal of Chemical Engineering, 81*, 2–16.

Klein, J. A., & Vaughen, B. K. (2017). *Process Safety: Key concepts and practical applications.* Boca Raton, FL: CRC Press/Taylor & Francis.

Kletz, T. A. (1978). What you don't have can't leak. *Chemistry & Industry, 6*, 287–292.

Kletz, T. A., & Amyotte, P. R. (2010). *Process plants: A handbook for inherently safer design* (2nd ed.). Boca Raton, FL: CRC Press/Taylor & Francis.

Kletz, T. A., & Amyotte, P. R. (2019). *What went wrong? Case histories of process plant disasters and how they could have been avoided* (6th ed.). Oxford, UK: Elsevier.

Mannan, S., & Lees, F. (2005). *Lees' loss prevention in the process industries* (3rd ed.). Oxford, UK: Elsevier/Butterworth-Heinemann.

Minerva. (2013). *Teaching module: Process safety management.* Minerva. Retrieved November 26, 2019, from https://safetymanagementeducation.com/teaching-resources/teaching-modules/.

MKOCPSC/ISC (Mary Kay O'Connor Process Safety Center/IChemE Safety Centre). (2017). *Process safety for the 21st century and beyond.* MKOCPSC/ISC. Retrieved November 26, 2019, from http://psc.tamu.edu/wpcontent/uploads/0001_17%20ISC_21st%20Century_Brochure%20-%20Final.pdf.

Murphy, M. R., & Borene, M. J. (2019). How to conduct a dust hazards analysis. *Process Safety Progress, 38*(2), 1–6.

NFPA (National Fire Protection Association). (2016). *NFPA 652—Standard on the fundamentals of combustible dust.* Quincy, MA: National Fire Protection Association.

NFPA (National Fire Protection Association). (2019). *NFPA 652—Standard on the fundamentals of combustible dust.* Quincy, MA: National Fire Protection Association.

NIOSH (National Institute for Occupational Safety and Health). (2015). *Hierarchy of controls.* NIOSH. Retrieved December 1, 2019, from https://www.cdc.gov/niosh/topics/hierarchy/default.html.

NRC (National Research Council). (2012). *The use and storage of methyl isocyanate (MIC) at bayer CropScience.* Washington, DC: National Academy of Sciences.

Parr Instrument Company. (2019). *Tubular reactors systems.* Parr Instrument Company. Retrieved January 10, 2020, from https://www.parrinst.com/products/tubular-reactor-systems/5400-continuous-flow-tubular-reactors/documents-2/.

Perry, J. A., Ozog, H., Murphy, M., & Stickles, R. P. (2009). Conducting process hazard analyses for dust-handling operations. *Chemical Engineering Progress, 105*(2), 28–35.

PRAB. (2020). *Disposing swarf safely.* Retrieved January 10, 2020, from https://www.prab.com/customer-solutions/waste-products/swarf/.

RUF US Inc. (2016). *Magnesium briquetting provides a safer solution for your magnesium scraps.* RUF. Retrieved January 10, 2020, from https://www.ruf-briquetter.com/materials/metals/magnesium-briquetting.

State of New Jersey. (2017). *Guidance for Toxic Catastrophe Prevention Act (TCPA): Inherently Safer Technology (IST) review*. State of New Jersey. Retrieved December 14, 2019, from https://www.nj.gov/dep/enforcement/tcpa/downloads/istguidance_rev2.pdf.

Vaughen, B. K., & Bloch, K. (2016). Use the bow tie diagram to help reduce process safety risks. *Chemical Engineering Progress*, *112*, 30–36.

Washington State Department of Labor and Industries. (2017). *CR-101 preproposal statement of inquiry, August 22 2017: Chapter 296-67 WAC, safety standards for process safety management of highly hazardous chemicals*. Washington State Department of Labor and Industries. Retrieved December 14, 2019, from https://lni.wa.gov/rulemaking-activity/AO17-20/1720CR101.pdf.

Washington State Department of Labor and Industries. (2018). *Process safety management advisory committee. October 10, 2018, PSM rulemaking meeting: Initial rule language presentation meeting*. Washington State Department of Labor and Industries. Retrieved December 14, 2019, from https://lni.wa.gov/safety-health/grants-committees-partnerships/advisory-committees/sh-comimttees-process-safety-management#meetings.

Further reading

Amyotte, P. R., Goraya, A. U., Hendershot, D. C., & Khan, F. I. (2006). Incorporation of inherent safety principles in process safety management, center for chemical process safety, American institute of chemical engineers. In *Proceedings of 21st annual international conference—Process Safety challenges in a global economy, Orlando, FL (April 23–27, 2006)* (pp. 175–207).

California Code of Regulations. Process safety management for petroleum refineries, Title 8, §5189.1.

Contra Costa County (CCC), California, Industrial safety ordinance (ISO), County Ordinance Chapter 450–8 Contra Costa County (CCC).

Mannan, M. S. (2012). Lees' loss prevention in the process industries, Vol. 2. (4th ed.). In Oxford, UK: Elsevier/Butterworth-Heinemann. Chapter 17.

New Jersey Administrative Code, Toxic catastrophe prevention Act (TCPA) regulations, N.J.A.C. Vol. 7:31.

RISO Richmond Industrial Safety Ordinance (RISO), Municipal Code, City of Richmond, California, Chapter 6.43.

Washington State Department of Labor and Industries. (2019). *Process safety management advisory committee. September 24, 2018, chapter 296-67 WAC process safety management for refineries, Part A: Discussion draft round 2*. Washington State Department of Labor and Industries. Retrieved December 14, 2019, from https://lni.wa.gov/safety-health/grants-committees-partnerships/advisory-committees/_psmdocs/PSM-DiscussionDraftRound2.pdf.

Kolibri 360 GmbH
Ferdinand-Jühlke-Straße 7
99095 Erfurt, Deutschland
produktsicherheit@kolibri360.de